PERCHLORATE

Environmental Occurrence, Interactions and Treatment

PERCHLORATE

Environmental Occurrence, Interactions and Treatment

edited by

Baohua Gu

Oak Ridge National Laboratory, Oak Ridge, Tennessee

John D. Coates

University of California, Berkeley, California

Springer

Library of Congress Control Number: 2005937702

ISBN-10: 0-387-31114-9   e-ISBN-10: 0-387-31113-0
ISBN-13: 978-0-387-31114-2

Printed on acid-free paper.

Printed in the United States of America.

9 8 7 6 5 4 3 2 1   SPIN 11347507

springer.com

CONTENTS

Chapter 1. Perchlorate: Challenges and Lessons

S. E. Cunniff, R. J. Cramer and H. E. Maupin

Chapter 2. The Chemistry of Perchlorate in the Environment

Gilbert M. Brown and Baohua Gu

Chapter 3. Occurrence and Formation of Non-Anthropogenic Perchlorate

W. Andrew Jackson, Todd Anderson, Greg Harvey, Greta Orris, Srinath Rajagopalan, and Namgoo Kang

Chapter 4. Alternative Causes of Wide-Spread, Low Concentration Perchlorate Impacts to Groundwater

Carol Aziz, Robert Borch, Paul Nicholson, and Evan Cox

Chapter 5. Stable Isotopic Composition of Chlorine and Oxygen in Synthetic and Natural Perchlorate

Neil C. Sturchio, J. K. Böhlke, Baohua Gu, Juske Horita, Gilbert M. Brown, Abelardo D. Beloso, Jr., Leslie J. Patterson, Paul B. Hatzinger, W. Andrew Jackson, and Jacimaria Batista

Chapter 6. Recent Developments in Perchlorate Detection

Pamela A. Mosier-Boss

Chapter 7. The Ecotoxicology of Perchlorate in the Environment

Philip N. Smith

Chapter 8. Perchlorate Toxicity and Risk Assessment

David R. Mattie, Joan Strawson, and Jay Zhao

Chapter 9. Using Biomonitoring to Assess Human Exposure to Perchlorate

Benjamin C. Blount and Liza Valentín-Blasini

Chapter 10. Recent Advances in Ion Exchange for Perchlorate Treatment, Recovery and Destruction

Baohua Gu and Gilbert M. Brown

Chapter 11. Field Demonstration using Highly Selective, Regenerable Ion Exchange and Perchlorate Destruction Technologies for Water Treatment

Baohua Gu and Gilbert M. Brown

Chapter 12. The Microbiology of Perchlorate Reduction and its Bioremediative Application

John D. Coates and Laurie A. Achenbach

Chapter 13. The Biochemistry and Genetics of Microbial Perchlorate Reduction

Laurie A. Achenbach, Kelly S. Bender, Yvonne Sun, and John D. Coates

Chapter 14. Field Demonstration of *In Situ* Perchlorate Bioremediation in Groundwater

P. B. Hatzinger, J. Diebold, C. A. Yates and R. J. Cramer

Chapter 15. Perchlorate Removal by Modified Activated Carbon

Robert Parette and Fred S. Cannon

Chapter 16. Titanium Catalyzed Perchlorate Reduction and Applications

Baohua Gu, Peter V. Bonnesen, Frederick V. Sloop, and Gilbert M. Brown

Chapter 17. Membrane and Other Treatment Technologies – Pros and Cons

Ping Zhou, Gilbert M. Brown, and Baohua Gu

PREFACE

Perchlorate (ClO_4^-) has been detected recently in groundwater, surface water, and soils and, more ominously, in plants, food products and human breast milk in many areas of the United States and the world. Because of its potential health affect on thyroid function by interfering with iodide uptake, the widespread occurrence of perchlorate in the environment has generated considerable interest in its contamination source, environmental interactions, toxicology, risk assessment, and remediation technologies. Most perchlorate is manufactured for use as a primary ingredient of solid rocket propellant and explosives. However, perchlorate is also used in pyrotechnic devices, such as fireworks, highway flares, gun powder, and air bags, and in a wide variety of industrial applications such as tanning and leather finishing, rubber manufacturing, and paint and enamel production. Naturally-occurring perchlorate is also known to exist, particularly in the hyperarid Atacama Desert in Chile. The widespread use and the presence of both natural and anthropogenic perchlorate thus have resulted in intense public debate and far-reaching ramifications, ranging from public health issues to liabilities that could be imposed by environmental cleanup needs.

The goal of this book is to provide the current state of science and technology with respect to the occurrences and potential sources of perchlorate contamination, its behavior, exposure pathways, and detection in the environment, toxicology and risk assessment, and recent advances in treatment technologies for removing perchlorate from contaminated soil and water. To this end, internationally recognized experts in each respective field of perchlorate research have contributed to this text to render a complete inter-disciplinary overview of the state of the science. The book is intended to serve as a comprehensive reference for environmental professionals, regulators, policy makers, scientists, engineers, and others interested in issues associated with perchlorate in the environment. The book consists of 17 chapters covering diverse subjects. The first six chapters describe the challenges and various sources of perchlorate contamination, its chemistry and detection in the environment, its natural occurrence and unique isotopic signatures that may be used for environmental forensics. Chapters 7 to 9 summarize our current understanding of perchlorate toxicology, risk assessment, and exposure pathways. The remaining chapters address recent advances in innovative treatment technologies for remediating perchlorate contaminated soil and water. In particular, significant advances in selective ion exchange and its regeneration technologies enable the treatment of large volumes of contaminated water with reduced costs. Furthermore, rapid advances in our understanding of the microbiology, biochemistry and genetics of perchlorate-reducing microorganisms offer great hope for eliminating perchlorate from contaminated environments in the future. Presented in this book several chapters are also devoted to field demonstration and case studies involving the use of highly-selective, regenerable ion exchange processes, in situ bioremediation strategies, and modified activated carbon technologies for perchlorate removal. Further attention is given to other treatment technologies, such as titanium-catalyzed reduction and membrane filtration, and to the pros and cons of various remedial options.

We wish to thank the authors for their contributions and for their cooperation during the preparation of this book. Special thanks are expressed to Juske Horita and Denise Parker for their expert review and editorial support. Finally, this book would not have been possible without the support of the U.S. Department of Energy, Department of Defense's Strategic Environmental Research and Development Program (SERDP) and the Environmental Security Technology Certification Program (ESTCP), and the Laboratory Directed Research and Development Fund of the Oak Ridge National Laboratory.

Baohua Gu, Oak Ridge, Tennessee

John D. Coates, Berkeley, California

Chapter 1

Perchlorate: Challenges and Lessons

S. E. Cunniff[1], R. J. Cramer[2] and H. E. Maupin[3] †

[1]Office of the Secretary of Defense, Arlington, VA 22202
[2]Indian Head Division, Naval Surface Warfare Center, Indian Head, MD 20640
[3]Naval Environmental Health Center, Portsmouth, VA 23708

HOW PERCHLORATE BECAME AN ISSUE

The U.S. Environmental Protection Agency (EPA) has been identifying toxic chemicals, establishing reference doses protective of public health, and regulating chemicals and their cleanup for decades. So what is it about perchlorate that made federal and state legislators, policy makers, and the press single it out over several other chemicals that possess potentially greater risks to public health? Many factors were involved:

- Strong desires by the public and some regulators to take immediate remedial action before regulatory standards had been promulgated.

- A collision of expectations over regulatory processes and cleanup responsibilities.

- Marked disagreement over the interpretation of science used in developing a draft reference dose.

- A narrow focus on environmental compliance, resulting in funding priorities being directed to regulatory compliance activities.

- Inability to determine accurately the sources of perchlorate or to distinguish between sources when multiple potential sources were present.

- Implementation and reporting problems with EPA Method 314 for the analysis of perchlorate in drinking water, i.e., the application to media for which it was not designed, and, the tendency for reporting false positives, without verification with more expensive methods (as required by EPA Method 314 due to its non-specific nature).

---

† The views expressed in this chapter represent those of the authors, not those of the department of the Navy nor the Department of Defense.

- Difficulties in communicating to the public complex scientific information, operating and policy challenges, and real differences in interpretations.

- Viewing the Department of Defense (DoD), NASA, and aerospace industry contractors – the users of large amounts of perchlorate – as a source of remediation funding for cash-strapped water districts.

In addition, five other "hot button" issues brought perchlorate in the environment to the attention of legislators, policy makers, environmental and public health officials and interest groups, the press, and, of course, the general public. These issues will each be described in greater detail below.

Drought and Water Demand in the Southwest

The southwestern United States is facing a drought at a time of declining water resources flexibility and a reduced administration support for water reclamation funding. Population growth, shifting and more complex demands, impaired water quality, new regulatory requirements, and decadal variations and long-term global climate changes affect water supply and demand. As our fresh water supplies continue to be stretched, the presence of any contaminant is an increasingly significant compounding factor.

Population in the southwestern and front-range states is predicted to climb from 91 million in 2000 to 125 million in 2025, creating a 89 million acre-feet (MAF) shortfall of water, a significant demand that exceeds the needed supply.[1] Drought further aggravates a challenged water supply system. By 2004, the Colorado River's Lake Mead was down 80 feet and Lake Powell was at an all-time low.[2] These two reservoirs provide a significant portion of the surface water supplies upon which the Los Angeles, Las Vegas, Phoenix, and Tucson areas depend. Based on dendrochronological and paleoecological studies, several publications predict that the southwestern United States is entering a prolonged dry period.[3,4] In addition, anthropogenic climatic changes pose water supply uncertainties. California water planners are now attempting to incorporate such impacts into their water strategic planning.[5,6]

The use of the Colorado River water by California is changing dramatically. In 2003, California agreed to reduce its use of Colorado River water by 800,000 acre-feet annually--a 20% reduction. With decreased access to Colorado River water, California's dependence on groundwater becomes greater and the contamination of groundwater more critical. In California, where much of the perchlorate contamination has been found, between 30 and 40 % of the water supply and 43% of the drinking water comes from groundwater supplies.[7]

Over the past decade, recycling, desalination, conjunctive use, water conservation, and water marketing have all been used to improve overall water supply reliability. Two of these options, recycling and conjunctive use, require the availability of unimpaired groundwater basins. In recycling processes, also known as reuse or reclamation, recycled water is often mixed with surface supplies in reservoirs or groundwater basins. In conjunctive use, which is joint management of surface and groundwater supplies, depleted groundwater basins are recharged and serve as underground reservoirs. Thus, the quality of groundwater becomes even more important to protect to ensure these basins and their supplies can be reliably used.

When city supplies are threatened by drought and contamination, and options like water recycling have limited financial support, few inexpensive options exist to obtain additional supplies. Congress initiated a federal water recycling assistance program, The Reclamation Wastewater and Groundwater Study and Facilities Act of 1992, Title XVI of Public Law 102-575 as amended, which provided a 25% federal cost-sharing for construction of treatment plants for impaired waters. Since its inception, the demand for this program has always far outstripped available funding. At the beginning of the 21st century, as the Administration looked for ways to balance the federal budget, it sought to reduce funding for federal support of water recycling. In seeming disagreement with the Administration, Congress repeatedly partially restored funding for water recycling. Thus from the perspective of urban water purveyors, federal funding for water recycling became less certain.

Critical Importance of Perchlorate

Perchlorate has been used for decades in the United States national defense and space programs. Ammonium perchlorate (AP) was selected as an oxidizer of choice for solid propellants because of its performance, safety, and ease of manufacturing and handling. It is critical to the safe and effective performance of systems ranging from tactical missiles such as Sidewinder, AMRAAM, and Tomahawk to strategic missiles such as Minutemen and all boosters for missile defense. It is a major component in almost all solid rocket and missile propellants. AP and, to a lesser extent, potassium perchlorate (KP), are found in thousands of military combat and training munitions. Perchlorate salts are an essential component in composite propellants, underwater explosives, and pyrotechnic compositions. These materials are used in rocket motors, flares, underwater munitions, gas generators, igniters, and in many other devices. Over 100 military rocket motors, as well as NASA's Space Shuttle Solid Rocket Boosters, use AP-based composite propellants. KP is used mostly in initiation systems, flares, smokes, and other pyrotechnic devices. All Navy underwater munitions, all tactical and strategic missiles that use composite propellants, and most of the pyrotechnic devices in the stockpile of combat and training munitions contain AP or KP.

Perchlorates began to be increasingly used in the late 1950s because they were more stable and significantly increased performance over traditional nitroglycerin nitrocellulose double-base propellants. They increased the range of tactical and strategic missiles, and their low sensitivity allowed missiles to be stored and operated over a wider temperature range. At present, there exist no suitable replacements for perchlorate as an oxidizer that will satisfy military performance, safety, logistics, and environmental requirements. However, investigations to find suitable perchlorate replacements are actively underway, and promising substitutes for use in some of the Army's training activities have been found.

Production of AP and KP in the U.S. and Allied countries is very large, and the cost of ammonium perchlorate salt is under $1.00 per pound. Candidate replacements currently in research and development are not available in large quantities, and costs may be on the order of $1,000 per pound or even more.

Regulation of perchlorate at very low levels threatened DoD's readiness and training activities as well as safety for storage and handling issues thus significant concerns were voiced about the potential regulatory impact to DoD mission accomplishment.

Regulatory Decision-Making Process

At the beginning of the 21st Century, members of the regulated community, including the DoD, saw a need for a more transparent, science-based process for making regulatory decisions. It was felt a process should be developed that supported decisions based on sound science, separated science and science policy decisions, and adequately vetted policy decisions. Congress and the Office of Management and Budget established two significant standards for federal science: (1) the Information Quality Act, Section 515 of Public Law 106-554, and, (2) the Information Quality Bulletin for Peer Review, December 15, 2004. These documents established higher standards for both the transparency and the quality of science in the science-based decision-making processes. Together, they called for more rigorous standards for introducing objectivity, integrity, reproducibility and utility in collecting and disseminating data, and they created new responsibilities for federal officials to explain decisions based on their use of that data. These combined requirements could have as significant an effect on transparency and public review of government decision-making as does the National Environmental Policy Act.

In the wake of these changes, the regulated community, including DoD, argued that before being asked to invest millions of dollars in clean up, regulatory levels should be supported by solid scientific evidence pointing decisively to an increased health benefit. They wanted assurance that perchlorate's Reference Dose (RfD), upon which regulatory standards would ultimately be developed, would be based on an independent peer review of the complete body of science and on informed policy evaluations. Since unnecessary and overly protective regulations would increase costs without providing additional safety, they asked for rigorous evaluations of the full societal costs and benefits. In fact, environmental requirements can induce significant operational and financial risk and therefore costs to all segments of society, for example, local government, industries, and individual property owners. Although Executive Order 12866 requires a Regulatory Impact Assessment that reflects conduct and disclosure of such cost and benefit comparisons of major rule makings, the listing of an RfD on EPA's Integrated Risk Information System (IRIS) is not viewed by EPA as a regulatory action requiring this review. Nevertheless, since IRIS values, such as an RfD, are often adopted at the state level as cleanup values, and since they often form the basis for policy guidance and decision making, it was imperative that the perchlorate RfD receive a systematic, unbiased review to provide sound information to decision-makers on public health benefits.

Policy ramifications of perchlorate regulations depend to a large degree on the ultimate regulatory level set by EPA. The potential impact of restrictions on perchlorate use on the DoD mission includes the loss of access to and use of military lands; decreased realism in training; the need for stable munitions to prevent unintended explosions; increased munitions storage requirements; redesign of weapons systems; changes to acquisition and logistics processes; loss of perchlorate sources in the United States; and, increased costs of cleanups, including industrial cleanups. Evaluation of these impacts is critical for informed decision making and public awareness.

The perchlorate issue has also been involved in the question of whether the establishment of uncertainty factors is strictly a scientific decision, as it is now considered, or a science policy decision. Uncertainty factors are assigned values based on the lack of data available when developing an RfD.

EPA defines the RfD: "an estimate with uncertainty spanning perhaps an order of magnitude of a daily oral exposure to the human population, including sensitive subgroups, that is likely to be without an appreciable risk of deleterious effects during a lifetime. It can be derived from a No Observed Adverse Effects Level, Lowest Observed Adverse Effects Level, or benchmark dose, with uncertainty factors generally applied to reflect limitations of the data used."[8]

Assigning uncertainty factors is somewhat subjective. A weight of 1, 3, or 10 is assigned to each of three factors. The addition of such factors is viewed by some as an injection of policy that changes the results of the scientific findings. Such a case argues that an evaluation of other societal risks should also be performed. Others argue that it is the policy maker – not the scientist – who ought to determine the level of uncertainty acceptable based on public health impacts and other societal costs.

Another very contentious issue that has serious ramifications is the question of the appropriate use of guidance during the interim period between development of an RfD and the ultimate regulation of a chemical. Guidance is often in place years before completion of the formal standard development process. For DoD, application of guidance as if it were regulation represents *de facto* regulation without the benefit of the full policy vetting process typical of regulation development.

Perceived Levels of Risk

Levels of risk that are generally acceptable to the public become unacceptable when the health of children, babies, and fetuses is threatened. The elevated risk to these groups from perchlorate, compounded by the fact that its risk was highly debated among very credible scientists, heightened the fear and concern of environmental activists, consumers, and public health advocates. Despite the January 2005 findings of the National Academy of Sciences, who was charged to review the science upon which EPA based its risk assessment of perchlorate, concern and debate over safe exposures is still covered by the press.

Communication Problems

One of the first articles printed about perchlorate in drinking water stated inaccurately that perchlorate causes cancer and hyperthyroidism. In fact, at the time, experts agreed that perchlorate does not cause cancer in humans, and in the past, perchlorate was used successfully to treat hyperthyroidism. Despite attempts to correct the press, these misleading statements continued to appear over a year. For the most part, key stakeholders were often not skilled enough at media relations to ensure that their positive messages regarding advances in the science and other accomplishments related to perchlorate were getting out.

Coverage of science by journalists often fails to include all sides of complex issues and tends to highlight controversy and division rather than demonstrate where strong agreement does exist. It is disconcerting to the public to hear that experts are in disagreement. Education of all stakeholders, including the media and the public, through clear, concise, and consistent explanations delivered through informational exchange forums and other effective outreach efforts, might have allayed some unnecessary fears and concerns.

FACING THE CHALLENGE OF EMERGING CONTAMINANTS

The Perchlorate Interagency Working Group (IWG), co-chaired by the Office of Science and Technology Policy and the Office of Management and Budget, was formed to facilitate coordination and interagency dialogue on issues surrounding perchlorate. EPA and other federal agencies sought a resolution of the science upon which EPA's draft risk assessment was based and, under the auspices of the IWG, a review by the National Academy of Sciences (NAS) was sought. During the 18-month NAS deliberations, DoD was found without an alternative level to use to identify cleanup priorities in its site-specific risk assessments.

During this time period, DoD established guidance specific to perchlorate environmental sampling in order to characterize potential perchlorate releases at its facilities. That policy, released on September 29, 2003, directed DoD facilities to sample sites having both suspected perchlorate releases and the potential for a complete exposure pathway. DoD also included perchlorate as one of four unregulated materials that it would investigate in its assessments of the impacts of the use of military munitions on the Department's operational testing and training ranges.[9]

DoD realized early on that regulation of perchlorate at very low levels could compromise military readiness and decrease its capability to perform its mission. DoD concluded from its experiences with the perchlorate risk assessment process that (i) better processes were needed to keep key parts of DoD informed of the implications of the evolving science, (ii) more robust ways were needed to become engaged in the development and review of science, and (iii) routes were needed to effectively introduce accurate information into interagency policy discussions. In short, DoD wanted to ensure that the full body of science was accounted for by federal or state regulators, that the science and regulatory processes were systematic and transparent, and that all policy makers were aware of the full ramifications of their decisions.

Chemical risk assessments based on medical and environmental toxicity data are the first input into the development of a risk assessment and life-cycle risk management investment strategy. Operational risk assessment – focused on military needs and a robust, holistic review of medical, environmental, occupational health, and other legal, financial, technical, and existing policy requirements – must occur enterprise-wide. To ensure that DoD's risk management polices are reviewed and revised by an integrated team, DoD brought together representatives from its acquisition, health affairs, research, readiness, outreach/public affairs, operations, and environmental safety and occupational health communities to develop and maintain a strategic, proactive stance on emerging contaminants. This approach was designed to accomplish the following objectives:

- Facilitate strategic policy formulation

- Assess the relevance and impact of regulation of emerging contaminants on military readiness and program execution

- Develop robust, consistent information and data management to ensure efficient and wise expenditures that support sound decision-making

- Target resources and staff effectively on strategically selected high-priority issues

These policies represent the first of several important steps DoD has taken toward transforming its "compliance only" approach to program management to a "going beyond compliance" approach that maximizes the capability of DoD assets. DoD Directive 4715.1E challenges the paradigm that environmental compliance is the end goal and stresses the importance of mission success.[10] By addressing emerging contaminants in a unified fashion, DoD is able to direct its resources where the benefits are greatest, which will ultimately decrease its financial, human health, and legal risks.

PERCHLORATE ENVIRONMENTAL SCIENCE AND TECHNOLOGY DEVELOPMENT

Identification of Sources of Perchlorate

It is generally recognized that past manufacturing processes, residues from open-burn and open-detonation demilitarization sites, and range residues from testing and training of perchlorate containing munitions are all potential sources of perchlorate in the environment attributable to DoD activities. The extent to which perchlorate is found at DoD properties is being addressed under current DoD perchlorate sampling and analysis policy. Perchlorate is also used in a wide variety of commercial chemical and industrial processes. A partial listing includes automobile air-bag initiators, water-gel and emulsion blasting agents, matches and fireworks, road flares, perchloric acid desiccation and digestion and metal etching, and photographic flash powder. It is also an incidental byproduct in the manufacture of sodium chlorate used in agricultural herbicides and defoliants.

It is difficult to assess the total contributions of commercial sources to the widespread presence of perchlorate in the environment when there is no control of perchlorate releases, as in the case of commercial fireworks, flares, and blasting agents. The magnitude and extent of these releases has been summarized in a Strategic Environmental Research and Development Program (SERDP), where perchlorate source contributions from Chilean nitrate-based fertilizers, fireworks, safety flares, blasting explosives, and electrochemically produced chlorine products are discussed in great detail.[11] SERDP is the Department of Defense's corporate environmental research and development program planned and executed in full partnership with the Department of Energy and the Environmental Protection Agency.

As a gas generator igniter, perchlorate is used widely in air bags, which are inflated by gaseous reaction products. Although igniter designs differ, normally 100 mg perchlorate salt is used per igniter. Several million igniters are manufactured in the U.S. alone for air-bag applications. The final fate of the 25,000 pounds per year of perchlorate used in gas generators is unknown.

Water-gel and emulsion blasting agents containing perchlorates are used in difficult rock blasting, underground and trenching, deep boreholes, and other applications that require extra energy over conventional agents. If applied where surface and groundwater can be affected, such uses could lead to a higher incidence of incomplete combustion and become a source of perchlorate contamination. Little data are available on releases of perchlorate from these sources.

In addition, a certain amount of perchlorate occurs naturally in the environment. The widespread commercial use and the natural occurrence of perchlorate make assigning responsibilities for perchlorate releases very difficult. To determine DoD responsibilities and vulnerabilities, a thorough understanding of the various sources is critical. Studies on the causes, distribution, fate, and, most important, methods to distinguish anthropogenic from natural perchlorate in the environment are in progress. Isotopic ratio analysis, a method discussed in its own chapter, is being used to differentiate perchlorate sources and to document perchlorate biodegradation.

Remediation Technologies

Low-cost and effective cleanup technologies are a key research area. DoD has invested approximately $40 million to develop a suite of perchlorate cleanup technologies including bioremediation, ion exchange, the use of novel adsorbants and membranes, all of which are described in greater detail in the following chapters.

Perchlorate Replacement Research

DoD is working to develop solid rocket propellants, pyrotechnics, and explosives that do not contain ammonium perchlorate as the oxidizer. Under study are ammonium dinitramide and other nitramines and nitramides and the use of minimum signature or double-base propellants in lieu of composite formulations. While alternative energetic oxidizers exist, significant cost, availability, and performance issues remain that have prevented their use in fielded weapon systems.

Each rocket or missile system has unique performance demands that require consideration when attempting to replace AP. Generally, DoD uses ammonium perchlorate as the oxidizer in most solid composite rocket/missile propellants. The SERDP has funded development work to eliminate toxic materials from solid rocket propellants and to find perchlorate alternatives.[12, 13]

Perchlorates are used as an oxidizing component in many incendiary mix and flash compositions. The formulations are used as markers to produces flash and smoke to designate an impact point and as igniters for flammable liquids and explosive trains. Over 7 million rounds utilizing these perchlorate-based formulations are used for training on Navy and Army ranges. Developing perchlorate-free incendiary and pyrotechnic technologies for projectiles is of great interest. The new technologies must meet or exceed current DoD performance and safety standards in areas such as toxicity, stability, sensitivity, ignitability, energy, and burning rate, and yet reduce the environmental impact of the materials in production and in use.

OUTREACH

The presence in the environment of contaminants that have unknown or little understood health and environmental affects will always be of concern. Sources of the material, responsible parties, regulatory level--all receive a high level of interest by the public, industry, states, and federal agencies through news articles, congressional task forces and reports, and speakers at national conferences. The constantly changing nature of issues, the multitude of new chemical listings, the frequent new policy developments, and the breadth of the services and other DoD component activities related to emerging contaminants require an extraordinary level of internal coordination. Information needs to be shared in a way that leverages internal resources and reduces duplication of effort. Similarly, effective outreach to the public and a continual dialogue with regulators and the technical community are necessary to establish avenues to exchange information among all stakeholders. Effective outreach and communication practices lead to many benefits, such as:

- Stakeholder education
- Greater awareness of real risks
- Full agreement on how to address a risk
- Consumer and stakeholder participation in decision-making
- Understanding and acceptance of decisions
- Consumer confidence

LESSONS LEARNED

The perchlorate experience illustrates the challenges DoD faces during the time a chemical is identified as a potential concern but is not yet regulated. As a consequence of vocal concern over contamination of public drinking water supplies by perchlorate, DoD was initially targeted as the principal source for all perchlorate contamination and thus became the focus of efforts to secure remedial actions. Communities and states quickly became dissatisfied with DoD's ability to respond to their concerns.

All aspects of the challenges that emerging contaminants present will require sound and creative technical attention. New analytical methods, field testing and sampling techniques, new tools for risk assessment, new methods for risk management, cleanup, recycling, pollution prevention, and environmental management are just the few of the areas that must be examined.

DoD faces increasing pressures to address substances used in the conduct of mission that are not currently subject to regulatory controls or are currently regulated but under consideration for more restrictive standards. When these substances are identified as potential contaminants or are being reviewed for stricter regulations, DoD faces challenges in determining how to direct its assets. DoD has a clear responsibility to protect public health and the environment, but it is often difficult to determine how to appropriately commit resources and initiate responses prior to the completion of the regulatory process.

Regulations developed to protect environmental and public health are based on peer-reviewed science and should be developed in a transparent manner that allows input by those agencies and individuals that will be affected by the regulation. This is necessary to ensure real public benefits are achieved that do not adversely affect DoD's ability to protect national security or require unnecessary expenditures. Oversight to ensure compliance with the requirements of the Information Quality Act, Administrative Procedure Act (APA), environmental statutes, and applicable Executive Orders should do much to improve the quality and overall time for development of regulatory standards.[14,15]

In the case of perchlorate, EPA's draft risk assessment suggested an RfD for perchlorate of 0.00003 mg/kg.[16] Although EPA released the draft RfD as a "do not quote or cite" document, the public posting nevertheless led state and federal regulatory offices to use the proposed RfD as the basis for their guidance and practice. These events call into question the practice of posting draft risk assessments prior to completion of independent peer review. Doing so risks misdirecting private and taxpayer dollars and may generate unsubstantiated concern in the general public.

Once good science is assured, transparent policy development will include full evaluation of public health benefits and can better sustain natural infrastructure as well as assess operational and financial risk. Scientific studies and analyses that comprise risk assessments should be clearly separated from the risk management procedures that take into account regulatory impact analyses. The following ideas are offered for further dialogue and consideration to assist agencies in moving forward to address emerging contaminants:

- Communicate regularly and often with established federal points of contact.

- Implement transparent processes.

- Use peer-reviewed science.

- Disclose science, scientific rationale, and processes used, as well as studies not used, to develop RfDs.

- Ensure peer review of proposed RfDs.

- Submit draft risk assessments to all interested federal agencies for interagency review and comment, prior to public release.

- Quickly elevate unresolved issues.

- Seek earlier engagement of all agencies in regulatory agenda.

- Discontinue public posting of draft Reference Doses or Cancer Slope Factors (RfD/SF) prior to peer review.

- Publish a separate risk management document to accompany the RfD/SF.

- Develop internal guidance on risk assessment and management in the absence of formal standards.

CONCLUSIONS

DoD experienced threats to range sustainability and effective military training, decreased access to resources, and potential increases in weapon system costs – all from a draft risk assessment that was the subject of significant scientific debate. The perchlorate experience highlights the need for process reform to ensure that agency actions on emerging contaminants, achieve real public benefits without unintended adverse impacts on national security.

Historical compliance-based risk management strategies have proven insufficient to balance compliance with operational mission requirements. A reactive approach that limits itself to complying with regulations and standards after they are developed leaves little to no room for nimbly addressing new, unregulated contaminants and may result in financial and operational costs that are not supported by commensurate human health or environmental benefit. Significant changes in risk assessment policy and practice, including the strategic investment and use of chemical and site risk assessments, are required to transform a regulatory compliance-based environmental and occupational health management strategy into an operational capability/asset management strategy. DoD's risk assessment policies require review and revision to ensure that DoD programs and service components cost-effectively assess the life-cycle risks posed by emerging chemicals to operational requirements. Such approaches can lead to both short- and long-term decreases in DoD's costs of operating responsibly with respect to environment, safety and occupational health and to minimize cleanup costs.

REFERENCES

1. Solley, W. B. Estimates of water use in the western United States in 1990 and water-use trends 1960-90: report to the Western Water Policy Review Advisory Commission, 1997. http://hdl.handle.net/1928/369

2. United States Bureau of Reclamation. http://www.usbr.gov/uc/crsp/charts/displaysites.jsp

3. Gutzler, D. S., Drought in New Mexico: history, causes, and future prospects. Decision-Makers Field Guide 2003. http://repository.unm.edu/bittream/1928/290/1/gutzler decisionmakers2003.pdf

4. Scoderi, L. A. Drought periodicity over the past 2000 years in the American southwest. Proceedings Geological Society of America XVI International Union for Quaternary Research Congress Poster 74-1, 2003.

5. Landers, J. 'Climate change to alter California's water supplies,' study says. *Civil Engineering* 72(8): 16-17, 2002.

6. Weare, B. C. 2002. "Global climate change will affect air, water in California." *California Agriculture* 56(3): 89-96.

7. Layperson's Guide to California Water. Water Education Foundation, 2003. http://www.water-ed.org/store/itemdetail.asp?id=

8. U.S. Environmental Protection Agency integrated /risk information system IRIS database for risk assessment. Glossary of IRIS terms. http://www.epa.gov/iris/gloss8.htm

9. Guidance for Fiscal Years 2006 – 2011 Sustainable Ranges Programs. June 26, 2003.

10. Department of Defense Directive 4715.1E. Environment, Safety, and Occupational Health (ESOH). March 19, 2005.

11. Strategic Environmental Research and Development Program CP-1429. Evaluation of Alternative Causes of Widespread, Low Concentration Perchlorate Impacts to Groundwater. 2005. http://www.serdp.org/research/Compliance.html

12. Strategic Environmental Research and Development Program CP-1403. Synthesis, Evaluation, and Formulation Studies in New Oxidizers as Alternatives to Ammonium Perchlorate in DoD Missile Propulsion Applications. 2004. http://www.serdp.org/research/Prevention.html

13. Strategic Environmental Research and Development Program CP-1404. Robust, Perchlorate-Free Propellants with Reduced Pollution. 2004. http://www.serdp.org/research/Prevention.html

14. Information Quality Act, Pub. L. No 106-544, §315. Administrative Procedure Act 5 U.S.C. § 551*et seq.*

15. Executive Order 12866 Regulatory Planning and Review, September 30, 1993.

16. United States Environmental Protection Agency. Perchlorate Environmental Contamination: Toxicological Review and Risk Characterization. Washington, DC 20460, January 16, 2002. http://cfpub.epa.gov/ncea/cfm/recordisplay.cfm

Chapter 2

The Chemistry of Perchlorate in the Environment

Gilbert M. Brown and Baohua Gu

Chemical Sciences and Environmental Sciences Divisions
Oak Ridge National Laboratory, Oak Ridge, TN 37831

INTRODUCTION

Current interest in the chemistry of perchlorate is primarily due to its presence as a contaminant in groundwater and drinking water. The negatively charged perchlorate ion is composed of one chlorine atom surrounded by four oxygen atoms arranged in tetrahedral geometry. The perchlorate anion (ClO_4^-) commonly originates as a contaminant in the environment from the improper disposal of solid salts of ammonium, potassium, or sodium perchlorate[1]. These salts are very soluble in water, and the perchlorate ion is kinetically inert to reduction and has little tendency to adsorb to mineral or organic surfaces. Therefore the perchlorate anion persists in groundwater, and its mobility in surface or groundwater is so high that perchlorate essentially moves with the flow of water (diffusion and convection controlled movement).

Compounds containing perchlorate include the oxidant in solid rocket fuel as well as that in fireworks, military ordinance, flares, airbags, and other applications where an energetic oxidant is required. Ammonium perchlorate is among the most important propellants because it has a high oxygen content and decomposes to the gaseous phases products water, HCl, N_2, and O_2, leaving no residue. Salts of perchlorate do not function well in solid-fueled rockets after the salts adsorb too much water, and improper disposal has led to environmental contamination. A report on the occurrence of perchlorate in drinking water by the American Water Works Association[2] indicates that perchlorate contamination is a national problem with significant concentrations being found in 26 states and Puerto Rico.

The presence of perchlorate in groundwater and drinking water is a potential health concern because perchlorate can impair proper functioning of the thyroid gland.[3] The thyroid gland produces hormones that are maintained within narrow concentration limits by an efficient regulatory mechanism. Hormones required for normal development of the central nervous system of fetuses and infants are secreted by the thyroid gland. These hormones are required for normal skeletal growth and development. In both infants and adults, thyroid gland hormones determine metabolic activity and affect many organ systems. Iodine is a component of the thyroid hormones, and one of the functions of the thyroid gland is to control iodide levels in the bloodstream. Perchlorate competitively inhibits the uptake of iodide ions by the thyroid; in fact, potassium perchlorate has been used to treat hyperthyroidism. In the discussion of the physical properties of the perchlorate ion that follow, it will be seen that the competitive inhibition is a consequence of the hydration energies of the two ions. A recent report by the National Research Council[3] concluded that the levels of perchlorate that have been found in groundwater are unlikely to affect a healthy adult. However, the situation may be different for infants, children, pregnant women, and people with pre-existing thyroid disorders. The National Research Council suggested a reference dose of 0.7 µg/kg of body weight, which is an order of magnitude lower than the dose with no observed effect on iodine uptake by the thyroid. The committee concluded that this reference dose would protect the health of even the most sensitive populations. This reference dose is elevated from the provisional reference dose of 0.1 µg/kg suggested by the U.S. EPA in 2002. This latter value prompted the State of California to set a maximum contaminant level (MCL) of 6 µg/L for perchlorate in drinking water. The higher reference dose translates to an MCL for perchlorate in drinking water of 24.5 µg/L.

The environmental chemistry community has an interest in the chemical and physical properties of perchlorate for the purpose of developing analytical methods, separation schemes, and destruction technologies. A number of reviews[4,5,6,7] of perchlorate chemistry from this perspective have been published in recent years as well as the proceedings of a symposium.[8] A recent review paper[9] discusses bioremediation approaches to perchlorate contaminated waters. This chapter will not discuss bioremediation other than to comment on the mechanism of reduction.

REDOX PROPERTIES OF CHLORINE COMPOUNDS

Perchloric acid is the highest oxidation state of the element chlorine. A Latimer diagram[10] for chlorine in acid solution (Figure 1) is a convenient way to summarize the redox potentials relating stable compounds of chlorine. The standard reduction potentials[11] in Figure 1 refer to 1 M acid solution at 25°C. The convention is that if a reaction combined from one or more half-reactions of complementary electron count has a positive change in reduction potential (ΔE), then the reaction as written is thermodynamically favorable.

The Gibbs free energy change for a combination of half-cell reactions at 25°C can be computed from the equation

$$\Delta G = -nF\Delta E$$

where ΔG is the Gibbs free energy change, n is the number of electrochemical equivalents in the half-cell reaction, and F is the Faraday constant equal to 9.6485×10^4 J V^{-1} mol^{-1}.

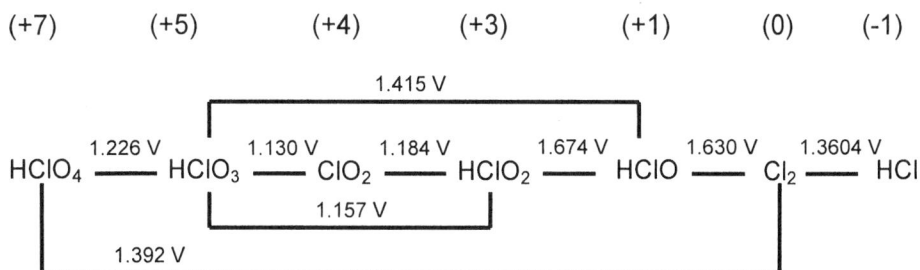

Figure 1. Latimer diagram giving the standard reduction potentials of the stable species of chlorine in aqueous acid solution at 25°C. The formal oxidation state of chlorine is given by the Arabic numeral above each species.

The Latimer diagram is particularly convenient for assessing the thermodynamic stability of individual species. If the potential to the right of a species shown in Figure 1 is more positive than the potential to the left, then the species is unstable to disproportionation. The perchlorate anion is formally in the +7 oxidation state of chlorine in which all valence electrons of chlorine have been lost. A striking observation is that all of the higher oxidation states of chlorine are relatively strong oxidizing agents with respect to the chloride ion.

The half-reactions relating perchlorate (ClO_4^-), chlorate (ClO_3^-), chlorite (ClO_2^-), hypochlorite (ClO^-), and chloride (Cl^-) can be combined to derive the thermodynamic potential for the eight-electron reduction of perchlorate to chloride

$$ClO_4^- + 8\,e^- + 8\,H^+ \rightarrow Cl^- + 4\,H_2O \qquad E = +1.388 \text{ V}$$

In comparison to the potential for oxidation of water to oxygen,[11]

$$O_2 + 4\,e^- + 4\,H^+ \rightarrow 2\,H_2O \qquad E = +1.272 \text{ V}$$

it is apparent that perchlorate is unstable with respect to the evolution of oxygen from water. Perchlorate reduction to chloride by this reaction does not occur spontaneously at ambient temperatures. In fact, concentrated perchloric acid is stable for many years. One of the authors has a bottle of high purity 72% perchloric acid that has lost none of its acid titer nor built up any chloride ion in a period of around 25 years. The failure of this reaction to occur spontaneously illustrates that reactions involving oxyanions of chlorine require oxygen atom transfer as well as electron transfer. For this process to occur, a mechanism must be available.

Frequently, the species of chlorine dissolved in solution are not in equilibrium, although some do exist in a rapidly established equilibrium. The reactions of HClO, Cl_2, and Cl^- occur rapidly, and in acid solution HClO reacts with Cl^- to produce Cl_2. Users of household bleach will realize that in a neutral or slightly alkaline solution, however, the situation is reversed and Cl_2 is the unstable species with respect to OCl^- and Cl^-.

Perchloric acid has a long history in analytical chemistry as an oxidant for destroying organic material by fuming at elevated temperature.[12] As will be noted later in this chapter, the active oxidizing agent in these reactions is molecular perchloric acid, not the perchlorate anion. The methodology for wet-ashing of organic materials for the determination of trace levels of metals is well-established.

The prevalence of perchlorate in drinking water has led to speculation that chlorine used to disinfect water is the source of perchlorate. A combination of individual half-reactions indicates that the disproportionation of Cl_2 to ClO_4^- and Cl^- is not favorable in acid solution. Gaseous ClO_2 has also been used as a disinfectant for water.[13] The Latimer diagram indicates that this species is unstable to the formation of chlorate and chlorite in solution. The formation of perchlorate by disproportionation of ClO_2, driven by the formation of HClO, is a thermodynamically favorable reaction:

$$2\ ClO_2\ +\ H_2O\ \rightarrow\ HClO_4\ +\ HClO \qquad \Delta G = -91.7\ kJ\ mol^{-1}.$$

However, even though this reaction is thermodynamically favorable, there is no clear evidence in the literature that the reaction occurs. A simple bimolecular reaction involving the transfer of two oxygen atoms has not been observed, and the mechanism of a process such as this will require intermediates that are not well understood. Since there is a fairly good database on the occurrence of perchlorate in drinking water,[2] it would be interesting to correlate reported perchlorate levels with the method of disinfection.

A draft report by the Massachusetts Department of Environmental Protection[14] presented evidence that perchlorate is spontaneously generated in both commercial strength (15%) hypochlorite solutions and in household bleach. The suggestion was made that sodium hypochlorite spontaneously decomposes to chlorate by the reaction

$$3 \text{ NaOCl}^- \rightarrow \text{ NaClO}_3^- + 2 \text{ NaCl}^- \text{ ,}$$

which is not a thermodynamically favorable reaction in acid solution, but which is thermodynamically favored in alkaline solution (ΔG = -159.8 kJ mol^{-1}). The spontaneous decomposition of chlorate to form perchlorate was claimed, i.e.,

$$4 \text{ NaClO}_3 \rightarrow 3 \text{ NaClO}_4 + \text{ NaCl} \text{ .}$$

This is of potential significance since 15% commercial bleach solutions are used by some municipalities for disinfection of drinking water. Although hypochlorite is made by an electrochemical process, commercial bleach solutions are made by bubbling chlorine gas into a NaOH solution, precluding the introduction of perchlorate as an electrochemical reaction byproduct. As previously noted, just because a reaction is thermodynamically favorable does not mean it occurs fast enough to be significant at ambient temperatures. A kinetic pathway with an activation energy accessible at ambient temperature is necessary, and the formation of perchlorate by this route seems dubious.

Salts of chlorate have been used as defoliants, leading to speculation that these could be sources of perchlorate in groundwater.[15] From the Latimer diagram in Figure 1, it is immediately apparent that the disproportionation of chlorate to perchlorate and chlorite in acid solution is not thermodynamically favorable. However, the potential for the HClO$_2$/HClO redox couple is quite positive, which provides the driving force for the net reaction

$$3 \text{ HClO}_3 \rightarrow 2 \text{ HClO}_4 + \text{ HClO} \qquad \Delta G = -73.1 \text{ kJ mol}^{-1}$$

to be thermodynamically favorable. The Latimer diagram indicates that chlorite is unstable to chlorate and hypochlorite in acid solution. However, chlorite is sufficiently stable kinetically that it is a readily identified and well-characterized species. The spontaneous formation of perchlorate and hypochlorite from chlorite,

$$3 \text{ HClO}_2 \rightarrow \text{ HClO}_4 + 2 \text{ HClO} \qquad \Delta G = -191.4 \text{ kJ mol}^{-1},$$

is thermodynamically favorable in acid. The mechanism (or even a detailed sequence of reactions) by which either of these reactions might occur is unknown. Although chlorate may not disproportionate to produce perchlorate at ambient temperatures, perchlorate could be present in chlorate as a byproduct from manufacturing, as will be discussed later.

GENERAL PHYSICAL AND CHEMICAL PROPERTIES OF PERCHLORIC ACID AND THE PERCHLORATE ANION

Perchloric acid is among the strongest acids, and this species is completely dissociated in water. Simply stated, an acid is a proton donor and a base is a proton acceptor (Bronsted theory). The perchlorate anion is the conjugate base of perchloric acid. Within this framework, acid strength may be defined as the tendency to give up a proton. Acid-base reactions occur when an acid reacts with the conjugate base of a weaker acid to transfer a proton. When acid-base reactions occur between acids of comparable strength, the reaction proceeds to a measurable equilibrium that allows the relative strength of the two acids to be measured. In Table 1, relative strengths of a series of inorganic acids are given. For the very strong acids, pK_a values are approximate but relative strengths of the various acids are accurately known. The conjugate base of perchloric acid, the ClO_4^- anion, has its negative charge distributed equally over four oxygen atoms, and this delocalization of the conjugate base contributes to the increased acidity of perchloric acid. Perchloric acid occupies a unique position in the development of

Table 1. Inorganic Acid pK_a Values.[a]

Acid	Conjugate Base	Approximate pK_a[b]
FSO_3H	FSO_3^-	<-12
$HClO_4$	ClO_4^-	-10
HI	I^-	-10
H_2SO_4	HSO_4^-	-10
HBr	Br^-	-9
HCl	Cl^-	-7
H_3O^+	H_2O	-1.74
HNO_3	NO_3^-	-1.4
HSO_4^-	SO_4^{-2}	1.99
HF	F^-	3.17

[a]Data from Smith, M. B. and March, J. *March's Advanced Organic Chemistry: Reactions, Mechanism, and Structure,* 5th Ed. New York: John Wiley and Sons, 2001, p 329.

[b]Values relative to water.

superacids.[16] This designation was given to a solution of perchloric acid in glacial acetic acid. Concentrated perchloric acid is commercially available as a 70 to 72% solution, which corresponds to $ClO_4 \cdot 2H_2O$. Anhydrous perchloric acid can be prepared by vacuum distillation of concentrated acid in the presence of a dehydrating agent such as $Mg(ClO_4)_2$. However, the anhydrous acid is stable only for a few days at low temperature, decomposing to give $HClO_4 \cdot H_2O$ (84.6% acid) and ClO_2.[16] Perchloric acid in the absence of sufficient water is a very powerful oxidizing agent and a potent dehydrating agent. Contact between organic material and anhydrous or concentrated perchloric acid can produce explosions.[12]

Inorganic chemists, studying metal ions in solution, and electrochemists have long prized perchlorate salts and perchloric acid for use as an inert electrolyte because perchlorate is generally a non-complexing anion that is a poor nucleophile and is also kinetically inert to oxidation and reduction. The poor coordinating ability and low nucleophilicity are consistent with the delocalization of the negative charge of the perchlorate anion over four oxygen atoms. Although perchlorate complexes of metal ions are known, the stability constants are quite low. Because of its poor coordinating ability, perchlorate is widely used as a counter-ion in studies of metal ion complexation. It is used to adjust the ionic strength in kinetic studies involving transition metals where other electrolytes might complex the metal ion.[4]

The ClO_4^- ion is kinetically inert to exchange of oxygen atoms with water. The exchange of ^{18}O-labeled water with $HClO_4$ of normal isotopic composition at a concentration of 9 mol/L is insignificant after 63 days at 100°C. Based on these measurements, the half-life for exchange at room temperature is estimated to be greater than 100 years.[17] By analogy to the other oxyanions of chlorine, the rate of exchange is anticipated to be at least first order in acid. Thus, the rate of exchange in neutral groundwater is expected to be negligible. This is anticipated because exchange can occur only by expansion of the coordination sphere of chlorine, a pathway expected to have a very high activation energy, or by protonation of the perchlorate anion, an unlikely possibility given the acidity of perchloric acid.

The hydration energy of perchlorate anions in water is among the lowest of common inorganic anions, a fact that has significant consequences for the separation and analysis of perchlorate. The hydration energy of an anion is the free energy for transfer of the unsolvated gas-phase anion to an aqueous solution, i.e.,

$$X_g^- \rightarrow X_{aq}^- .$$

The standard free energy for this process correlates remarkably well with the Born equation, which gives the energy to charge a sphere of radius R in a medium of dielectric constant ε as

$$\Delta G^{\circ}_{Born} = \frac{Bz^2}{R(1-1/\varepsilon)}$$

where z is the charge on the ion and B is a constant equal to -69.47 kJ nm mol^{-1} at 25°C.[18] Although this equation represents the zeroth-order approximation to the computation of hydration energy, previous workers have empirically improved the fit by adjusting the radius by a constant and adding a small correction term. For singly-charged ions such as perchlorate, the hydration energy is predicted to decrease as the size of the ion increases. This trend is shown very clearly in the data in Table 2 for a series of inorganic anions commonly found in groundwater.

Table 2. Hydration Energies of Common Inorganic Anions[a]

Anion, X$^-$	Thermochemical radius, nm	ΔG_h obs, kJ mol^{-1}
F$^-$	0.126	-465
OH$^-$	0.133	-430
HCO$_3^-$	0.156	-335
Cl$^-$	0.172	-340
Br$^-$	0.188	-315
NO$_3^-$	0.196	-300
H$_2$PO$_4^-$	0.200	-465
I$^-$	0.210	-275
ClO$_4^-$	0.240	-205
CO$_3^{-2}$	0.178	-1315
SO$_4^{2-}$	0.230	-1295
PO$_4^{3-}$	0.238	-2765

[a]Data from Moyer and Bonnesen.[18]

One of the most effective processes for the removal of perchlorate contamination from ground and surface waters is sorption on solid phase anion exchange resins. Commercially available anion exchange resins typically are based on cross-linked polystyrene or polyacrylate beads with quarternary ammonium exchange sites. The process of transferring an anion from aqueous solution to the organic resin surface can be understood by an electrostatic model of anion transfer, which is discussed in some detail in Chapter 10 and by Diamond and Whitley[19] and by Moyer and Bonnesen.[18] Perchlorate is more strongly sorbed on these organic resins than other anions commonly found in groundwater. For singly-charged anions, this is the expected result based on the hydration energies. A generalization is that anions with small hydration energies are selectively transferred from an aqueous to a nonaqueous phase. Optimizing resin design for the selective sorption of perchlorate from groundwater is discussed in Chapter 10.

The thyroid gland in mammals appears to function as an ion exchanger with a preference for large, poorly-hydrated anions. It seems clear from the relative hydration energies of perchlorate, iodide ion, and pertechnetate and perrhenate anions (TcO_4^- and ReO_4^-, respectively, which have hydration energies very close to that of perchlorate), that these anions have common characteristics in this regard. Radioactive perrhenate and pertechnetate preferentially concentrate in the thyroid, and these anions are similar in size and charge density to perchlorate and iodide. The thyroid is not the only naturally occurring anion exchanger known. Others are noted in the section on perchlorate fate in the environment.

Another consequence of perchlorate's low hydration energy is the ease with which this ion can be separated from more hydrophilic ions for analysis by ion chromatography. Ion chromatography with detection by conductivity change is destined to be the workhorse for potable water analysis in the US for the foreseeable future.[6] It is a fair statement that the discovery of groundwater contamination with low concentrations of perchlorate can be directly attributed to developments in ion chromatography. Optimization of the sorbant phase is obviously important, but in general, more hydrophilic ions are eluted before less hydrophilic ions.

INDUSTRIAL PREPARATION OF PERCHLORATE

Perchlorates are produced in a two-step electrochemical process in which sodium chloride is first oxidized to sodium chlorate, and then sodium chlorate is oxidized to sodium perchlorate.[20] The potential for oxidation of chlorate to perchlorate

$$ClO_4^- + 2e^- + 2H^+ \rightarrow ClO_3^- + H_2O \quad E = +1.226V$$

is close to the potential for oxidation of water to oxygen

$$O_2 + 4 e^- + 4 H^+ \rightarrow 2 H_2 O \qquad E = +1.272 \text{ V} .$$

The production process is operated at high voltage in an undivided cell with an anode of either PbO_2 or Pt-coated material. The cell tank is operated as the cathode, which is typically bronze or 316 stainless steel. Sodium dichromate is added to the cell to prevent undesirable reduction of chlorate or perchlorate. Despite the potentially competitive evolution of oxygen, the efficiency at a Pt anode is 90 to 97%. The production of perchlorate (or even the capacity for production) in the United States is difficult to estimate because perchlorates are classified as strategic materials. Immediately following World War II, capacity was thought to be about 18,000 tons per year.

In contrast, the capacity for production of chlorate, an intermediate in the production of perchlorate is well known. For a number of years, the Electrochemical Society has published an annual report on the electrolytic industries, and sodium chlorate production is a major category. The two most recent reports are cited here.[21,22] Sodium chlorate is produced by electrolysis of an acidic brine solution followed by crystallization to isolate the product. Over 95% of the sodium chlorate produced is used in the pulp and paper industry where it is converted to ClO_2 for use as a bleaching agent. The pulp and paper industry is under environmental pressure to convert to elemental-chlorine-free processes. The capacity for sodium chlorate in North America was 2.1 million tons in 2000, and the world capacity in the same year was estimated to be 3 million tons. For use in the paper industry, sodium chlorate is converted to ClO_2 by reduction with hydrogen peroxide in the presence of sulfuric acid. The similarities in potential for oxidation of brine to chlorate or to perchlorate have led to speculation that commercial grade sodium chlorate is contaminated with perchlorate.[15] It has been reported that laboratory grade sodium chlorate contains 0.2% perchlorate on a weight basis, while the analytical reagent grade chemical has 200 to 900 ppm.[23] Specifications given in authoritative reference works[24] indicate that technical grade sodium chlorate is 99.5% pure, and the American Chemical Society analytical reagent specifications do not mention perchlorate as a significant impurity in sodium chlorate. It seems unlikely that perchlorate is present at more than a hundred ppm or so even in technical grade material. Nonetheless, given the large quantities of sodium chlorate used annually by the pulp and paper industry, sodium chlorate cannot be ignored as a possible source of perchlorate contamination in the environment. Salts of chlorate are used as defoliants, and this is also a possible source for the introduction of perchlorate into the environment.

CHEMICAL REDUCTION OF PERCHLORATE

As mentioned previously, perchlorate reduction in aqueous solution is strongly favored thermodynamically, i.e.,

$$ClO_4^- + 2e^- + 2H^+ \rightarrow ClO_3^- + H_2O \qquad E = +1.226V,$$

yet the kinetics of the reduction reactions indicate the rates of these processes are inhibited. Earley,[25] Espenson[26] and others[27,28] have commented on the unusual lack of reactivity of perchlorate. Chemical reactions that lead to the reduction of perchlorate are of great interest to those seeking economical and efficient methods for the remediation of contaminated groundwater. Perchlorate does not react at an appreciable rate at ambient temperature with reducing agents commonly used in environmental remediation such as sulfite, dithionite, or zero-valent iron. Because of the large kinetic barriers, chemical reduction of perchlorate engenders great academic interest as well. Of notable scientific interest is the fact that some weaker reducing agents react with perchlorate at measurable rates, whereas stronger reducing agents fail to react at all. The fundamental issue is can the reaction rate be predicted, even if only to the precision of several orders of magnitude, from the properties of the reactants.

Taube[28] has noted that there are no examples of the removal of oxygen from perchlorate by typical nucleophilic reducing agents, and all of the well-characterized examples of perchlorate reduction under ambient conditions involve metal ions. The perchlorate anion is resistant to the solvated electron in both water and ammonia. This stability in the presence of such a strong and kinetically active reducing agent is understandable because perchlorate has no low-lying unfilled molecular orbitals, and thus it cannot readily take an electron to form the ClO_4^{-2} ion. Considering again the nucleophile as a reducing agent, the electrophile in this case is the chlorine center of perchlorate. For reaction with a nucleophilic reducing agent, the coordination environment of the tetrahedrally bound chlorine center must expand to five-coordination, and this seems to be an extremely high energy intermediate. However, perchlorate can be reduced if the reductant can accept or stabilize an O^{2-} group as a part of the reaction. Most evidence suggests that perchlorate is reduced only when atom transfer accompanies a change in the oxidation state of chlorine, i.e., when an oxygen atom is transferred to the reducing agent in a mechanism involving bonding of both chlorine and the reducing species to the transferring oxygen atom in the activated complex. In the case of perchlorate reduction, atom transfer can occur in both one- and two-electron processes as

$$ClO_4^- + e^- \rightarrow ClO_3^- + O^{2-}$$

$$ClO_4^- + 2 e^- \rightarrow ClO_3^- + O^{2-} .$$

There are well-defined examples of each process that have been described in the literature. Fairly detailed kinetic studies have been conducted involving the reduction of perchlorate by aquo ions of Ti(III),[29,30] Ru(II),[31] V(II), and V(III).[32] Detailed studies of reactions between perchlorate and complex metal ions, including an HEDTA chelate of Ti(III)[33] and the Ru(II) complex $Ru(NH_3)_5(H_2O)^{2+}$, have also been conducted.[34] Among the aquo metal ions, the order of reactivity is Ru(II) > Ti(III) > V(III) ≈ V(II) > Cr(II) ≈ Eu(II) ≈ 0. The aquo ions of Cr(II) and Eu(II) do not appear to be reduced by ClO_4^- at measurable rates (k <10^{-8} M^{-1} s^{-1} at 25°C), despite the fact that both ions are thermodynamically powerful reducing agents. These facts led Taube to postulate that metal ions for which the transferring oxygen atom is stabilized by multiple bond formation with the metal ion − that is by formation of an "yl" complex* − are the most kinetically active. For certain metal ions there is an abrupt change in the acidity of water bound to the metal ion with an increase in oxidation state. For metal ions that form "yl" complexes in higher oxidation states, the disassociation constant of the protons bound to water are such that both are lost, and a strong multiple bond between oxygen and the metal ion forms. It is this bonding that influences the rate of perchlorate reduction. The aquo ions of Cr(III), Eu(III), and V(III) do not form "yl" complexes at normal pH values, but Ti(IV), V(IV), and Ru(IV) are thought to exist either predominantly in this form or within a readily established equilibrium.

The reaction of the Ti(III) aquo ion with perchlorate is a particularly well-defined example of a one-electron redox process since titanium exists only in oxidation states (III) and (IV) in aqueous solution. The rate expression for this reaction is given by

$$\text{rate} = -\frac{d[Ti(III)]}{dt} = (k + k'[H^+])[Ti(III)][ClO_4^-]$$

where k = 1.9 x 10^{-4} $M^{-1}s^{-1}$ and k' = 1.25 x 10^{-4} $M^{-2}s^{-1}$. Taube[28] suggested that the acid dependent term in the rate expression may be spurious due to the use of Na^+ ion rather than Li^+ to replace H^+ while maintaining constant ionic strength. The reduction of perchlorate by Ru(II) could occur by either a one-electron or a two-electron redox process. The two-electron pathway gives Ru(IV) as an intermediate, which would react rapidly in a subsequent step with Ru(II) to give the observed Ru(III) product. Other complexes of

---

* yl complexes are M=O complexes such as $V(IV)O^{2+}$, referred to as the vanadyl ion.

Ru(IV) are well-characterized as oxo complexes with multiple bonds between the metal ion and oxygen. It is interesting that both V(II) and V(III) react with perchlorate at roughly the same rate. If the product of both reactions is the V(IV) oxo complex (vanadyl ion), then at first blush it would seem that the V(II) reductant, which is the stronger reducing agent of the two, would react faster in a two-electron process. However, experimental evidence suggests the two-electron pathway offers no advantage over the one-electron process (The first pK_a of V(III) is roughly three,[35] and the formation of an oxo complex under reaction conditions is precluded).

To put these reactions in perspective with regard to processing perchlorate-contaminated groundwater, Espenson[26] tabulated half-lives for the reaction of 10^{-3} M ClO_4^- with aqua metal ions at 25°C. The most rapidly reacting aqua metal ions among those discussed above are Ru(II) and Ti(III), which have calculated half-lives of 87 and 1470 hr, respectively. Thus, use of these metal ions in a direct or catalyzed process does not seem to offer a feasible method for treating groundwater even moderately contaminated with perchlorate.

A very clear example of a two-electron redox process has been reported by Espenson and coworkers,[36,37] The unusual organometallic oxide methyltrioxorhenium(VII), CH_3ReO_3, has been extensively investigated by Espenson and his group,[38,39] mostly as a catalyst for the oxidation of organic substrates by hydrogen peroxide. This compound and its reduced form, methylrhenium(V) dioxide, undergo facile oxygen atom transfer reactions. The reduced Re(V) form has been shown to be a strong two-electron reducing agent able to abstract an oxygen atom from many substrates including amine oxides, sulfoxides, epoxides, halogenated oxyanions (XO_n^- including perchlorate), and certain metal-oxo species such as VO^{2+}. The coordination geometry of methyltrioxorhenium is tetrahedral, but the exact composition of methylrhenium dioxide is unknown. In aqueous solution, two water molecules appear to be bound to the Re center, and the structure is possibly that shown in Figure 2. Although the oxidation state of rhenium formally changes during these reactions, Espenson and coworkers[38] characterize the reactions as atom transfer reactions rather than electron transfer reactions.

The stoichiometry for the net reaction of perchlorate with methylrhenium dioxide is given by the equation

$$4\,CH_3ReO_2 \;+\; ClO_4^- \;\rightarrow\; 4\,CH_3ReO_3 \;+\; Cl^-.$$

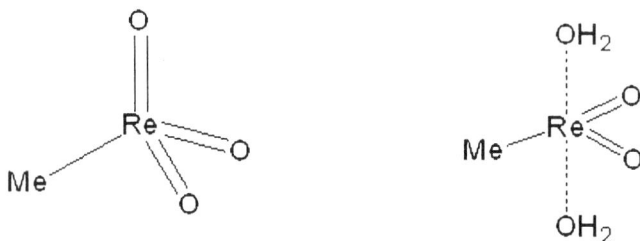

Figure 2. Structure of methylrhenium(VII) trioxide (left) and methylrhenium(V) dioxide (right).

An independent kinetic determination has shown that chlorate reacts with the Re(V) species nearly 10^4 times faster than with perchlorate. The eight-electron reduction of perchlorate is postulated to occur by a series of four step-wise two-electron reduction processes. The rate expression for the reaction of methylrhenium dioxide with perchlorate is deceptively simple

$$\text{rate} = -\frac{d[\text{ClO}_4^-]}{dt} = k[\text{CH}_3\text{ReO}_2][\text{ClO}_4^-]$$

where $k = 7.2$ $M^{-1}s^{-1}$ and the first of four sequential reactions is the rate-determining two-electron transfer step. This Re(V) species reacts with perchlorate more than a thousand times faster than any other reducing agent observed to date. Espenson[26] relates this to practical applications by noting that the half-life of 10^{-3} M perchlorate in the presence of this Re(V) species has been reduced from hours for the aquo metal ions to a few minutes.

Abu-Omar and coworkers[40,41] have studied the catalytic activity of Re(V) complexes in perchlorate reduction reactions. They used inexpensive reducing agents that might be useable in processing contaminated groundwater. They also investigated the catalytic activity of rhenium coordinated by oxazoline, particularly as applied to the net reaction between perchlorate and organic thioethers. The Re complex acts as a catalyst by oxygen atom transfer from a Re(VII) oxo complex to the organic thioether, producing the active Re(V) species. This complex reduces perchlorate by an oxygen atom transfer process that regenerates the Re(VII) species. Abu-Omar[41] indicates that oxazoline complexes of Re are superior to the methylrhenium trioxide – methylrhenium dioxide catalyst pair for perchlorate reduction because a stronger reducing agent is required to reduce the methyl complex to Re(V) and furthermore, that reaction becomes rate-limiting in the catalytic process. Methylrhenium dioxide also undergoes a polymerization process that renders it inactive.

A preponderance of the evidence suggests that reduction of perchlorate proceeds by an atom transfer reaction in which the transferring oxygen atom is bound to both chlorine and the reducing center in the activated complex. The only well-documented reactions occur at metal centers. Evidence for both one-electron and two-electron mechanisms has been presented. Taube[28] has suggested that rate enhancements occur when the oxidized product is stabilized by multiple bonding between the transferred oxo group and the metal center (formation of an –yl complex). Earley[25,33] has argued that the rate-determining factor is interaction of polarizable d-orbitals of the metal ion with a chlorine-centered orbital. Reactants in which the symmetry of the orbital donating the electron is appropriate to overlap with the accepting p-orbital (metal t_{2g} orbitals) will show a rate enhancement. Earley[25,33] points out that effective interaction of the orbitals will not occur unless perchlorate begins to distort in the activated complex in the direction of a more planar geometry, thereby lowering the energy of these orbitals. Therefore, distortion of perchlorate from strict tetrahedral geometry is a precondition for rapid reduction. The ideas of Taube and Earley are not particularly at odds for the identification of reactive metal ions, since those which form stable –yl ions are the same that Earley identified. The acid-dependence of the kinetics of perchlorate reduction by the Ti(III)HEDTA complex is somewhat unusual because it is second-order with respect to the concentration of acid. Liu, Wagner and Earley[33] offer this as a route by which perchlorate distorts from tetrahedral geometry, and they propose a mechanism in which coordinated perchlorate is protonated in the activated complex, a possibility that seems unreasonable given the weak basic character of the perchlorate anion. The kinetics for reduction of perchlorate by Re(V) complexes are completely independent of acid, and any mechanism that assumes protonation of the perchlorate anion as a precondition to reaction is untenable.

The instability of anhydrous perchloric acid and the explosive nature of ammonium perchlorate, in contrast with the kinetic inertness of perchlorate anions in aqueous solution, become understandable when the reaction pathways are considered. Anhydrous perchloric acid becomes kinetically active because a proton bound to oxygen is a legitimate way to stabilize the transfer of an oxygen atom. Ammonium perchlorate, like all ammonium salts of mineral acids, exists as molecular ammonia and perchloric acid when vaporized. Under these circumstances perchloric acid is predicted to be kinetically active.

ELECTROCHEMICAL REDUCTION OF PERCHLORATE

Electrochemical methods of effecting chemical reactions are often cited as potentially "green technologies" because the waste volume from electrochemical processes is usually minimal. Thus, direct electrochemical reduction of perchlorate and reduction via a pathway involving catalysis by a soluble species are potentially attractive routes for remediation of perchlorate-contaminated groundwater. As noted previously in this chapter, the reactivity of perchlorate toward reductants in solution is very low despite the strong thermodynamic driving force for its reduction. Direct electrochemical reduction rates for perchlorate are equally anomalous. Although perchlorate has been utilized as a non-reactive innocent electrolyte, the amount of experimental evidence for perchlorate reduction has been increasing. The literature dealing with electrochemical reduction of perchlorate was recently reviewed by Horáyni[42] and by Láng and Horáyni.[43]

Perchlorate is sufficiently electrochemically inactive in nonaqueous solvents, such that the alkali metals of corresponding alkali metal salts of perchlorates, used in large excess over the reactant of interest as a supporting electrolyte, are reduced before the perchlorate ion.[44] However, there are definitive examples of the electrochemical reduction of perchlorate in aqueous solution. The electrochemical reduction of perchlorate has been studied, using a number of electrode materials, including Pt,[45,46] WC,[47] Ir,[48,49] Re,[50] Rh,[51,52] Ru,[53] Tc,[54] Sn,[55] Co and Fe,[56] Ti[57] and Al.[58] Following Láng and Horáyni,[43] these electrochemical reactions may be classified as those occurring at noble metal electrode materials (Pt, Ir, Rh, and Ru), at non-noble electrodes (WC, Re, and Tc), and as perchlorate reduction occurring in conjunction with the corrosion of metals (Co, Fe, Ti, and Al).

The reduction of perchlorate at noble metal electrodes has been studied the most thoroughly, and until the 1990s, it was generally thought that perchlorate was not reactive at these electrodes. From a theoretical point of view, these are perhaps the most interesting results, and several generalizations can be made. Cyclic voltammetry and other transient potential control techniques indicate that current for the reduction of perchlorate is observed only at positive potentials following a negative-potential process involving hydrogen evolution. The reaction requires prior adsorption of perchlorate anion at the electrode surface. Chloride ion reduction products, as well as other anions such as sulfate, inhibit the reaction. The reduction of perchlorate at non-noble electrodes does not have a well-defined pattern. At WC electrodes, perchlorate reduction occurs along with hydrogen evolution, and the process can be described as electrochemically induced hydrogenation. Horányi and co-workers observed perchlorate at electrodeposited Tc[54] and Re[50] layers. Reduction was inhibited due to the formation of chloride ion, leading to speculation that adsorption of perchlorate is required for reduction to occur. The reduction of perchlorate at Sn[55] electrodes depends on the electrode surface properties and appears to require a surface oxide of Sn(II). The proposed mechanism involves reduction of ClO_4^- to surface-bound SnOHCl followed by reduction of this species to Sn liberating the Cl^- anion. Recent as well as older studies of the corrosion of some metals in perchlorate-containing acidic solutions indicate that the reduction of perchlorate cannot be assumed insignificant. The reduction of perchlorate occurs during the formation and breakdown of passive oxide layers on Al,[58] and Cl^- was definitively observed as a reduction product. Brown[57] observed the reduction of perchlorate during active dissolution of Ti metal. Recent observations on the corrosion of Fe and Co metals[56] in perchorate-containing solutions indicate that perchlorate reduction occurrs concurrently with metal dissolution. The open circuit potential at an actively corroding electrode occurs where the current for metal dissolution is equal to the current for hydrogen evolution. In perchlorate-containing solutions, the current for the cathodic process must include the current required for perchlorate reduction. This latter result has some implications for the processing of perchlorate-contaminated groundwater by finely-divided Fe metal. Although it appears that the reaction may be too slow to be of practical value, the fact that Fe dissolution occurs concurrently with perchlorate reduction and hydrogen evolution should be considered.

Analogous to the reduction of perchlorate in homogeneous solution, the reduction of perchlorate at an electrode (or any surface) must involve oxygen atom transfer. A mechanism for perchlorate reduction at a Ti electrode was proposed by Brown.[57] The basic principles of this mechanism were generalized by Lang and Horanyi,[43] and applied to the reduction of perchlorate at a Rh electrode. The concept of an atom transfer reaction at an electrode is not as well-defined as it is in solution, but for reactions at a normally oxygen- (or hydroxyl-) covered electrode, the first step is adsorption at an oxygen vacancy. Assume for the purpose of discussion that the surface of a generalized metal, Me, is covered with an air- and water-formed surface oxide that will be described as Me(I)OH. In further analogy with the solution chemistry, perchlorate will not react at the electrode at an appreciable rate if the surface is not capable of stabilizing the surface oxide, thus

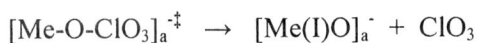

$$[Me(I)(OH)]_a + e^- + H^+ \rightarrow [Me(0)]_a + H_2O$$

$$[Me(0)]_a + ClO_4^- \rightarrow [Me\text{-}O\text{-}ClO_3]_a^{-\ddagger}$$

$$[Me\text{-}O\text{-}ClO_3]_a^{-\ddagger} \rightarrow [Me(I)O]_a^- + ClO_3$$

$$[Me(I)O]_a^- + H^+ \rightarrow [Me(I)(OH)]_a$$

where subscript "a" indicates a surface-stabilized adsorbed species, and the symbol ‡ indicates the activated complex. In this formulation, the surface species $[Me(0)]_a$ represents an oxygen vacancy. The oxidation state of the predominant surface species varies with the metal species, and adjustment of the model should be made accordingly. Several concrete predictions arise from this model that can be tested experimentally. If perchlorate reduction occurs by a one-electron mechanism as shown in the four equations above, then the potential dependence of the current (Tafel slope) is predicted to be 120 mV per decade of current. If a two-electron process is indicated, then the Tafel slope is predicted to be 60 mV per decade at 25°C. The pH-dependence of the current will be governed by the pH-dependence of the predominant species on the electrode surface. Perchlorate is a very poor nucleophile, and other anions in solution are predicted to displace it from the oxygen vacancy.

The prognosis for remediation of groundwater or drinking water contaminated with low concentrations of perchlorate by direct electrochemical reduction of perchlorate is not favorable. Direct electrochemical reduction is slow enough to be controlled by reaction kinetics, thereby requiring a large surface area electrode. The reaction will proceed only with a large overpotential, and the reduction of protons to hydrogen, as well as other reactions, will invariably be competing reactions that are also promoted by high surface area. An electrochemical reaction catalyzed by a solution redox couple is also possible, but this process is controlled by the kinetics of the solution processes, which were shown to be kinetically sluggish in the previous section. However, the process discussed by Gu et al. in Chapter 16 may be feasible. In this reaction, perchlorate is removed from solution by sorption on an ion exchange resin, and a reducing agent generated in an electrochemical process is circulated through the resin to reduce the sorbed perchlorate.[59]

FORMATION OF PERCHLORATE IN THE ENVIRONMENT

Perchlorate is known to occur naturally within extensive caliche deposits of the Atacama Desert in Chile. These deposits have been extensively mined for $NaNO_3$ and were the major source of nitrogen fertilizer in the world prior to 1930. Some ore processing is performed to improve $NaNO_3$ purity, yielding a product containing from 0.07 to 0.2 weight percent perchlorate. To gauge the magnitude of these deposits consider that Ericksen[60] indicates about 25 million metric tons of $NaNO_3$ were produced from this area between 1830 and 1980. Although the deposits have the appearance of evaporate minerals, the ^{14}N-^{15}N and ^{16}O-^{17}O-^{18}O ratios have values characteristic of mass-independent fractionation, suggesting an atmospheric origin of the nitrate in which ozone was the oxidant. As discussed in Chapter 5, the perchlorate recovered from these naturally occurring ores also has an isotopic signature that indicates atmospheric ozone was involved in their formation. These factors point to a long term atmospheric deposition process as the source of nitrate and perchlorate in Chilean caliche.

It is now well-established that atmospheric ozone carries a unique isotopic signature.[61] Reactions with ozone occur in both the upper troposphere and the lower stratosphere. The air in the troposphere cools with increasing altitude above the earth's surface until a minimum temperature is reached at the tropopause, which is the boundary between the troposphere and the stratosphere at an altitude of approximately 10 km. The temperature increases with altitude in the stratosphere due to excess kinetic energy released by solar-driven photochemical reactions, extending to an altitude of approximately 50 km. Reactions involving ozone are the primary heat-generating reactions in this layer. The maximum concentration of O_3 occurs between 15 and 25 km above the surface of the earth, depending on latitude and season.[62] Ozone is formed in the atmosphere by a three-body reaction in which oxygen atoms react with oxygen molecules to form a vibrationally excited ozone molecule that may either dissociate back to reactants or collide with an inert molecule M to form stable ozone:

$$O\cdot \ + \ O_2 \ + \ M \ \rightarrow \ O_3 \ + \ M \ .$$

Photolysis of ozone to oxygen atoms and oxygen molecules also occurs:

$$O_3 \ + \ h\nu \ \rightarrow \ O\cdot \ + \ O_2 \ .$$

The principle chlorine-containing species in the upper troposphere and lower stratosphere are hydrochloric acid, chlorine monoxide, and hypochlorous acid.[63] The species $ClONO_2$ is prominently mentioned in the literature as well. The chemistry of chlorine in the atmosphere has been extensively studied in the past two decades, in particular the influence of chlorine compounds on ozone depletion. Chlorofluorocarbon refrigerants are relatively unreactive in the troposphere and insoluble in water. Thus, they are neither oxidized in the lower atmosphere nor removed by rain. Photochemical reactions of these compounds in the stratosphere are a source of chlorine radicals that are responsible for catalytic cycles that lead to decreased levels of ozone, particularly in the polar regions. One of the catalytic cycles leading to ozone loss is the reaction sequence

$$Cl\cdot + O_3 \rightarrow ClO\cdot + O_2$$

$$ClO\cdot + O\cdot \rightarrow Cl\cdot + O_2$$

The combination of these two reactions represents the catalytic reaction of oxygen atoms with ozone to form two oxygen molecules, i.e.,

$$O_3 + O\cdot \rightarrow {}^1O_2 + {}^3O_2$$

Although anthropogenic sources of chlorine may significantly influence the chemistry of chlorine in the present day atmosphere, they cannot be responsible for the naturally occurring perchlorate in Chilean desert deposits.

The chemistry of ozone and NO_2 is important in the troposphere, but the hydroxyl radical is generally acknowledged as the most important free radical in the lower atmosphere and appears to drive the oxidation chemistry of most species emitted into the atmosphere during the day.[64] The chemistry of halogens derived from sea water has been reviewed recently by Finlayson-Pitts.[65] Sea salt is a major source of halogens in the atmosphere of marine areas due to airborne particles generated by wave action. A number of reactions have been suggested that lead to chlorine atoms from NaCl aerosols, such as

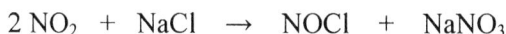

$$2\,NO_2 + NaCl \rightarrow NOCl + NaNO_3$$

in which the nitrosylchloride species is known to photochemically dissociate to form NO and chlorine atoms. The nitric acid reacts with sea salt aerosols to form HCl that can react with hydroxyl radicals as

$$HCl + OH\cdot \rightarrow Cl\cdot + H_2O,$$

although this reaction is thought to be too slow to serve as a significant source of chlorine atoms. A more likely source of atomic chlorine is the reaction of hydroxyl radicals with solid NaCl crystallites to produce molecular chlorine, which is then photolyzed to chlorine atoms. Chlorine atoms rapidly react with ozone to form chlorine monoxide as follows

$$Cl\cdot + O_3 \rightarrow ClO\cdot + O_2 ,$$

and chlorine atoms also react with hydroxyl radicals to form hypochlorous acid, HOCl:

$$Cl\cdot + OH\cdot \rightarrow HOCl.$$

Another reaction of importance is that of chlorine monoxide with NO_2:

$$ClO\cdot \; + \; NO_2 \;\; \rightarrow \;\; ClONO_2 \;.$$

HOCl, ClO, and $ClONO_2$ along with HCl all have sufficiently long lifetimes and mobility to be transported to the stratosphere. Volcanic activity can be another significant source of chlorine compounds in the troposphere and lower stratosphere.

The Atacama Desert is extremely dry, receiving on average less than 3 mm of rain annually. The nitrate mineral beds lie in the rain shadow between the coastal range (altitude 1 to 3 km) to the west and the Andes Mountains to the east. One of the reasons the extensive nitrate beds exist is that this area has not had significant rainfall for millions of years. Michalski et al.[66] summarize the evidence for a photochemical process involving ozone leading to the generation of nitrate. They conclude that the extensive deposits are the result of atmospheric deposition of the solid phase produced by gas to particle conversion. These deposits could have accumulated from atmospheric deposition over a period of 200,000 to 2 million years under hyper-arid conditions such as those currently found in this area. The perchlorate in these nitrate beds must have been formed at the same time under similar conditions.

Isotopic ratios in the Chilean nitrate deposits suggest that perchlorate formed in the atmosphere by a process involving ozone as the oxidant. Older literature[67] suggests that $HClO_4$ may be a sink for atmospheric chlorine since it is not as susceptible to photodisassociation as are lower oxidation states of chlorine. Researchers proposed that $HClO_4$ was formed by reaction of hydroxyl radicals with ClO_3 :

$$ClO_3\cdot \; + \; OH\cdot \;\; \rightarrow \;\; HClO_4$$

although proponents of this theory acknowledged that the reaction had never been observed even in the laboratory. Evidence for the possible presence of perchloric acid in the stratosphere was provided by balloon-borne FTIR spectroscopy at an altitude of 15 to 20 km in September 1993.[68] A significant fraction of the inorganic chlorine could not be accounted for by the sum of the species HCl, $ClONO_2$, and HOCl. The reaction of ClO in sulfuric acid aerosols to form perchloric acid was proposed to account for the remaining reservoir of inorganic chlorine. Direct observation of perchloric acid in the stratosphere by laser ionization mass spectroscopy was reported, and the maximum concentration occurred at 19 km, which generally corresponds to the altitude at which ozone and ClO have concentration maxima.

The photochemistry and chemistry of chlorine oxides in the atmosphere has been studied in some detail,[63] and it is difficult to produce and maintain substantial quantities of either ClO_2 or ClO_3. There is little evidence for the latter species in the atmosphere, and the former is rapidly photolyzed to ClO and atomic oxygen. In fact, chlorine oxide reacts with oxygen atoms to form chlorine atoms and O_2 as follows

$$ClO\cdot + O\cdot \rightarrow Cl\cdot + O_2 .$$

Chlorine atoms react with ozone to form ClO, thereby imparting the "ozone signature" to the oxygen atom. Chlorine dioxide can be generated in the atmosphere by reaction of ClO with BrO (generated by the reaction of ozone with bromine atoms) as

$$ClO\cdot + BrO\cdot \rightarrow ClO_2\cdot + Br\cdot$$

in a reaction that imparts to both oxygen atoms of ClO_2 the "ozone signature." The characteristic photochemical reaction of ClO_2 in the gas phase is photodisassociation to form ClO and oxygen atoms. However, the photochemistry in the condensed phase is different. In solution or a matrix, the photolysis products are chlorine atoms and oxygen molecules. Further reaction to produce higher oxidation states of chlorine appears to be possible only if the ClO_2 is sorbed to an ice crystal, sulfuric or nitric acid aerosol, or volcanic aerosol particle of some sort. Chlorine dioxide has a favorable partition coefficient to equilibrate into the condensed phase. We have previously noted that ClO_2 is unstable to disproportionation to $HClO_3$ and $HClO_2$. Furthermore, ClO_2 is thermodynamically unstable to disproportionation to $HClO_4$ and HOCl, although a reaction pathway leading to these products is difficult to envision. In aqueous solution, ozone is known to react rapidly with ClO_2 to form $HClO_3$. Thus, there appear to be ample precedents in the literature to oxidize chlorine atoms to $HClO_3$ by processes involving ozone as the source of the oxygen atoms. Ice crystals and acid aerosols containing water are fairly abundant in the atmosphere at the tropopause. Any reaction producing perchloric acid from chloric acid would be purely speculative, but for the sake of completeness, we will suggest ozone or oxygen atoms as the oxidant that takes $HClO_3$ to $HClO_4$. This reaction sequence gives rise to several predictions that could be tested in the laboratory. The conversion of HCl to perchlorate should be catalyzed by bromine and also by the presence of sulfuric acid aerosols.

FATE OF PERCHLORATE IN THE ENVIRONMENT

Despite the strongly oxidizing nature of perchlorate, it is known to be stable and nonreactive in aqueous systems. The high stability of ClO_4^- in water is due to its large kinetic barrier to reduction as well as its reluctance to bind to surfaces (poor nucleophile, poor coordinating ability). Simply lowering the Eh of water to below -200 mV does not induce abiotic reduction of the perchlorate ion.[4,5] Perchlorate has reportedly been reduced by some transitional metal ions, but concentrations of these reduced transition metal ions in natural waters and geologic media are perhaps too low to impact the fate of ClO_4^- in the environment. Although the reduction of ClO_4^- by Fe(II) has been reported at elevated temperatures,[69] it does not occur at ambient temperatures and near-neutral pH conditions.[70] However, perchlorate can be reduced on iron surfaces where mixed valence iron-oxyhydroxides and Fe(II) co-exist.[71] The reaction was found to accelerate under acidic pH conditions but was inhibited by the presence of soluble chloride ion. It seems clear from the results of electrochemical experiments that competitive adsorption at an active site is the mechanism of inhibition, as discussed in an earlier section. Recent studies have also revealed that nano-size iron powder significantly enhances the reduction rate of ClO_4^-, particularly at elevated temperatures[72]. The average size of iron nanoparticles used was about 57 nm, and they were synthesized in the laboratory by the reduction of ferrous ion using sodium borohydride ($NaBH_4$). Nevertheless, no measureable reduction of ClO_4^- occurs using bulk iron filings, microscale iron powders, or Fe(II) ion alone so that the role of iron or iron-containing minerals in the natural attenuation of ClO_4^- is also likely to be insignificant in the subsurface environment.

Perchlorate is poorly retained or sorbed by sediment minerals in the subsurface because of its negative charge and its noncomplexing nature with metal ions. Its large ionic size and low charge density reduce its affinity for metal cations and make it highly soluble and thus exceedingly mobile in natural aqueous systems. As such, perchlorate sorption to soils and sediments has not been widely observed nor extensively studied. The most comprehensive study of perchlorate retention and mobility in soils came from Urbansky and Brown,[73] who reported the adsorption and release of ClO_4^- in a variety of soils and minerals at varying perchlorate concentrations. Their findings support the widely-accepted view that perchlorate does not appreciably sorb to soils or minerals and that its mobility and fate are largely influenced by hydrologic and biologic factors. However, sorption of perchlorate appears to occur in soils with appreciable anion-exchange capacities (AEC). These soils act like anion exchangers, and perchlorate is sorbed by the replacement of other anions such as chloride, sulfate and nitrate. Because of the chemical similarities between ClO_4^-, iodide (I^-), and radioactive pertechnetate (TcO_4^-) anions, the sorption behavior of I^- and TcO_4^- may also be used to infer the mobility of perchlorate in soil. A review of the literature indicates that only highly weathered soils dominated by iron and aluminum-oxyhydroxides and kaolinite are capable of exhibiting a significant AEC and, hence, adsorption capacity for I^- and TcO_4^- anions, which increases as pH decreases below the point of zero charge.[74] Ticknor and Cho[75] reported no detectable iodide sorption in the pH range of 7.5 to 8 on a wide variety of minerals such as calcite, chlorite, epidote, goethite, gypsum, hematite, kaolinite, bentonite, muscovite, and quartz. Whitehead[76] found that although freshly precipitated Fe(III) and Al(III) oxides could sorb a substantial amount of I^- at pH below 5.5, sorption decreased to zero as the pH increased to about 7.0. Routson et al.[77] studied the sorption of TcO_4^- on two highly weathered soils and found that TcO_4^- was poorly sorbed by the soils. Sisson et al.[78] observed that a sandy loam soil was unable to sorb TcO_4^- even under acidic pH conditions. Therefore, unless appreciable anion exchange capacity is confirmed, sorption and retention of perchlorate in soil sediments will likely be negligible.

The most significant factors affecting perchlorate fate in the subsurface are likely to be dilution as it migrates from the source, biological uptake, and possibly degradation under anaerobic conditions. Like many other contaminants, perchlorate plume concentrations tend to decrease significantly from the source area due to dilution. Flowers and Hunt[79] modeled the density-driven flow of a highly concentrated perchlorate brine with concentrations as high as 200 g/L (close to the solubility of NH_4ClO_4) exhibiting an estimated density of 1.1 kg/L. They indicated that release from the source would be mass-transfer limited under density-driven flow conditions with a time-scale of approximately 100 years. However, it is noted that at relatively high ClO_4^- concentrations, perchlorate may "salt out" or precipitate with cations such as potassium because potassium perchlorate has a relatively low solubility (~10 g/L at 20°C), thus decreasing its migration potential.

A widely studied mechanism for natural attenuation of perchlorate in the subsurface environment is anaerobic microbial degradation, which is described in detail in Chapters 12 to 14. For example, in a study of the potential of natural wetland systems to remove ClO_4^-, Tan et al.[80] reported that soil microbial degradation played a more important role in perchlorate attenuation than plant uptake and transformation. Indigenous soil microorganisms also were found to be responsible for rapid degradation of perchlorate under saturated near-surface conditions in a study of the fate of perchlorate in streambed sediments.[81] In the unsaturated zone, microbial reduction of perchlorate may occur if electron donors such as acetate and hydrogen are added.[82] It has been reported that the addition of acetate and hydrogen greatly enhance the reduction rate of perchlorate, although a longer lag period is observed for hydrogen (41 d) than for acetate (14 d). The detailed biochemistry of the microbial reduction of perchlorate has not been extensively studied. Based on the previous discussion of perchlorate reduction, the detailed mechanism will surely involve metal ion catalysis.

Perchlorate uptake by plants, trees, forage and edible vegetation is known to occur and thus represents another natural attenuation and redistribution pathway.[83] In fact, phytoremediation of perchlorate has been proposed as a technology to remove ClO_4^- from contaminated soils and groundwater, particularly for shallow contamination.[84] In a field-scale study of perchlorate accumulation in aquatic and terrestrial plants, Tan et al.[85] reported a large potential for ClO_4^- accumulation in plant species such as smartweed, watercress, and 40 trees such as ash, chinaberry, elm, willow, mulberry, and hackberry. Concentrations of accumulated perchlorate in plant tissues was as much as 100 times higher than the bulk water perchlorate concentration. Furthermore, trees located closer to the stream had higher ClO_4^- accumulations than trees located farther away, likely caused by a relatively high ClO_4^- concentration closer to the stream. The bioaccumulation or concentration factor for ClO_4^- was even greater for wheat stems and heads (230 to 260) and alfalfa (380).[30] Because of its high solubility and mobility, perchlorate follows water movement within the plant and tends to accumulate more in leaves, where transpiration occurs, rather than in fruits. For example, perchlorate concentrations in soybean leaves grown in the greenhouse were found to be significantly higher than in the soybean seeds, and concentrations in tomato leaves were also higher than in the fruit[30]. There is little evidence of perchlorate degradation or biotransformation in plants. The only evidence of ClO_4^- reduction in poplar tree tissue was reported by van Aken and Schnoor[29] using radio-labeled $^{36}ClO_4^-$. Their results indicated that 27% of ClO_4^- was translocated to the leaves, while 67% remained in the solution. Analysis of the radioactivity recovered in solution showed 68% as nontransformed ClO_4^- and about 32% as chloride-36, suggesting ClO_4^- degradation in the rhizosphere. Most of the radioactivity recovered in leaf extracts was nontransformed ClO_4^-, while small amounts of reduced metabolites were identified in the following percentages: Cl^- (1.6), ClO_2^- (2.4), ClO_3^- (4.8), and an unidentified organic compound (1.4). These studies provide evidence of possible perchlorate reduction and transformation inside plant tissues, although the overall impact on natural attenuation of perchlorate in the environment is yet to be determined. Plants which naturally contain anion exchangers should be investigated for perchlorate retention.

CONCLUSIONS

The persistence of perchlorate in groundwater can be understood from its chemical and physical properties and its chemical reactivity. Although perchlorate is a powerful oxidizing agent, its notorious lack of reactivity can be understood from the requirement that reduction involves oxygen atom transfer. Because perchlorate is relatively unreactive, remediation schemes involving direct chemical or electrochemical reduction are not effective. The low hydration energy of perchlorate anions favors sorption on organic anion exchange resins. Remediation methods involving collection on such resins prior to reduction or some other means of disposal are feasible. Biological systems that naturally reduce and degrade perchlorate are also a potentially practical means of remediating perchlorate-contaminated groundwater in a cost-effective manner.

REFERENCES

1. Hogue, C. "Rocket-Fueled River," *Chem. Eng. News* 2003, *81*, 37–46.

2. Brandhuber, P and Clark, S. "Perchlorate Occurrence Mapping," American Water Works Association: Washington, DC, 2005, report available at the web site www.awwa.org/Advacocy/PerchlorateOccurrenceReportFinalb02092005.pdf

3. National Research Council. *Health Implications of Perchlorate Ingestion.* The National Academies Press: Washington, DC, 2005.

4. Urbansky E.T. Perchlorate chemistry: implications for analysis and remediation. Bioremediation J. 1998, 2: 81-95.

5. Motzer, W. E., Perchlorate: Problems, Detection, and Solutions, Environ. Forensics, 2, 2001, 301-311.

6. Urbansky, E. T., Quantitation of Perchlorate Ion: Practices and advances Applied to the analysis of Common Matrices, Crit. Rev. Anal Chem, 30, 2000, 311-343.

7. Urbansky, E. T. and Schrock, M. R., Issues in Managing the Risks Associated with Perchlorate in Drinking Water, J. Environ. Mgmt, 1999, 56:79-95.

8. Urbansky, E. T., Ed. *Perchlorate in the Environment*; Kluwer Academic/Plenum Publishers: New York, 2000.

9. Hatzinger, P., Perchlorate Biodegradation for Water Treatment, Environ. Sci. Technol., 2005, 39:239A-247A.

10. Latimer, W. E., *Oxidation Potentials,* 2nd Ed, New York: Prentice Hall, 1952.

11. Bratsch, S. D. Standard Electrode Potentials and Temperature Coefficients in Water at 298.15 °K, J. Phys Chem Ref Data, 1989, 18: 1-21.

12. Shilt, A.A., *Perchloric Acid and Perchlorates*, Columbus, OH: GF Smith Chemical Company, 1979.

13. Gordon, G., Is all Chlorine Dioxide Created Equal?, J. Am. Water Works Assoc, 2001, 163-174.

14. The Occurrence and Sources of Perchlorate in Massachusetts, draft report by Massachusetts Department of Environmental Protection, August, 2005, http://www.mass.gov/dep/

15. Jackson, W. A., Anandam, S. K., Anderson, T., Lehman, T., Rainwater, K., Rajagopalan, S., Ridley, M., and Tock, R. Ground Water Monitoring and Remediation, 2005, 25:137-149.

16. Olah, G. A., G. K. Surya Prakash, and J. Sommer, *Superacids*, New York: John Wiley and Sons, 1985.

17. Hoering, T. C., Ishimori, F. T., and McDonald, H. O., The Oxygen Exchange Between Oxy-anions and Water. 11. Chlorite, Chlorate and Perchlorate Ions, J. Am Chem Soc, 1958, 80:3876-3879.

18. Moyer, B. A. and P. V. Bonnesen, "Physical Factors in Anion Separation," in *Supramolecular Chemistry of Anions*, A. Bianchi, K. Bowman-James, and E. Garcia-Espana, ed, New York: Wiley-VCH, Inc, 1979, p 1-44.

19. Diamond, R. M., Whitney, D. C. In *Ion Exchange, Vol. 1*; Marinsky, J. A., Ed.; Marcel Dekker: New York, 1966, pp 277-351.

20. Mendiratta, S. K., Dotson, R. L., and Brooker, R. T., Perchloric acid and Perchlorates, in Kirk-Othmer Encyclopedia of Science and Technology, 4th Ed, New York: John Wiley and Sons, 1996, Vol 18, p157-170.

21. Arora, P. and Srinivasan, V., Report on the Electrolytic Industries for the Year 2001, J. Electrochem. Soc., 2002, 149:K1-K29.

22. Srinivasan, V. and Lipp, L., Report on the Electrolyric Industries for the Year 2002, J. Electrochem. Soc., 2003, 150:K15-K38.

23. Burns, D. T., Chimpalee, N., and Harriott, M., Flow-Injection Extraction-Spectrophotometric Determination of Perchlorate with Brilliant Green, Anal. Chim. Acta., 1989, 217:177-181.

24. Encyclopedia of Chemical Processing and Design, McKetta, J. J. and Weismantel, G. E., Ed, New York: Marcel Dekker, 1995, Vol 51, p180

25. Earley, J. E., Tofan, D. C., Amadei, G. A., "Reduction of Perchlorate Ion by Titanous Ion in Ethanolic Solution," In *Perchlorate in the Environment*; Urbansky, E. T., Ed.; New York Kluwer/Plenum, 2000, pp 89-98.

26. Espenson, J. H., "The Problem and Perversity of Perchlorate," In *Perchlorate in the Environment*; Urbansky, E. T., Ed.; New York: Kluwer/Plenum, 2000, pp 1-7.

27. Urbansky, E.T. Perchlorate chemistry: implications for analysis and remediation. Bioremediation J 1998; 2: 81-95.

28. Taube, H."Observations on Atom Transfer Reactions" In *Mechanistic Aspects of Inorganic Reactions*; Rorabacher, D. B., Endicott, J. F., Eds.; ACS Symposium Series No. 198, 1982, pp 151.

29. Duke, F. R. and Quinney, P. R. The Kinetics of Reduction of Perchlorate Ion by Ti(III) in Dilute Solution, J. Am Chem Soc, 1954, 76:3800-3803.

30. Cope, V. W., Miller, R. G., and Fraser, R. T. M., Titanium (III) as a Reductant in Electron-Transfer Reactions, J. Chem Soc (A), 1967, 301-306.

31. Kallen, T. W. and Earley, J. E., Reduction of Perchlorate by Aquoruthenium(II), Inorg Chem, 1971, 10:1152-1155.

32. King, W. R and Garner, C. S., Kinetics of the Oxidation of Vanadium(II) and Vanadium(III) Ions by Perchlorate Ion, J. Phys. Chem, 1954, 58:29-33.

33. Liu, B.-Y, Wagner, P.A., and Earley, J. E., Reduction of Perchlorate Ion by (N-(Hydroxyethyl)ethylenediaminetriacetato)aquotitanium(III), Inorg. Chem., 1984, 23:3418-3420.

34. Endicott, J. F. and Taube, H., J. Am. Chem. Soc., 84, 1962, 4984; Endicott, J. F. and Taube, H., Inorg Chem., 4, 1965, 437.

35. Baes, C. F. and Mesmer, R. E. *Hydrolysis of Cations*, New York: Wiley-Interscience, 1973, p 199.

36. Abu-Omar, M. M. and Espenson, J. H., Facile Abstraction of Successive Oxygen Atoms from Perchlorate Ions by Methylrhenium Dioxide, Inorg. Chem., 1995, 34:6239-6240.

37. Abu-Omar, M. M., Appelman, E. H., and Espenson, J. H., Oxygen-Transfer Reactions of Methylrhenium Oxides, Inorg. Chem., 1996, 35:7751-7757.

38. Espenson, J. H. and Abu-Omar, M. M., in *Electron Transfer Reactions: Inorganic, Organic, and Biological Applications,* Isied, S. S., ed, Advances in Chemistry 253, American Chemical Society: Washington, DC, 1997, 99-134.

39. Espenson, J. H. Atom-Transfer Reactions Catalyzed by Methyltrioxorhenium(VII)— Mechanisms and Applications, Chem. Commun, 1999, 479-488.

40. Abu-Omar, M. M., McPherson, L. D., Arias, J., and Bereau, V. M., Clean and Efficient Catalytic Reduction of Perchlorate, Angew. Chem. Int. Ed., 2000, 39:4310-4313.

41. Abu-Omar, M. M., Effective and Catalytic Reduction of Perchlorate by Atom-Transfer Reaction - Kinetics and Mechanism, Comments on Inorganic Chemistry, 2003, 24:15-37.

42. Horanyi, G. "Electrosorption Studies in Electrocatalysis," in *A Specialist Periodical Report*, Vol 12, Spivey, J. J. (ed), The Royal Society of Chemistry: Cambridge, 1996, pp 254-301.

43. Láng, G. G. and Horayni, G., Some Interesting Aspects of the Catalytic and Electrocatalytic Reduction of Perchlorate Ions, J. Electroanal. Chem., 2003, 552:197-211.

44. Adams, R. N., *Electrochemistry at Solid Electrodes,* Marcel Dekker: New York, 1969, p 29.

45. Horanyi, G. and G. Vertes. Catalytic and electrochemical reduction of perchlorate ions on platinum in aqueous solution. J. Electroanal. Chem. 64 (1975) 252-254.

46. Bakos, I., G. Horanyi, An experimental study of the relationship between platinization and the shape of the voltammetric curves obtained at Pt/Pt electrodes in 0.5M H_2SO_4. J. Electroanal. Chem. 1992, 332:147-154. G. Horanyi, and I. Bakos, Experimental evidence demonstrating the occurrence of reduction processes of ClO_4^- ions in an acid medium at platinized platinum electrodes. J. Electroanal. Chem. 1992, 331:727-737.

47. Horanyi, G. and G. Vertes, Reduction of perchlorate ions by molecular hydrogen in the presence of tungsten carbide. Inorg. Nucl. Chem. Lett. 1974, 10:767-770.

48. Sanchez Cruz, M., M.J. Gonzalez Tejera, and M.C. Villamanan, Reduction electrochimique de l'ion ClO4- sur l'electrode d'iridium. Electrochim. Acta. 1985, 30:1563-1569.

49. Go´mez, R. and M.J. Weaver, Electrochemical infrared studies of monocrystalline iridium surfaces. Part 2: Carbon monoxide and nitric oxide adsorption on Ir(110). Langmuir. 1998, 14:2525-2534.

50. Bakos, I., G. Horanyi, S. Szabo, E.M. Rizmayer, Electrocatalytic reduction of ClO4- ions at an electrodeposited Re layer. J. Electroanal. Chem. 1993, 359:241-252.

51. Horanyi, G., E.M. Rizmayer, A radiotracer study of the adsorption of Cl⁻ ions on rhodized electrodes. J. Electroanal. Chem. 1986, 198:379-391. C.K. Rhee, M. Wasberg, G. Horanyi, A. Wieckowski, Strong anion/surface interactions: perchlorate reduction on Rh(100) electrode studied by voltammetry. J. Electroanal. Chem. 1990, 291:281-287.

52. Rhee, C.K., M. Wasberg, P. Zelenay, and P. Wieckowski, Reduction of perchlorate on rhodium and its specificity to surface crystallographic orientation. Catalysis Lett. 1991, 10: 149-164. Clavilier, J., M. Wasberg, M. Petit, and L.H. Klein, Detailed analysis of the voltammetry of Rh(111) in perchloric acid solution. J. Electroanal. Chem. 1994, 374:123-131.

53. Colom, F. and M.J. Gonzalez-Tejera, Reduction of perchlorate ion on ruthenium electrodes in aqueous solutions. J. Electroanal. Chem. 1985, 190:243-255.

54. Horanyi, G. and I. Bakos, Combined radiometric and electrochemical study of the behavior of Tc (VII) ions at gold and platinized surfaces in acidic media. J. Appl. Electrochem. 1993, 23: 547-552.

55. Almeida, C.M.V.B., Giannnetti, B. F., Rabockai, Electrochemical Study of Perchlorate Reduction at Tin Electrodes, J. Electroanal. Chem., 1997, 422:185-189.

56. Lang, G., Inzelt, G., Vrabecz, A., and Horanyi, G., Electrochemical Aspects of Some Features Connected with the Behavior of Iron Group Metals in Aqueous Perchloric Acid/Perchlorate Media, J. Eleectroanal Chem, 2005, 582:249-257.

57. Brown, G. M., The Reduction of Chlorate and Perchlorate Ions at an Active Titanium Electrode, J. Electroanal Chem, 1986, 198:319-330.

58. Painot, J. and J. Augustynski, Etude potentiostatique et spectroscopique de l'aluminum recouvert par une couche d'oxyde: effet de differents anions. Electrochim. Acta. 1975, 20, 747-752.

59. Gu, B., Brown, G. M. Regeneration of anion exchange resins by catalyzed electrochemical reduction. 2002:US Patent No. 6,358,396.

60. Ericksen, G.E. The Chilean nitrate deposits. Amer Scientist 1983; 71: 366-374.

61. Brenninkmeijer, C. A. M., Janssen, C. , Kaiser, J., Rockmann, T., Rhee, T. S., and Assonov, S. S., Isotope Effects in the Chemistry of Atmospheric Trace Compounds, Chem Rev, 2003, 103:5125-5161.

62. Thrush, B. A., The Chemistry of the Stratosphere, Rep. Prog. Phys, 1988, 51:1341-1371.

63. Bedjanian, Y. and Poulet, G., Kinetics of Halogen Oxide Radicals in the Stratosphere, Chem. Rev., 2003, 103:4639-4655.

64. Monks, P. S., Gas-phase Radical Chemistry in the Troposphere, Chem. Soc. Rev., 2005, 34:376-395.

65. Finlayson-Pitts, B. J., The Tropospheric Chemistry of Sea Salt: A Molecular-Level View of the Chemistry of NaCl and NaBr, Chem. Rev., 2003, 103: 4801-4822.

66. Michalski G., Böhlke J.K., and Thiemens M.. Long term atmospheric deposition as the source of nitrate and other salts in the Atacama Desert, Chile: New evidence from mass-

independent oxygen isotopic compositions. Geochim Cosmochim Acta 2004; 68: 4023-4038.

67. Simonaitis, R., and Heicklen, J. Perchloric Acid: A Possible Sink for Stratospheric Chlorine, Planet. Space Sci. 1975, 23:1567-1569.

68. Jaeglé, L., Yung, Y. L., Toon, G. C., Sen, B., and Blavier, J.-F. Balloon observation of organic and inorganic chlorine in the stratosphere: the role of HClO4 production on sulfate aerosols, Geophys. Res.Lett,. 1996, 23:1749-1752.

69. Gu, B., Dong, W., Brown, G. M., Cole, D. R. Complete degradation of perchlorate in ferric chloride and hydrochloric acid under controlled temperature and pressure. *Environ. Sci. Technol.*, 2003; *37*:2291-2295.

70. Moore, A. M., De Leon, C. H., Young, T. M. Rate and extent of aqueous perchlorate removal by iron surfaces. *Environ. Sci. Technol.*, 2003; *37*:3189-3198.

71. Moore, A. M., Young, T. M. Chloride interactions with iron surfaces: implications for perchlorate and nitrate remediation using permeable reactive barriers. *J. Environ. Eng.*, 2005; *131*:924-933.

72. Cao, J. S., Elliott, D., Zhang, W. X. Perchlorate reduction by nanoscale iron particles. *J. Nanopart. Res.*, 2005; *7*:499-506.

73. Urbansky, E. T., Brown, S. K. Perchlorate retention and mobility in soils. *J. Environ. Mon.*, 2003; *5*:455-462.

74. Gu, B., Schulz, R. K. Anion retention in soil: possible application to reduce migration of buried technetium and iodine. *NUREG/CR-5464. U.S. Nuclear Regulatory Commission*, 1991:Washington, DC 20555.

75. Ticknor, K. V., Cho, Y. H. Interaction of iodide and iodate with granitic fracture-filling minerals. *J. Radioanal. Nucl. Chem.*, 1990; *140*:75-90.

76. Whitehead, D. C. The sorption of iodide by soil components. *J. Sci. Food Agric.*, 1974; *25*:73-79.

77. Routson, R. C., Jansen, G., Robinson, A. V. Am-241, Np-237 and Tc-99 sorption on two United States subsoils from differing weathering intensity areas. *Health Physics*, 1977; *36*:21-30.

78. Sisson, D. H., MacLean, S. C., Schulz, R. K., Borg, R. J. A preliminary study of the migration of technetium in soil under hydrous conditions. *National Technical Service, UCRL-52832*, 1979:U.S. Department of Commerce.

79. Flowers, T. C., Hunt, J. R. In *Perchlorate in the Environment*; Urbansky, E. T., Ed.; Kluwer/Plenum: New York, 2000, pp 177-188.

80. Tan, K., Jackson, W. A., Anderson, T. A., Pardue, J. H. Fate of perchlorate-contaminated water in upflow wetlands. *Wat. Res.*, 2004; *38*:4173-4185.

81. Tan, K., Anderson, T. A., Jackson, W. A. Temporal and spatial variation of perchlorate in streambed sediments: results from in-situ dialysis samplers. *Environ. Poll.*, 2005; *136*:283-291.

82. Nozawa-Inoue, M., Scow, K. M., Rolston, D. E. Reduction of perchlorate and nitrate by microbial communities in vadose soil. *Appl. Environ. Microbiol.*, 2005; *71*:3928-3934.

83. Tan, K., Anderson, T. A., Jackson, W. A. Temporal and spatial variation of perchlorate in streambed sediments: results from in-situ dialysis samplers. *Environ. Poll.*, 2005; *136*:283-291.; Nozawa-Inoue, M., Scow, K. M., Rolston, D. E. Reduction of perchlorate

and nitrate by microbial communities in vadose soil. *Appl. Environ. Microbiol.*, 2005; *71*:3928-3934. Urbansky, E. T., Magnuson, M. L., Kelty, C. A., Brown, S. K. Perchlorate uptake by salt cedar (Tamarix ramosissima) in the Las Vegas Wash riparian ecosystem. *Sci. Total Environ.*, 2000; *256*:227-232.

84. van Aken, B., Schnoor, J. L. Evidence of perchlorate (ClO_4^-) reduction in plant tissues (poplar tree) using radio-labeled (ClO_4^-)-Cl-36. *Environ. Sci. Technol.*, 2002; *36*:2783-2788.

85. Tan, K., Anderson, T. A., Jones, M. W., Smith, P. N., Jackson, W. A. Accumulation of perchlorate in aquatic and terrestrial plants at a field scale. *J. Environ. Qual.*, 2004; *33*:1638-1646.

Chapter 3

Occurrence and Formation of Non-Anthropogenic Perchlorate

W. Andrew Jackson,[*] Todd Anderson, Greg Harvey, Greta Orris, Srinath Rajagopalan, and Namgoo Kang

[*] *Dept. of Civil Engineering, Texas Tech University, Lubbock, TX 79409*

INTRODUCTION

The issue of naturally-occurring perchlorate is one that has environmental, agricultural, and economic implications. Natural perchlorate was first identified in Chilean nitrates over 100 years ago.[1] Ericksen also reported a trace of perchlorate in the "blister" caliche of a nitrate deposit in southeastern California[2]. However, certainly up to the very recent time period the natural occurrence of perchlorate was thought largely to be strictly confined to the Atacama dessert and its well studied nitrate deposits, with the exception of some now rejected reports[3] and other unconfirmed findings.[4] Interestingly, one of the first major hypotheses and now verified mechanism[5] for the occurrence of perchlorate was atmospheric production and deposition over extremely long time periods. While it is certainly true that the Atacama is one of the driest places on Earth and has been for the last 6-15 million years[5] other extremely arid areas exist, so it is reasonable that other arid areas should also posses some amount of perchlorate. Arid areas in this country, (e.g. the Death Valley region of the Mojave Desert in California) and other deserts throughout the world have been noted to contain natural surficial nitrate-rich salts and deep subsurface soils acting as repositories for nitrate and chloride. Stable isotope studies of these natural surficial nitrate rich salts by the United States Geological Survey (USGS)[6] support the hypothesis that these long term nitrate accumulations must be of atmospheric origin. The combined oxygen and nitrogen isotope signatures of the nitrate in these deposits are significantly different from other natural and anthropogenic sources of nitrate.

The processes that form and accumulate nitrate may offer clues to the formation of natural perchlorate. It is possible that the ultimate source of natural perchlorate could be aerosols formed in the atmosphere. Chlorine, along with other halogens, participates in numerous photochemical reactions in the atmosphere, and it is known that various oxidized forms of chlorine, including ClO_4^-, are present in aerosols in trace amounts.[7] In fact, if recently reported precipitation data are applicable globally[8] the question is not what areas contain natural perchlorate but rather what areas concentrate naturally deposited perchlorate to environmentally relevant concentrations.

The issue of documenting perchlorate's natural occurrence has been certainly problematic given older analytical methodologies, the typically low levels of perchlorate occurrence (commonly a few tens of ppb or less), the inconsistent results of various investigators, and reservations about the existence of naturally-occurring perchlorate in areas other than the Atacama Desert. In addition, at least in the U.S. where interest in perchlorate is highest, the high profile industrial occurrences in the American Southwest (prime potential area for accumulation of perchlorate) have probably served to overshadow any anomalous detections potentially due to natural sources. This combined with the extreme difficulty in proving the origin of perchlorate found in the environment have led to a rather sparse data set on its potential occurrence. With the growing interest in its occurrence and production from natural mechanisms as well as the important isotopic work reported by Bao and Gu,[9] Sturchio et al.,[10] and Michalski et al.[5] to help determine its source, the overall importance as both an environmental pollutant and component of basic geochemical cycling may soon be at hand. This chapter is focused on documenting both the few known occurrences of perchlorate believed to be of natural origin as well as a discussion of the current theories on possible production mechanisms and the overall importance of naturally occurring perchlorate in the environment.

OCCURRENCES

Perchlorate Salt Deposits of the Atacama Desert

Reliable sources of nitrate before the 20th century were scarce and coveted. The discovery of massive deposits of sodium nitrate in the hyper-arid coastal Atacama Desert was a historically important event and these deposits had a significant impact on the world socio-political scene for over a century. These commercially and strategically valuable deposits of nitrate in the Atacama Desert have been the object of both contention and scientific curiosity. As early as the 1850's the genesis of the Chilean nitrate deposits was of interest to geologists. At one time or another, various theories involving seaweed, bird guano, bacteria, or electricity were in vogue with the scientific community.[11] Ultimately most of these theories could not be supported by the available evidence and were abandoned. As recently as the 1980's researchers such as George Ericksen of the USGS were investigating and speculating over the origin of these nitrate deposits. Along with the nitrate deposits, early chemists discovered chromates, iodate, lithium, and perchlorate. Until the advent of stable oxygen and chlorine isotope technology, the origin of the various constituents of Chilean caliche Blanco remained a mystery. Recent studies have afforded more insight and evidence to support the view that the nitrate and perchlorate and most likely the chromate found in the Atacama Desert are formed atmospherically.[5, 6, 9]

The Atacama Desert is one of the most arid regions in the world with less than 2 mm of rain annually[5]. The nitrate deposits are known to occur in the two northern provinces of Tarapaca and Antofagasta in Chile. The deposits occur in a band up to 30 km wide and for approximately 700 km in length or over an area of about 21,000 sq.km.[11] Most of the deposits are at altitudes of 1000 – 2000 m, but some may be at altitudes of 3000 – 4000 m. The local distribution of the deposits are believed to be a function of remobilization of existing salts by rainfall, eolian transport, capillary action, sedimentation, discharge from groundwater, and hydrothermal changes.[5]

Table 1. Average content of ionic saline constituents in mined caliche[12]

Ion	Percentage
SO_4^{2-}	10
Na^+	6.9
NO_3^-	6.3
Cl^-	4.6
Ca^{2+}	1.8
K^+	0.7
Mg^{2+}	0.5
$B(OH)_4^-$	0.5
IO_3^-	0.06
ClO_4^-	0.03

The average Chilean caliche ore reportedly contains 6.3% nitrate and 0.03% perchlorate by weight.[12] The nitrate deposits also contain iodate, chromate, dichromate, sulfate, and chloride. The chemical composition of typical nitrate deposits is difficult to estimate as there are great local and regional variations in the ore composition. Best estimates were obtained from ore processing facilities in Chile based on large tonnage of the processed deposits from 1932 to 1967 (Table 1).[12] These constituents represent only the water soluble portion of the ore, which accounts for about 30 percent of the ore with insoluble salt debris and saline minerals. The bulk anionic composition of the ore is reasonably consistent with an atmospheric source of nitrate, sulfate, and chloride after considering the insoluble sulfates, wind erosion, and leaching of salts to groundwater. The molar ratio of iodate to nitrate in the atmosphere is similar to that found in the ore deposits, further supporting an atmospheric source[5]. Recent research has more definitively shown through stable isotopic techniques that the origin of much of the nitrate in the Atacama as well as sulfate is indeed atmospheric.[5]

The sources of the more exotic "ates" (chromate and perchlorate) are less understood, while the atmospheric production of iodate is fairly well characterized.[13] Ericksen[11] speculated that photochemical reactions between chlorine and ozone may be the source of perchlorate in the Atacama, citing work by Simonaitis and Heicklen[14] as supporting this mechanism. Very recently perchlorate derived from Atacama nitrate ore has been shown to have significant $\Delta^{17}O$ value, a clear indication of atmospheric production involving ozone.[9]

Total sodium nitrate in commercial grade ore (> 7 percent sodium nitrate) was estimated at 200 billion kilograms,[15] or 33 billion kilograms as N. The poorer grade sodium nitrate may be several times more than the commercial grade with a final estimate of total nitrate in both high and low grade ore of 100 billion kilograms of N or ~ 443 billion kilograms as nitrate[5]. Assuming that the ratio of perchlorate to nitrate (0.0048) reported by Grossling and Ericksen[12] is consistent between ore grades (highly speculative) the total estimated mass of perchlorate in the Atacama deposits is approximately 2 billion kilograms. The hyper-arid conditions of the Atacama are estimated to have existed for 6-15 million years. Assuming a hyper-arid period of 10 million years and a depositional area of 21,000 km^2, this would equate to a depositional flux of 9.5 $\mu g/m^2$-year. In comparison, the nitrate depositional flux using this same analysis would equate to 2.1 $mg\text{-}NO_3/m^2$-year although Michalski et al.[5] report an estimated total flux of 73 $mg\text{-}NO_3/m^2$-year. Flux estimates are highly speculative as the total depositional area may be larger than the current area of deposits and there is little known of both the quantity of low grade ores and composition of these ores. Regardless, these estimates suggest the potential for significant accumulation of atmospheric perchlorate in other arid areas of the world which have existed for significant periods of time if the depositional flux can be assumed to translate to other global positions.

Chile started exporting the nitrate saltpeter on a large scale in 1830. The nitrate was purchased by France, United States, England, and later by Germany, Italy and other European countries. Most of the Chilean nitrate processing was done by foreign capital (mainly England and United States).

The Chilean nitrate was used as fertilizers and in production of smokeless gunpowder and explosives. About 75 million metric tons was exported from Chile from 1830 to 1930 with an exponential increase in exports peaking at 1920 for World War I (Figure 1).[16] However, the synthesis of ammonia by the Haber-Bosch method in 1918, developed by the Germans during World War I, became a serious competitor of Chilean nitrate. After controlling the entire market in the late nineteenth and early twentieth century, Chilean nitrate started losing its market share to synthetic nitrogen. By 1950, Chilean nitrate accounted for only 15% of world fixed nitrogen and by 1980, it was only 0.14% of the world market for fixed nitrogen.[11] According to U.S. Bureau of Mines Mineral Yearbook 1938 – 1993,[17] total Chilean nitrate imported to the U.S. from 1938 to 1993 was 20 million metric tons (Figure 2). Renner anecdotally quotes a 1999 annual U.S. use figure of 68,000 metric tons Chilean nitrate[18]. However, according to other sources,[19] the total use of $NaNO_3$ (Chilean and synthetic) in the U.S. had already dropped to 31,500 metric tons by 1994.

A study by Urbansky et al.[20] found certain lots of $NaNO_3$ fertilizer to contain over 0.1% ClO_4^-. Further, an earlier study[21] reported 0.15 - 0.26% perchlorate in Chilean nitrate deposits. However, a recent letter to the EPA from SQM North America, the sole provider of Chilean nitrate, indicated that SQM had modified its refining process to produce fertilizer containing less than 0.01% perchlorate.[22] Using a historical percentage value of 0.1% for perchlorate in imported Chilean $NaNO_3$ would result in a total imported perchlorate mass of 16.1 million kilograms only using the import values from 1938 to 1968, the time period with the best historical information. This value would be a conservative estimate as production of Chilean nitrate peaked in 1930. Other occurrences of suspected natural perchlorate are much more sporadic and significantly less well understood.

Perchlorate in Other Salt Deposits

As has been discussed, for many decades Chilean nitrates of the Atacama desert have remained the best known and most significant source of natural nitrates, perchlorates and chromates and for more than 100 years, geologists commonly believed the presence of perchlorate, iodates, and chromates in Chilean nitrate deposits to be a unique occurrence, with the origin and mechanism of accumulation of these deposits a matter of speculation. In the late 1990's, reports were surfacing of findings of perchlorate in fertilizers. Although, with the exception of perchlorate in Chilean nitrate fertilizers, these reports were not substantiated; the initial reports led to a USAF-initiated study with the USGS to look for possible natural perchlorate in geologic materials from environments with similar characteristics to the Chilean nitrates.

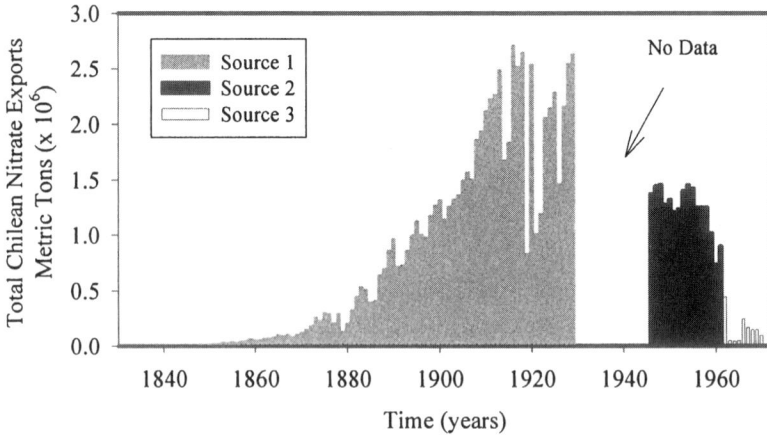

Figure 1 Total Chilean Nitrate Exported from Chile: Source 1[16] ; Source 2[17.] (Converted to sodium nitrate values from N valuues); Sourec 3[17]

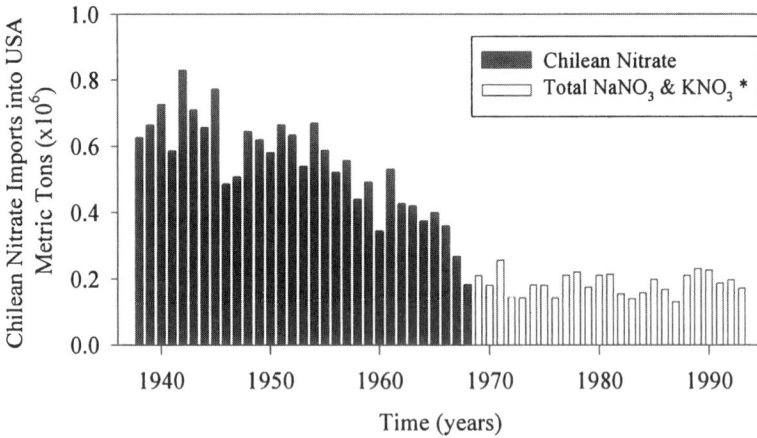

Figure 2 Toal Chilean Nitrate imported ino USA[17]
* Total Sodium Nitrate and Pottasium Nitrate imported into USA reported from 1969 (Specific value for Chilean Nitrate Unknown)

One of the first discoveries of this effort was that Ericksen[2] had reported finding a "trace" of perchlorate in nitrate-rich caliche blister from southeastern California.   Not only did this support the idea that natural perchlorate might occur outside of Chilean nitrate fields, but it also indicated perchlorate could be present in amounts as high as tens of parts per million since the analytical method was relatively insensitive.   Ericksen[15, 23] had earlier reported the association of nitrate deposits with saline minerals and deposits, and arid climatic conditions. He also suggested that the perchlorate was most likely present as potassium perchlorate or in the crystal structure of other minerals. Van Moort[24] had indicated that there is a tendency for perchlorate to be more concentrated in the basal parts of the nitrate deposits and that preservation was dependent on a largely arid environment. Based on this and other information, evaporite and evaporite-related samples of various ages were acquired from North and South America by Orris et al.[25] (Figure 3). The geologic samples were selected to maximize the chances of finding perchlorate given the limited understanding that was based on the Chilean deposits and to represent a variety of geologic ages. Where possible, samples were collected from just above impermeable or relatively impermeable horizons or were representative of late-stage evaporitic processes. Other geologic samples, such as limestone and phosphate rock were also tested for perchlorate, although they were not expected to contain perchlorate.

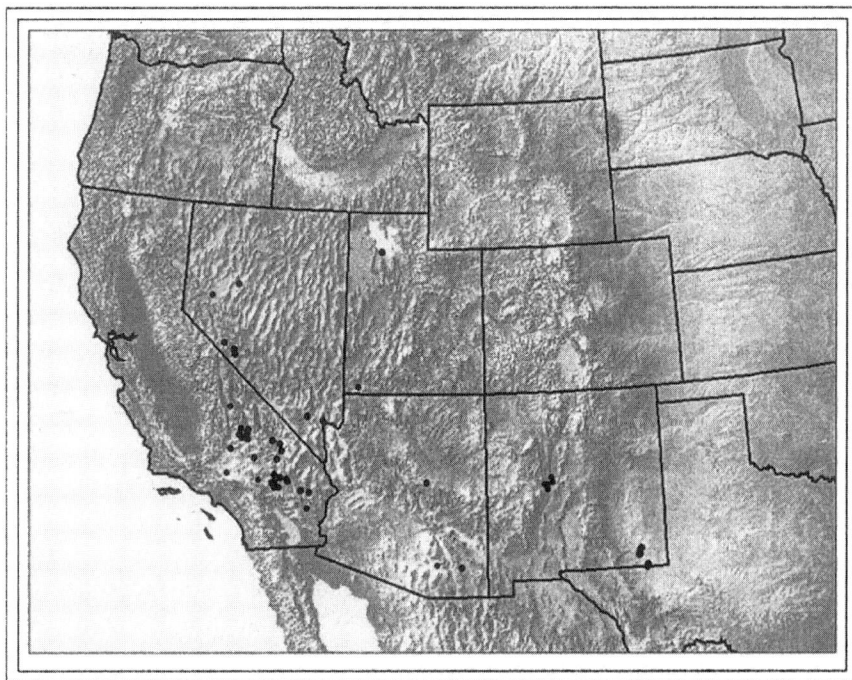

Figure 3. USGS distribution of sampling points for surficial perchlorate in the American Southwest.

Perchlorate was also identified in approximately one-half of the evaporite samples analyzed; although the initial results could be questioned as ion chromatography is non-specific for perchlorate and has increased unreliability for samples with high total dissolved solids. In addition, there were indications that the method of sample processing could influence results. However, the presence of perchlorate was confirmed by other methods, duplicate analyses gave similar results for a given sample (ranging from 0 to hundreds of parts per million), and there was no correlation of TDS or salinity with detectable perchlorate within the evaporite samples collected. While uncertainties about the exact quantity of perchlorate remained, perchlorate was probably present within many of the samples and some initial observations could be made:

- The distribution of perchlorate is highly variable.
- Perchlorate was not typically associated with sulfate minerals.
- Perchlorate seems to be affiliated in post-halite mineralization in evaporite sequences.
- Stratigraphic control and the age of the geologic system appear to be important.

Subsequently, reanalysis of most of the original samples, as well as new samples, by IC-MS-MS has shown that all evaporite samples (including those in which no perchlorate was initially found) containing K- and (or) Mg- salts contain detectable perchlorate, typically in the few to tens of ppb. Perchlorate has not been detected to date in samples of pure sulfates or halite; but is variably present in the soils and clay sediments associated with playas. The IC-MS-MS has allowed reliable determination of perchlorate at the ppb level even in the presence of high TDS. While a full understanding of the presence of perchlorate in evaporites and soils from arid areas is lacking, this study highlights the potential for the widespread occurrence of perchlorate and further supports both the mechanisms of atmospheric production as well as the important role of evaporative concentration in producing environmentally significant perchlorate concentrations.

The Southern High Plains, Texas, and New Mexico

In addition to the occurrence of perchlorate in evaporites, a more regional occurrence of perchlorate in the saturated and unsaturated zone has been reported in a large area of the panhandle of Texas and extreme eastern edge of New Mexico.[26, 27] As part of the Unregulated Contaminant Monitoring Rule a number of Public Water System (PWS) wells were tested in the Spring of 2002 in West Texas. Interestingly, perchlorate was detected in a number of samples. Following confirmation of these results and the subsequent testing of additional wells in the area with continued presence of perchlorate, a more detailed investigation was commenced.[26] The investigated area was approximately 155,000 km,[2] although the extent of the final impacted area was not determined (Figure 4). This is approximately twice the size of the area containing the known deposits of the Atacama Dessert in Chile.[11] The area is composed of both the southern reaches of the high plains as well as the low rolling hills to the south and east of the caprock escarpment. The majority of the study area is underlain by the High Plains Aquifer System, commonly referred to as the Ogallala. Other major aquifers in the study area include the Edwards-Trinity (plateau) aquifer which exists both independently in the southern portions of the study area as well as underlying parts of the region dominated by the Ogallala, the Cenzoic Pecos Alluvium, the Seymour, and underlying most of the study area the Dockum Aquifer (Figure 4). The High Plains portion of the study area is dominated by agricultural activity which largely started in the 1930s. In addition, substantial oil and gas exploration has and currently occurs throughout the area.

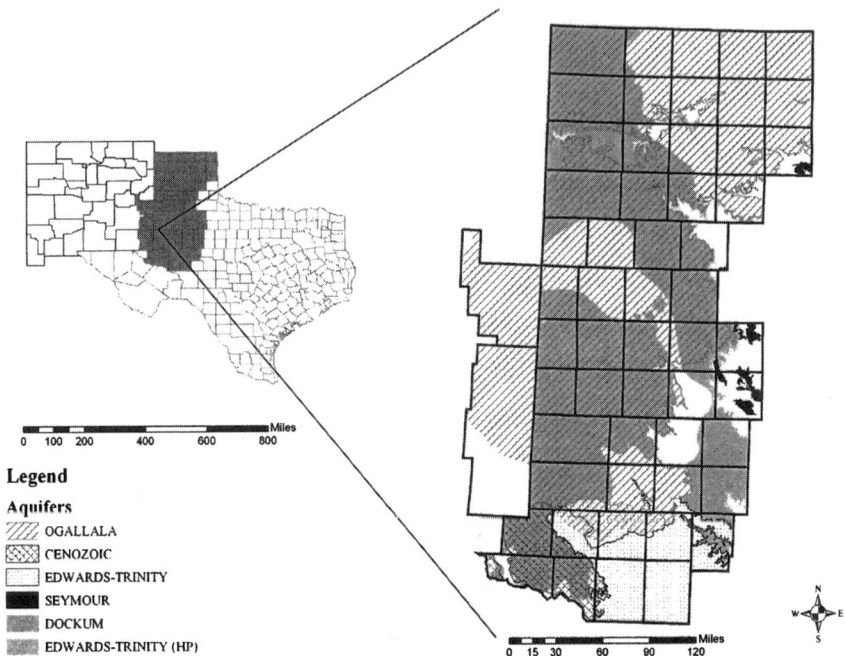

Figure 4. Overview of study area and major aquifers.

The distribution of perchlorate in the saturated zone (Figure 5) was produced by interpolating 773 wells which consisted of Public Water System, Private, Agricultural, and USGS wells.[26] Detailed information on the distribution of perchlorate in most areas is very poor with relatively few wells to represent large areas. However, it is clear that perchlorate is widely distributed across the study area with highest concentrations located in the southern portions of the area overlying the Ogallala aquifer, and areas least impacted in the extreme southeast (Figure 4). While much of the groundwater was impacted by perchlorate, large areas had relatively low perchlorate concentrations often below 1 µg/L. All major aquifers in the area were impacted to some extent although with the exception of the Ogallala and to a lesser extent the Cenzoic, relatively few wells were sampled in other systems due to the limited use of the aquifers. Total mass of perchlorate in the saturated zone of Ogallala aquifer is estimated to be approximately 1.6×10^6 kg.

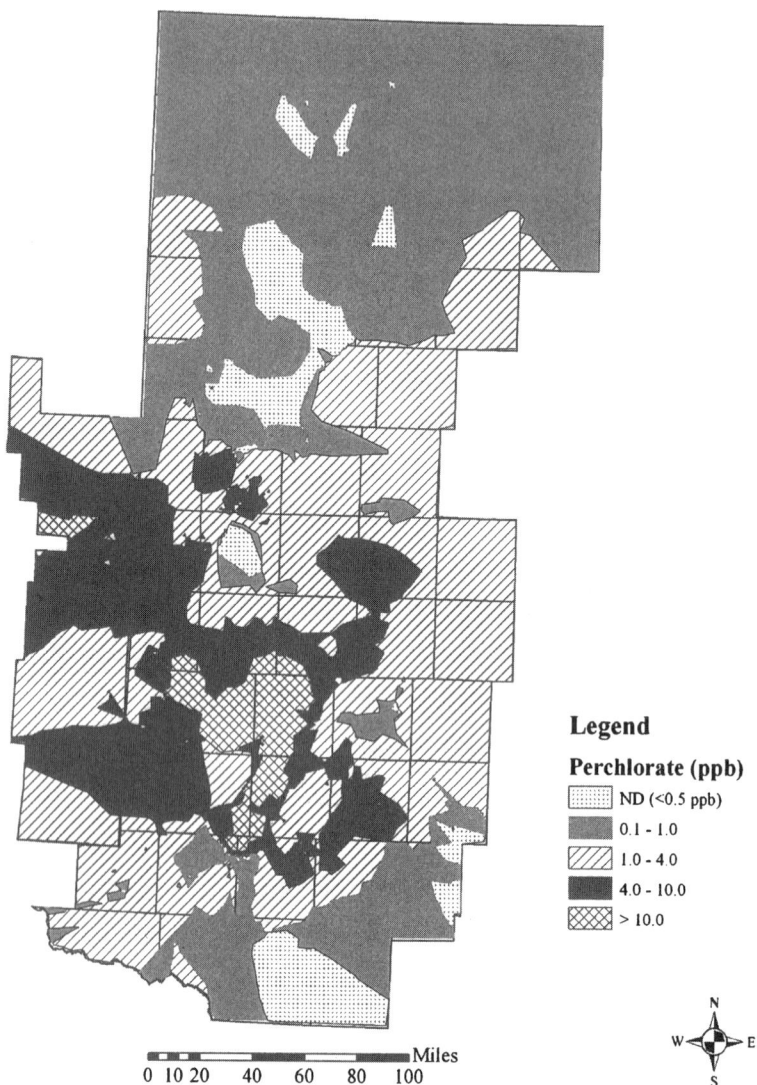

Figure 5. Distribution of perchlorate in the Southern High Plains. Distribution is an interpolation of well data using ARCGIS from 773 wells.[26]

Vertical distribution of perchlorate within the High Plains Aquifer System was typically highest at the surface of the aquifer and decreased with depth.[26] In addition, at least for some deeper intervals the age of the water was estimated to be pre-1950s by tritium dating. A strong inverse relationship was found between both total saturated thickness and depth to water and perchlorate concentration for the overall combined data set which led the authors to conclude that the source of the perchlorate was a surface source and that time of infiltration and dilution of infiltrating water were both important factors in the final concentration of perchlorate. Supporting this conclusion were results from several locations in which unsaturated sediments were extracted and analyzed for perchlorate.[28] Two of the sites had no previous history of crop production or oil or gas exploration, while the other sites were in areas of previous known crop production. Results reported indicate that in the two non impacted sites distribution of perchlorate was similar to chloride distribution with profiles stereotypical of atmospherically deposited species. Profiles at the other sites were more variable with no clear patterns. Total mass of perchlorate at each site varied but ranged from 8.5 - 73 mg/m^2 with an average of 48 mg/m^2. Total perchlorate in the unsaturated zone was estimated at approximately 7.5×10^6 kg.

Numerous sources were proposed to account for the perchlorate in the study area including Defense sites, flares, historic use of Chilean Nitrate, chlorate defoliants, among others.[26] Detailed studies have shown that of these only Chilean nitrate could have been potentially responsible, however a number of lines of evidence were reported to make this extremely unlikely including the total mass of Chilean nitrate required (~48% of the total imported into the United State from 1935-1970), detailed analysis of the co-occurrence of land use and perchlorate occurrence, as well as the distribution of perchlorate in the unsaturated sediments and age of the impacted groundwater. The conclusions of the study were that while no direct evidence exists, the only source capable of explaining the distribution over such a wide area and vertically within the sediments was a natural atmospheric source with deposition over long time periods. In addition, the high correlation ($r^2 = 0.44$) between iodate and perchlorate in groundwater supported the atmospheric origin of the perchlorate[8]. The relationship between groundwater perchlorate concentration and agriculture activity was solely proposed to be due to flushing of accumulated perchlorate in the unsaturated zone into the saturated zone and potentially additional concentration due to increased evaporative processes.

Other Occurrences

A number of sporadic reports exist which support a more widespread distribution of natural perchlorate. Initial samples were analyzed by ion chromatography in the late 1990's at Wright-Patterson Air Force Base. This preliminary study found perchlorate could be detected by ion chromatography in fish meal (10 ppb), blood meal (5 ppb), and kelp meal (300 ppm).[25] In a very recent study of the groundwater in New Mexico, water from wells and springs sampled on the Pajarito Plateau, Espanola Basin, and the Rio Grande North of Taos were found to have an average perchlorate concentration of 0.24 µg/l and precipitation samples collected had trace (0.006-0.017 µg/l) concentrations of perchlorate as well.[29] In another study, upper Colorado River water collected near the upper reaches of the water shed contained low amounts of perchlorate, generally < 0. 2 µg/l.[30] Another intriguing result has been reported in which kelp meal samples were positive for perchlorate. Samples of *Chondrus crispus* from the Western North Atlantic and *Gigartina* from the Pacific Ocean near Chile were analyzed by a modified EPA 8321 IC-MS-MS methodology. The IC-MS-MS results showed that *Chondrus crispus* samples from the Atlantic contained between 20-80 ppb and the *Gigartina* sample from Chile contained less than 10 ppb (Greg Harvey).[31]

POTENTIAL MECHANISM OF FORMATION

As previously mentioned, a significant body of data now shows that the perchlorate found in the Atacama Dessert of northern Chile was formed atmospherically. Other evidence of atmospheric production include the detection of perchlorate in the atmosphere[7], as well as more recent reports of perchlorate in precipitation[8]. Perchlorate has been formed in experiments potentially relevant to atmospheric conditions including molecular chlorine (Cl_2) and ozone (O_3) in the presence of sulfuric acid (H_2SO_4).[32] Although surface oxidative weathering may also be a secondary source of perchlorate [33], no current field evidence exists to support its contribution but a number of reported experiments confirm the potential for its contribution to the pool of naturally produced perchlorate.

Atmospheric Mechanisms of Perchlorate Formation

Atmospheric production of perchlorate has been proposed by Grossling and Ericksen[12] but was formally posed as a realistic pathway by Simmonatis and Heiklen,[14] who recognized that the possibility of perchlorate formation in the stratosphere could be an important sink to explain the significant deficit in the inorganic chlorine budget. A number of potential mechanisms can be formulated theoretically of which very few have been shown to produce perchlorate experimentally. Regardless, all of the mechanisms require some source of chlorine and some source of energy (UV or lightening) in order to produce energetic oxidant species such as O_3, hydrogen peroxide (H_2O_2), hydroxyl radical ($\cdot$OH), perhydroxyl radical ($HO_2\cdot$), superoxide radical ($O_2^-\cdot$), ground oxygen atom ($O(^3P)$), excited oxygen atom ($O(^1D)$), etc. which may directly or indirectly participate in oxidizing atomic chlorine ($Cl\cdot$) to perchlorate. In addition to these criteria, environmental factors in the atmosphere that may also play important roles in the production of perchlorate include reaction pH,[34, 35] moisture content (usually expressed in relative humidity)[36, 37], concentrations of reactants such as $Cl\cdot$, O_3, H_2O_2, $\cdot OH$[38], transition metals such as dissolved iron (Fe), manganese (Mn), copper (Cu),[39,40, 41] and reaction phases such as gas, liquid, gas-liquid interface, and solid crystal.[42-46] Lastly, the atmospheric phase (homogenous vs. heterogeneous) in which critical chlorine oxidation reactions occur may also be of importance.

Homogeneous Mechanisms of Perchlorate Production - It is well known that atmospheric oxidants such as ozone and reactive HO_x radicals readily react with inorganic chlorine species. While little is known about the final oxidation steps leading to perchlorate, a large volume of literature is available on the oxidation of chlorine. The photochemical reaction of ozone in the ocean surface layer is likely to be a significant source of Cl_2 from aqueous sodium chloride (NaCl) and genuine sea-salt particles.[47,48] Subsequent photolysis of volatile Cl_2 or its hydrolysis product, HOCl, in the aqueous phase will generate $Cl\cdot$ or $ClO\cdot$.[47,49,50] In the stratosphere, these reactive inorganic chlorine radicals can be readily converted into the predominant sink such as hydrochloric acid (HCl) and minor sinks such as hypochlorous acid (HOCl), chlorine nitrate ($ClNO_2$), and Cl_2 depending on environmental conditions.[51]

Model I: The first document dealing with perchlorate formation in the atmosphere appeared three decades ago.[14] The researchers speculated that formation of perchlorate rather than HCl could be a more effective sink for inorganic chlorine radicals ($Cl\cdot$) in the stratosphere via combined oxidation steps. They speculatively proposed that oxidation of $Cl\cdot$ may be proceeded via airborne oxidants including ozone and reactive free radicals such as $\cdot$OH, $HO_2\cdot$, and possibly $\cdot$O. The key mechanism includes:

$$Cl\cdot + O_3 + M \rightarrow ClO_3\cdot + M$$
$$ClO_3\cdot + \cdot OH \rightarrow HOClO_3$$

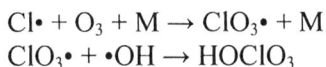

Although this mechanism is comprised of simple steps, disadvantages are implicit such that perchlorate formation would be kinetically slow as non-selectively reactive $\cdot$OH competes with various reactants other than $ClO_3\cdot$ and depends directly on the availability of $ClO_3\cdot$ production.

Model II: An alternative mechanism has been proposed by Prasad and Lee.[52] The key mechanism involves oxidant species such as molecular oxygen (O_2), O_3, and $HO_2\cdot$. It was suggested that $ClO\cdot O_2$, a reaction intermediate, would be formed as a weakly bound complex. The $ClO\cdot O_2$ can enhance the rate of the formation of $ClO\cdot O_3$, which could then react with $HO_2\cdot$ to produce perchlorate:

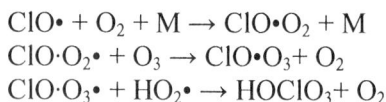

$$ClO\cdot + O_2 + M \rightarrow ClO\cdot O_2 + M$$
$$ClO\cdot O_2\cdot + O_3 \rightarrow ClO\cdot O_3 + O_2$$
$$ClO\cdot O_3\cdot + HO_2\cdot \rightarrow HOClO_3 + O_2$$

They claimed that this mechanism has the advantage that perchlorate formation does not depend on $ClO_3\cdot$ and $\cdot OH$. It was also suggested that the reaction between $ClO_3\cdot$ and $\cdot OH$ can produce $HO_2\cdot$ and $OClO\cdot$ via an intermediate, $[HOOClO_2]^*$, and therefore impact the chain reactions and partitioning of chlorine species.

$$ClO_3\cdot + \cdot OH \rightarrow [HOOClO_2]^* \rightarrow HO_2\cdot + OClO\cdot$$
$$OClO\cdot + \cdot OH \rightarrow HOClO_2 \rightarrow HOCl + O_2$$

Heterogeneous Mechanisms of Perchlorate Production - Martin et al.[32] first reported that the heterogeneous reactions of $Cl\cdot$ and $ClO_2\cdot$ on H_2SO_4-coated Pyrex produced HCl as a major product with small quantities of perchlorate as a minor product:

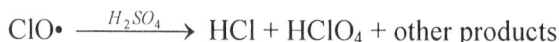

$$ClO\cdot \xrightarrow{\ H_2SO_4\ } HCl + HClO_4 + \text{other products}$$

Trace amounts of perchlorate were detected in authentic sulfate aerosols from the upper troposphere and lower stratosphere.[7,53] However, this mechanism was proposed especially under volcanic aerosol conditions. No robust mechanism for sulfate aerosol-associated perchlorate formation has yet been proposed.

Relation of Proposed Mechanisms to Isotopic Data

Once formed, perchlorate is stable against UV-light.[54] Its oxygen atoms will not exchange with water even at $100^{\circ}C$ in 1 M sulfuric acid over 63 days.[55] Therefore, Bao and Gu[9] proposed that measurement of oxygen isotope composition can provide clues to the origin and subsequent alteration of perchlorate in the environment. They found that natural perchlorate showed a unique oxygen isotope signature significantly different from anthropogenic perchlorate in terms of $d^{18}O$ (-17.3 ~ - 19.9 for various man-made perchlorate vs. -4.5 ~ -9.0 for natural sources) and $\Delta^{17}O$ (-0.06 ~ - 0.20 for various man-made perchlorate vs. 8.8 ~ 9.6 for natural sources). A non-zero

$\Delta^{17}O$ value is a strong indication of incorporation of O derived from O_3 produced through mass independent photochemical reactions.

They noted that water (or •OH), O_2, and O_3 as three sources of oxygen atoms may contribute to the final oxygen isotopic composition of atmospheric perchlorate. Although there exist highly varying ranges for $\delta^{18}O$ values with respect to temperature, pressure, and altitude, they formulated the approximate relationship using the reported mean values of $\delta^{18}O$ for O_3, O_2, and •OH (100‰, 23.5‰, and -165‰, respectively) and -4.6 to -9.0‰ for perchlorate oxygen:

$$1\delta^{18}O\ (O_3) + 2\ \delta^{18}O\ (O_2) + 1\delta^{18}O\ (\bullet OH) \approx \delta^{18}O\ (ClO_4^-)$$

They confirmed that up to one O_3 and a single •OH need to be incorporated in the product perchlorate during the formation process and therefore the oxidation of chlorine by ozone to produce perchlorate could be a sink for atmospheric chlorine.

Surface Oxidative Weathering

Heterogeneous photocatalysis may be a secondary source of perchlorate formation in desert environments. For example, illumination of TiO_2, a frequently observed mineral oxide even in soils, produces photo-excited electrons (e^-) in conduction band and holes (h^+) in electron-depleted valence band.[56]

$$TiO_2 \xrightarrow{\ hv\ } TiO_2\,(e^- + h^+)$$

These can migrate to the oxide surface and undergo half-cell reactions that are part of a catalytic cycle. The half-cell reactions comprise two primary pathways for substrate oxidation. Adsorbed substrate (S_{ads}) can undergo direct oxidation in the valence-band hole:

$$TiO_2\,(h^+) + S_{ads} \rightarrow TiO_2 + S\bullet^+_{ads}$$

In addition, electron transfer from high concentrations of adsorbed solvent molecules such as H_2O_{ads} and OH^-_{ads} generate surface-bound hydroxyl radicals ($\bullet OH_{ads}$), which can be transported to bulk aqueous solution. This indirect oxidation mechanism is known to be of greater importance as long as water molecules are present:[57]

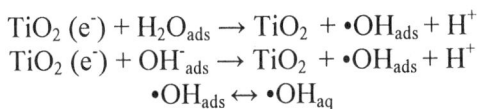

$$TiO_2\,(e^-) + H_2O_{ads} \rightarrow TiO_2 + \bullet OH_{ads} + H^+$$
$$TiO_2\,(e^-) + OH^-_{ads} \rightarrow TiO_2 + \bullet OH_{ads} + H^+$$
$$\bullet OH_{ads} \leftrightarrow \bullet OH_{aq}$$

$$S_{ads} + \bullet OH_{ads} \rightarrow S\text{-}OH\bullet^+_{ads}$$
$$S_{aq} + \bullet OH_{aq} \rightarrow S\text{-}OH\bullet^+_{aq}$$

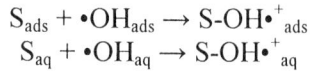

It is well known that the photolysis of nitrite (NO_2^-) and nitrate (NO_3^-) solutions generate $\bullet OH$.[58] The photolysis of nitrite in the wavelength of 200 – 400 nm undergoes the formation of nitrogen oxide ($NO\bullet$) and oxide radical ($O\bullet^-$), which protonates to form $\bullet OH$ at pH < 12 ($pK_a = 11.9$):

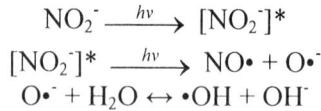

$$NO_2^- \xrightarrow{\ hv\ } [NO_2^-]^*$$
$$[NO_2^-]^* \xrightarrow{\ hv\ } NO\bullet + O\bullet^-$$
$$O\bullet^- + H_2O \leftrightarrow \bullet OH + OH^-$$

Similar to that of NO_2^-, the photolysis of NO_3^- proceeds via two primary photolytic pathways at illumination wavelength above 280 nm although there has been some debate on the mechanisms:

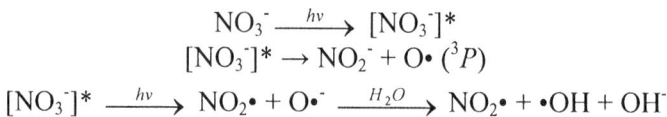

$$NO_3^- \xrightarrow{\ hv\ } [NO_3^-]^*$$
$$[NO_3^-]^* \rightarrow NO_2^- + O\bullet \ (^3P)$$
$$[NO_3^-]^* \xrightarrow{\ hv\ } NO_2\bullet + O\bullet^- \xrightarrow{\ H_2O\ } NO_2\bullet + \bullet OH + OH^-$$

Recently, Miller et al.[33] reported that perchlorate was produced from titanium dioxide (TiO_2) or nitrate-photocatalyzed oxidation of dry chloride salts. They proposed that $\bullet OH$ may be primarily responsible for small quantities of perchlorate (up to 16 ng/g soil) that have been found in desert soil based on previous findings that desert soils can generate $\bullet OH$ under irradiation.

CONCLUSIONS

While relatively few studies have been conducted investigating the occurrence of natural non-Chilean perchlorate it is likely that perchlorate is much more widespread than previously thought albeit at relatively low concentrations. While no systematic study has been conducted it is also reasonable to conclude that concentrations of naturally produced perchlorate will be highest in environments that have been arid for long periods of time 1 x 10^3 to 1 x 10^6 years. The arid environment serves both to concentrate the perchlorate as well as to prevent anaerobic conditions in which perchlorate could be microbially degraded. In very arid areas the perchlorate may be stored in the subsurface unsaturated zone much as nitrate has been reported[59]. In other areas in which conditions allow for infiltration into groundwater, perchlorate may be present although if significant outflow exists or if anaerobic conditions exist the perchlorate would not be expected to accumulate significantly. Interestingly, areas which may be most impacted are those in which irrigated agriculture has been established in semi-arid or arid areas. The irrigation can serve to both mobilize stored perchlorate (and other salts) allowing its infiltration into groundwater as well as increase the effect of evaporative concentration.

Much more difficult to assess is the overall importance of naturally produced perchlorate in the environment. Certainly there may be significant legal and economic consequences if some occurrences can be attributed to non-industrial sources, although it may prove very difficult to determine when low level perchlorate concentrations are due to natural background. From a strictly health based view the source of the perchlorate is irrelevant but from a practical view there may be significant issues if no "deep pockets" are available to fund treatment programs. Scientifically understanding perchlorate's role in the global biogeochemical cycle will certainly be of consequence. Other more speculative issues include the role of natural perchlorate in the evolution of bacterial enzymes capable of utilizing perchlorate (perhaps explaining their ubiquitous nature), and perhaps an explanation for the perchlorate in vegetation, dairy milk, and even human breast milk from areas with no known sources of perchlorate (FDA, 2004). [60]

Plants are known to concentrate perchlorate in leafy vegetation.[61-63] Using the lower end of concentration factors (~300) reported by Jackson et al. [63] and an average perchlorate concentration in precipitation[8] of 0.2 μg/L, would lead to a background concentration of perchlorate in leafy vegetation of 60 μg/kg. Interestingly, in a national survey of lettuce types conducted by the Food and Drug Administration average concentrations of perchlorate in different lettuce types ranged from 7-12 μg/kg which could theoretically be attributed solely to atmospherically deposited perchlorate. Of course these calculations are highly speculative as no information is available on the irrigation source, type of irrigation, lettuce processing or other important variables that could significantly effect these calculations. While no such simple calculations can be conducted to explain the presence of perchlorate in dairy milk (average concentration = 5.76 μg/L) [60] certainly a concentration factor may exist due to the presence of the sodium iodide symporter in the mammary gland. Significantly more studies will need to be conducted to determine regional background groundwater and soil perchlorate concentrations, confirm plant concentration factors, and explore the transport of ingested perchlorate into both dairy and human milk.

REFERENCES

1. Schumacher, J. *Perchlorates: Their Properties, Manufacture and Uses*, New York: Reinhold Publishing Corporation, 1960.

2. Ericksen, G.E., Hosterman, J.W., St. Amand., P. Chemistry, mineralogy and origin of the clay-hill nitrate deposits, Amargosa River Valley, Death Valley region, California, U.S.A. *Chem. Geol.* 1988; 67:85-102.

3. Baas Becking, L.G.M., Haldane, A.D., Izard, D. Perchlorate, an important constituent of seawater. *Nature* 1958; 182:645-7.

4. Loach, K.W. Estimation of low concentrations of perchlorate in natural materials. *Nature* 1962; 195:794-5.

5. Michalski, G., Böhlke, J.K., Thiemens, M. Long term atmospheric deposition as the source of nitrate and other salts in the Atacama Desert, Chile: new evidence from mass-independent oxygen isotopic compositions. *Geochim. Cosmochim. Acta* 2004; 68:4023-38.

6. Böhlke, J.K., Ericksen, G.E., Revesz, K. Stable isotope evidence for an atmospheric origin of desert nitrate deposits in northern Chile and southern California, USA. *Chem. Geol.* 1997; 136:135-52.

7. Murphy, D..M., Thomson, D.S. Halogen ions and NO^+ in the mass spectra of aerosols in the upper troposphere and lower stratosphere. *Geophys. Res. Lett.* 2000; 27:3217-20.

8. Dasgupta, P.K., Martinelango, P.K., Jackson, W.A., Anderson, T.A., Tian, K., Tock, R.W., Rajagopalan, S. The origin of naturally occurring perchlorate: the role of atmospheric processes. *Environ. Sci. Technol.* 2005; 39:1569-75.

9. Bao, H., Gu., B. Natural perchlorate has a unique oxygen isotope signature. *Environ. Sci. Technol.* 2004; 38:5073-7.

10. Sturchio, N.C., Hatzinger, P.B., Arkins, M.D., Suh, C., Heraty, L.J. Chlorine isotope fractionation during microbial reduction of perchlorate. *Environ. Sci. Technol.* 2003; 37:3859-63.

11. Ericksen, G.E. The Chilean nitrate deposits. *Am. Scientist* 1983; 71:366-74.

12. Grossling, B.F., Ericksen, G.E. Computer Studies of the composition of Chilean nitrate ores: Data reduction, basic statistics, and correlation analysis. USGS Open File Series, no. 1519, 1971.

13. Vogt, R. "Reactive Halogen Compounds in the Atmosphere." In *Iodine Compounds in the Atmosphere,* Fabian, P., Singh, O.N., eds. Berlin Heidelberg: Springer-Verlag, 1999.

14. Simonaitis, R., Heicklen, J. Perchloric acid: a possible sink for stratospheric chlorine. *Planet. Space Sci.* 1975; 23:1567-9.

15. Ericksen, G.E. Geology of salt deposits and the salt industry of northern Chile. USGS Open File Series, no. 689. 1963

16. Asociación de Productores de Salitre de Chile (1930*). Industria del Salitre de Chile 1830 - 1930*. Valparaíso: Sociedad Imprenta y Litografía Universo. Retrieved January 2005 from: http://www.albumdesierto.cl/ingles/2estadis.htm

17. US Bureau of Mines, *Mineral Yearbook 1938 – 1993*. Retrieved from: http://libtext.library.wisc.edu/cgi-bin/EcoNatRes/EcoNatRes-idx?type=browse&scope= ECONATRES.MINERALSYEARBK

18. Renner, R. Study finding perchlorate in fertilizer rattles industry. *Environ. Sci. Technol.* 1999; 33:394A-5A.

19. Beaton, J.D. *Fertilizer Use- A Historical Perspective.* Retrieved February 2005 from: http://www.back-to-basics.net/efu/pdfs/history.pdf

20. Urbansky, E. T., Collette, T.W., Robarge, W.P., Hall, W.L., Skillen, J.M., Kane, P.F. EPA/600/R-01/049. *Survey of Fertilizers and Related Materials for Perchlorate*: Final Report. Cincinnati, OH: U.S. Environmental Protection Agency, July 2001b.

21. Tollenaar, H., Martin, C. Perchlorate in Chilean nitrate as the cause of leaf rugosity in soybean plant in Chile. *Phytopathology* 1972; 62:1164-6.

22. Urbansky, E.T., Brown, S.K., Magnuson, M.L., Kelty, C.A. Perchlorate levels in samples of sodium nitrate fertilizer derived from Chilean caliche. *Environ. Pollut.* 2001a; 112:299-302.

23. Ericksen, G. E. Geology and origin of the Chilean nitrate deposits, 1981; U.S. Geological Survey Professional Paper 1188, 37.

24. Van Moort, J.C. Procesos naturals de enriquecimiento de iones nitrato, sulfato, perchlorato, iodato, borato, perclordo y chromato en los claiches del norte de Chile, in IV Congreso Geologico Chileno: Antofagasta, Chile, Universidad del Norte Chile, 1985; 4:3.674-702.

25. Orris, G.J., Harvey, G.J., Tsui, D.T., Eldridge, J.E. *Preliminary analyses for perchlorate in selected natural materials and their derivative products.* USGS Open File Report 03-314. Retrieved 2003 from: http://geopubs.wr.usgs.gov/open-file/of03-314/OF03-314.pdf

26. Jackson, W.A., Anderson, T.A., Lehman, T., Rainwater, K.A., Rajagopalan, S., Ridley, M., Tock, W.R. Distribution and potential sources of perchlorate in the high plains region of Texas. Final Report to the Texas Commission on Environmental Quality. 2004.

27. Jackson, W.A., Anandam, S., Anderson, T.A., Lehman, T., Rainwater, K.A., Rajagopalan, S., Ridley, M., Tock, W.R. Perchlorate occurrence in the Texas southern high plains aquifer system. *Groundwater Monitoring and Remediation* 2005a; 25:137-49.

28. Patil, L. B. *Vertical Distribution of Perchlorate in the Unsaturated Zone*, Master's Thesis, Texas Tech University, Lubbock, Texas, 2005.

29. Dale, M.R., Granzow, K.P., Yanicak, S.M., Englert, D.E., Longmire, P.A., Counce, D.A., Trace Perchlorate in Ground Waters of the Pajarito Plateau, Española Basin and the Rio Grande North of Taos, New Mexico. *Los Alamos National Laboratory Hydrogeologic Characterization Program Quarterly Meeting, 2004 October 25; New Mexico Environment Department, DOE Oversight Bureau* (poster presentation).

30. Roefer, P., Zikmund, K., Snyder, S. *Low level perchlorate sampling results in the Colorado river system, and Lake Mead.* Southern Nevada Water Authority Report, 2004.

31. Personal Communication with Greg Harvey

32. Martin, L. R., Wren, A. G., Wun, M. Chlorine atom and ClO wall reaction products. *Int. J. Chem. Kinet.* 1979; XI:543-57.

33. Miller, G., Kempley, R., Awadh, G., Richman, K. Photo-oxidation of chloride to perchlorate in the presence of titanium dioxide and nitrate. *Abstract of Papers of the American Chemical Society, August 22, U92-U92 056-AGRO Part 1.* 2004.

34. Pszenny, A.A.P., Moldanová, J., Keene, W.C., Sander, R., Maben, J.R., Martinez, M., Crutzen, P.J. Halogen cycling and aerosol pH in the Hawaiian marine boundary layer. *Atmos. Chem. Phys.* 2004; 4:147-68.

35. Keene, W.C., Sander, R., Pszenny, A.A.P., Vogt, R., Crutzen, P.J., Galloway, J.N. Aerosol pH in the marine boundary layer: A review and model evaluation. *J. Aerosol Sci.* 1998; 29: 339-56.

36. NRC. *Global Tropospheric Chemistry: A Plan for Action.* Washington, D.C.: National Academy Press, 1984.

37. Buckley, P.T., Birks, J.W. Evaluation of visible-light photolysis of ozone-water cluster molecules as a source of atmospheric hydroxyl radical and hydrogen peroxide. *Atmospheric Environ.* 1995; 29: 2409-15.

38. Faust, B.C. Photochemistry of clouds, fogs, and aerosols. *Environ. Sci. Technol.* 1994; 28: 217A-22A.

39. Arakaki, T., Faust, B.C. Sources, sinks, and mechanisms of hydroxyl radical ($^.$OH) photoproduction and consumption in authentic acidic continental cloud waters from Whiteface Mountain, New York: The role of the Fe(r) (r = II, III) photochemical cycle. *J. Geophys. Res.* 1998; 103:3487-504.

40. Behra, P., Sigg, L. Evidence of redox cylcling of iron in atmospheric water droplets. *Nature* 1990; 344:419-21.

41. Graedel, T.E., Mandich, M.L. Kinetic model studies of atmospheric droplet chemistry 2. Homogeneous transition metal chemistry in raindrops. *J. Geophys. Res.* 1986; 91:5205-21.

42. Roeselová, M., Jungwirth, P., Tobias, D.J., Gerber, R.B. Impact, trapping, and accomodation of hydroxyl radical and ozone at aqueous salt aerosol surfaces. A molecular dynamics study. *J. Phys. Chem.* 2003; 107: 12690-9.

43. Knipping, E.M., Dabdub, D. Modeling Cl_2 formation from aqueous NaCl particles: evidence for interfacial reactions and importance of Cl_2 decomposition in alkaline solution. *J. Geophys. Res.* 2002, 107:4360, ACH 8:1-30.

44. Keene, W.C., Khalil, M.A.K., Erickson III, D.J., McCulloch, A., Graedel, T.E., Lobert, J.M., Aucott, M.L., Gong, S.L., Harper, D.B., Kleiman, G., Midgley, P., Moore, R.M., Seuzaret, C., Sturges, W.T., Benkovitz, C. M., Koropalov, V., Barrie, L.A., Li, Y.F. Composite global emissions of reactive chlorine from anthropogenic and natural sources: reactive chlorine emission inventory. *J. Geophys. Res.* 1999; 104: 8429-40.

45. Bizjak, M., Grgic, I., Hudnik, V. The role of aerosol composition in the chemical processes in the atmosphere. *Chemosphere* 1999; 38:1233-1240.

46. Dentener, F.J., Carmichael, G.R., Zhang, Y., Lelieveld, J., Crutzen, P.J. Role of mineral aerosol as a reactive surface in the global troposphere. *J. Geophys. Res.* 1996; 101:22869-89.

47. Oum, K.W., Lakin, M.J., DeHaan, D.O., Brauers, T., Finlayson-Pitts, B.J. Formation of molecular chlorine from the photolysis of ozone and aqueous sea-salt particles. *Science* 1998; 279:74-7.

48. Behnke, W., Zetzsch, C. Heterogeneous formation of chlorine atoms from various aerosols in the presence of O_3 and HCl. *J. Aerosol Sci.* 1989; 20:1167-70.

49. Wongdontri-Stuper, W., Jayanty, R.K.M., Simonaitis, R., Heicklen, J. The Cl_2 photosensitized decomposition of O_3: The reactions of ClO and OClO with O_3. *J. Photochem.* 1979; 10:163-86.

50. Nowell, L.H., Hoigné, J. Photolysis of aqueous chlorine at sunlight and ultraviolet wavelengths - II. Hydroxyl radical production. *Wat. Res.* 1992; 26:599-605.

51. Francisco, J.S. *Ab initio* characterization of $HOClO_3$ and HO_4Cl: Implication for atmospheric chemistry. *J. Phys. Chem.* 1995; 99:13422-5.

52. Prasad, S.S., Lee, T.J. Atmospheric chemistry of the reaction $ClO + O_2 \leftrightarrow ClO \cdot O_2$ Where it stands, what needs to be done, and why? *J. Geophys. Res.* 1994; 99:8225-8230.

53. Jaeglé, L., Yung, Y.L., Toon, G.C., Sen, B., Blavier, J.-F. Balloon observation of organic and inorganic chlorine in the stratosphere: the role of $HClO_4$ production on sulfate aerosols. *Geophys. Res. Lett.* 1996; 23:1749-52.

54. Huie, R.E., Peterson, N.C. The photolysis of concentrated perchloric acid solutions. *J. Photochem.* 1983; 21:31-4.

55. Hoering, T.C., Ishimori, F.T., McDonald, H.O. The oxygen exchange between oxyanions and water. II. Chlorite, chlorate and perchlorate ions. *J. Am. Chem. Soc.* 1958; 80:3876-9.

56. Hoffmann, M.R., Martin, S.T., Choi, W., Bahnemann, D.W. Environmental applications of semiconductor photocatalysis. *Chem. Rev.* 1995; 95:69-96.

57. Legrini, O., Oliveros, E., Braun, A.M. Photochemical processes for water treatment. *Chem. Rev.* 1993; 93:671-98.

58. Mack, J., Bolton, J. R. Photochemistry of nitrite and nitrate in aqueous solution: a review. *J. Photochem. Photobiol.* 1999; 128:1-13.

59. Walvoord, M.A., Phillips, F.M., Stonestrom, D.A., Evans, R.D., Hartsough, P.C., Newman, B.D., Striegl, R.G. A reservoir of nitrate beneath desert soils. *Science* 2003; 302:1021-4.

60. FDA (2004). *Exploratory Data on Perchlorate in Food.* Retrieved March 2005 from: http://www.cfsan.fda.gov/~dms/clo4data.html

61. Tan, K, Anderson, T.A., Jones, M.W., Smith, P.N., Jackson, W.A. Uptake of perchlorate in aquatic and terrestrial plants at field scale, *J. Environ. Qual.* 2004; 33:1638-46.

62. Yu, L., Cãnas, J.E., Cobb, G.P., Jackson, W.A., Anderson, T.A. Uptake of Perchlorate in Terrestrial Plants. *Ecotoxicol. Environ. Safety* 2004; 58:44-9.

63. Jackson, W.A., Joseph, P.C., Patil, L.B., Tan, K., Smith, P.N., Yu, L., Anderson, T.A. Perchlorate Accumulation in Forage and Edible Vegetation. *J. Agri. Food Chem.* 2005b; 53:369-73.

Chapter 4

Alternative Causes of Wide-Spread, Low Concentration Perchlorate Impacts to Groundwater

Carol Aziz, Robert Borch, Paul Nicholson, and Evan Cox
GeoSyntec Consultants, Guelph, ON, Canada

INTRODUCTION

The frequency of detection of perchlorate in groundwater and drinking water supplies has been steadily increasing since its initial identification as a chemical of concern in 1997. It is currently estimated that perchlorate is present in groundwater in at least 30 states and affects the drinking water supplies of more than 20 million people in the southwestern United States (U.S.). The source of perchlorate in water supplies has typically been attributed to U.S. Department of Defense (DOD), National Aeronautics & Space Administration (NASA) and/or defense contractor facilities that have used ammonium perchlorate (AP) in rocket and missile propellants.

As a result of its high profile and its addition to the Unregulated Contaminant Monitoring Rule (UCMR List 1), which requires perchlorate analysis by large public water suppliers and selected small water utilities, most public water supplies are now being routinely analyzed for perchlorate. Through monitoring activities, perchlorate has been detected at low levels (typically less than 50 µg/L) in a significant number of areas without apparent military sources.

While natural sources or formation mechanisms for perchlorate may explain its presence in some cases,[1,2] widespread, low concentration perchlorate impacts in groundwater can apparently also result from a variety of non-military-based inputs as well, potentially including:

 i) storage, handling and use of Chilean nitrate-based fertilizers containing perchlorate;
 ii) manufacturing, storage, handling, use and/or disposal of fireworks containing perchlorate;
 iii) manufacturing, storage, handling, use and/or disposal of road flares containing perchlorate;

iv) manufacturing, storage, handling, use and/or disposal of explosives or pyrotechnics containing perchlorate; and/or

v) manufacture, storage, handling and use of electrochemically-prepared (ECP) chlorine products, primarily those that contain chlorate or were manufactured from chlorate feedstocks.

The potential impacts of these non-military perchlorate products and processes on the environment are discussed in the following sections.

CHILEAN NITRATE FERTILIZERS

Research by the U.S. Environmental Protection Agency (EPA) has confirmed that perchlorate is present in nitrate-based fertilizers manufactured from naturally-occurring caliche deposits mined from the Atacama Desert region of Chile.[3,4] Historical agronomic literature indicates that Chilean nitrate fertilizers were widely used in specific agricultural practices in the early to mid 1900s.[5,6,7] Past import statistics for Chilean nitrate and historical agronomic guidelines for sodium nitrate application for various crops (discussed below) indicate that significant quantities of perchlorate may have been unknowingly applied to agricultural soils over many decades from the early to mid 1900s. While the use of Chilean nitrate fertilizers steadily declined since about the 1930s, there is evidence of continued use through to the present day. For example, imports of fertilizer grade sodium nitrate supplied 27% and 6% of the total nitrogen used as fertilizer in 1939 and 1954,[8] respectively. Since 2002, it is estimated that some 75,000 tons of Chilean nitrate fertilizer have been used annually in the U.S.

This section summarizes pertinent information related to the import and use of Chilean nitrate fertilizers and explores the potential for present-day perchlorate impacts to groundwater from historical and on-going Chilean nitrate fertilizer uses for specific agricultural practices.

Chilean Nitrate Imports

Between 1909 to 1918 and 1925 to 1929, the U.S. imported approximately 7,500,000 and 5,300,000 tons of Chilean,[5,6] respectively, for a total of approximately 13,000,000 tons of Chilean nitrate. If we assume (based on these estimates) that approximately 1 million tons of Chilean nitrate were imported annually during 1919 through 1924, then approximately 19 millions tons of Chilean nitrate fertilizer were likely imported into the U.S. between 1909 and 1929.

During this period, it is estimated that between 49 and 70% of the imported Chilean nitrate was used as fertilizer, with an average of approximately 65%.[9] The percentage of Chilean nitrate used for fertilizer reportedly fluctuated based on its demand for use in explosives manufacturing. Assuming an average perchlorate concentration of about 0.2% in the Chilean nitrate[3] and that 65% of the imported Chilean nitrate (about 12 million tons) was used as fertilizer, then approximately 49 million pounds of perchlorate is likely to have been applied to agricultural soils during this time period.

Chilean nitrate fertilizer is still produced by SQM Corporation and makes up 0.14% of the total annual U.S. fertilizer application[3]. It is sold commercially as Bulldog Soda and is primarily used in a few niche markets and specialty products. Currently, world production is 900,000 tons/year of which 75,000 tons are sold to U.S. farmers for use on cotton, tobacco, and fruit crops.[3,10] SQM reports that the perchlorate concentration in Chilean nitrate fertilizer has been reduced through changes in the refinement processes since 2002. The current perchlorate concentration is reported as 0.01%,[4] which is more than an order of magnitude improvement compared to historic perchlorate contents. However, this amount still represents the potential introduction of more than 15,000 pounds of perchlorate annually to agricultural soils, the fate of which is not well understood.

Use of Chilean Nitrate Fertilizers

A wide variety of agricultural publications document that Chilean nitrate was a common nitrate fertilizer in the U.S. during the first half of the 20th century. For example, in its 1938 Yearbook, the U.S. Department of Agriculture (USDA) stated that "sodium nitrate and ammonium sulfate are undoubtedly the most widely used nitrogen fertilizers at the present time."[11] Similarly, the USDA Fertilizer Consumption and Trends in Usage report[7] identified Nitrate of Soda as the second most consumed fertilizer during its reporting period. While the use of Chilean nitrate fertilizers steadily declined since about the 1930s, there is evidence of continued use through to the present day. The following section discusses the use of Chilean nitrate fertilizer specifically related to the production of cotton, tobacco, and fruit, three crops for which Chilean nitrate use has been documented.

Cotton

Chilean nitrate fertilizer was often used to fertilize cotton and provided the necessary nitrogen for high yield crops.[12] It was typically used in delayed applications (side dressings). The application of nitrate of soda to cotton is dependant on soil quality and the corresponding amount of nitrogen available for plant uptake. Typical delayed application rates of nitrogen for cotton were 18 to 30 pounds per acre.[12] This application rate is equivalent to 110 to 190 pounds per acre of nitrate of soda, which is approximately 16% nitrogen,[13] or approximately 0.2-0.3 lb of perchlorate per acre.

Between 1909 and 1929, Texas was the largest cotton producing state, harvesting approximately 283 million acres of cotton over a twenty year period. However, only 7% of the acreage in Texas required fertilizer application.[12] By comparison, southeastern states such as North Carolina, South Carolina, Georgia, and Alabama harvested lower quantities of cotton, but the fertilizer requirement for these soils was much greater.[12] For example, during this time period, Georgia, Alabama, South Carolina and North Carolina typically fertilized 91 to 97% of the total cotton acreage.[12] While the contribution of Chilean nitrate to fertilization of the cotton acreage is not clearly defined, in 1928, Chilean nitrate accounted for approximately 35% of total nitrogen fertilizer used that year on a nitrogen basis.[6]

Georgia, Alabama, South Carolina, and North Carolina were heavily dependent on the use of Chilean Nitrate fertilizer, consuming between 63% to 75% of the total Chilean nitrate used domestically.[7] Based on the 1909 to 1929 import statistics (about 12 million tons of Chilean nitrate as fertilizer), a consumption rate of 63% to 75% for these states would represent the use of 7.6 to 9.0 million tons of Chilean nitrate, which in turn would represent the potential application of 30 to 36 million pounds of perchlorate to agricultural soils (all crops) in these states over the 1909 to 1929 time frame.

Tobacco

Chilean nitrate fertilizer was commonly used in the U.S. as a source of nitrogen for tobacco plants.[14] From 1909 to 1929, Kentucky was the largest producer of tobacco and harvested 10,000,000 acres. North Carolina was the second highest producer of tobacco, harvesting over 9,000,000 acres.[15]

Fertilizer application rates for tobacco vary with the season and soil quality; however, application rates of 30 to 40 pounds of nitrogen per acre were typically recommended.[16] To obtain this amount of nitrogen from nitrate of soda (16% nitrogen), approximately 185 to 250 pounds of nitrate of soda would have been applied per acre of tobacco. This range of application rates is similar to the application rates of nitrate of soda used today for certain tobacco crops (i.e., 3-5 lb/100 yd^2 or 195-325 lb/acre[17]). Prior to 2002, this Chilean nitrate fertilizer application rate would correspond to a perchlorate application rate of approximately 0.4 to 0.5 lb per acre.

Fruit

The historic use of Chilean nitrate fertilizers has been reported for fruit trees in California, with an accepted fertilization rate between 100 and 200 pounds per acre as nitrogen. This translates to application rates ranging between 625 and 1250 pounds per acre of sodium nitrate (16% nitrogen). For simplicity, if the average application rate is assumed to be 1000 pounds per acre per year of Chilean nitrate as suggested by Collings[18] in the textbook *Commercial Fertilizers*, then 2 pounds of perchlorate per acre per year may have potentially been applied to fruit orchard soils in some parts of California. Furthermore, between 1923 and 1960, 305,614 tons of Chilean Sodium Nitrate fertilizer were reported to have been used in California, according to data compiled by the California Department of Food and Agriculture. Assuming a perchlorate concentration of 0.2%, application of this mass of Chilean nitrate fertilizer would have resulted in the application of over 1.2 million pounds of perchlorate to agricultural soils/crops in California during this timeframe.

Potential to Impact Groundwater

While significant quantities of Chilean nitrate have historically been used to fertilize various crops, it is difficult to predict the fate and persistence of the applied perchlorate. The behavior of perchlorate in agricultural settings has not been investigated in detail, and several crucial aspects of perchlorate behavior in such settings (e.g., plant uptake, biodegradation, mobility in relation to soil factors, etc) are not well documented. However, nitrate (the principal component of the Chilean nitrate fertilizer) and perchlorate share important chemical features, and many aspects of the large body of literature concerning nitrate contamination of groundwater due to fertilizer use can be applied directly to understanding the potential for perchlorate contamination of groundwater through the same mechanism. The important aspects of the relationship between nitrate and perchlorate are summarized as follows:

Nitrate and perchlorate are present in the potential source material, Chilean nitrate fertilizer.

Nitrate (NO_3^-) and perchlorate (ClO_4^-) are both negatively charged ions and, as such, are highly mobile in soils. Soil particles are predominately negatively charged, and, therefore, electrostatic repulsion prevents adsorption.

Sodium nitrate and sodium perchlorate, the predominant forms of these constituents in Chilean nitrate fertilizer, are both highly soluble in water (1.8

and 4.4 pounds per gallon, respectively), and thus there are no solubility constraints on the flushing of these compounds from soil into groundwater.

Once in the vadose zone and groundwater, both nitrate and perchlorate are environmentally persistent and are not subject to chemical or biological breakdown under common groundwater conditions. The biological reduction of both nitrate and perchlorate requires the presence of organic matter, which can serve as electron donors, and anoxic conditions.

While the use of Chilean nitrate fertilizers containing perchlorate was most intense prior to 1950, the potential exists that impacts from these practices are only now being discovered in public water supplies. For example, Hudson et al[19] determined that water produced from 59 of 176 public water supply wells in the Los Angeles Basin was in excess of 50 years old. Bohlke[20] presents data for four representative surficial aquifers in the eastern U.S. with mean ages of 27-50 years, with some fraction of the groundwater being older. Fogg et al[21] and Weissman et al[22] discuss the significance of the dispersion of groundwater ages with regard to breakthrough time and persistence of agricultural pollutants, noting that in areas with deep alluvial aquifers, the observed nitrate pollution may be the result of agricultural practices more than 50 years previously. Given that perchlorate was a component of Chilean nitrate-based fertilizers, the hypothesis may be true for perchlorate.

The available nitrate literature reviewed for this chapter indicates that it is possible that low level perchlorate impacts to groundwater in some areas may be the result of historic use of Chilean nitrate fertilizers. Additional evaluation of soils and groundwater in common crop areas discussed in this section seems warranted to evaluate whether historical fertilizer practices can be expected to be the cause of low concentration perchlorate impacts to groundwater in some agricultural areas and watersheds.

FIREWORKS

Fireworks are widely used by both pyrotechnic professionals and individual consumers for celebratory displays. Perchlorate is known to be a component of many pyrotechnics, and as such, the manufacturing, storage, handling, use and disposal of these products have the potential for introduction of perchlorate into the environment. Many pyrotechnic displays are launched near or over surface waters, presumably for visual impact and safety reasons, increasing the potential for perchlorate impacts to water sources. The following sections discuss the potential for perchlorate to impact the environment.

Perchlorate in Fireworks

Perchlorate is a major component of fireworks and is used primarily as an oxidizing agent. It decomposes at moderate-to-high temperatures, liberating oxygen gas. Because oxidizers must be low in hygroscopicity, potassium salts are preferred over sodium salts. Potassium perchlorate has gradually replaced potassium chlorate as the principal oxidizer in civilian pyrotechnics because of its superior safety record. Potassium perchlorate produces mixtures that are less sensitive to heat, friction, and impact than those made with potassium chlorate, because of its higher melting point and less-exothermic decomposition.[23] Potassium perchlorate can be used to produce colored flames, noise, and light as summarized in Table 1. Ammonium perchlorate is also used in some fireworks formulations. Another potential source of perchlorate is from the potassium nitrate in the black powder used in the lift charge. Potassium nitrate made from Chilean nitrate can contain perchlorate, as has been well documented for sodium nitrate fertilizers.

Fireworks Consumption/Market

In 2003, 221 million pounds of fireworks were consumed in the U.S. This represents almost a 10-fold increase in consumption since 1976, as shown in Figure 1. The demand for fireworks is expected to increase, due to an upsurge of patriotism and an increase in the number of states permitting consumer fireworks. It is now legal to sell consumer fireworks in 43 states plus the District of Columbia.[24]

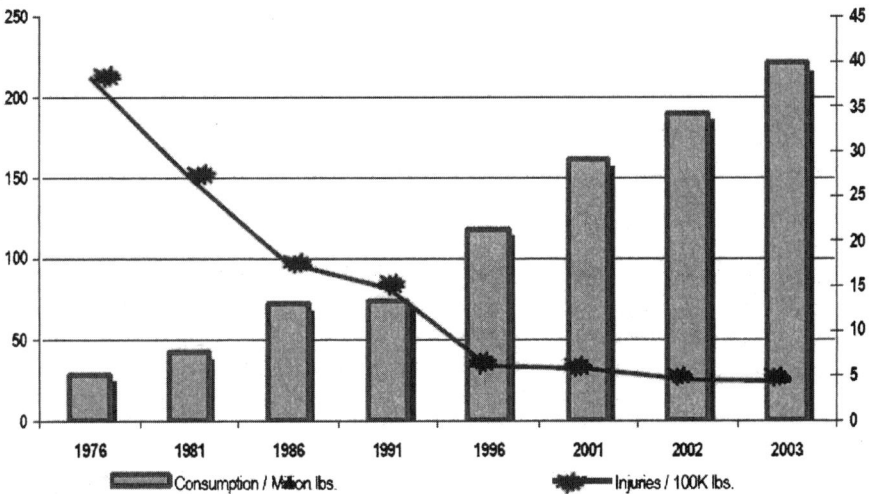

Figure 1: Fireworks Consumption in the United States from 1976-2003[26]

Table 1: Perchlorate Content and Effects in Fireworks

Purpose/Effect	Composition (% by Wt)	
White Light	Potassium Perchlorate	64
	Antimony, Sb	13
	Gum	10
	Potassium Nitrate	13
White Sparks	Potassium Perchlorate	42.1
	Titanium	42.1
	Dextrine	15.8
White Sparks "water fall"	Potassium Perchlorate	50
	"Bright" Aluminum Powder	25
	"Flitter" Aluminum, 30-80 mesh	12.5
	"Flitter" Aluminum, 5-30 mesh	12.5
Red Torch	Ammonium Perchlorate	70
	Strontium Carbonate	10
	Wood Meal (slow fuel)	20
Red Fireworks	Potassium percholrate	67
	Strontium Carbonate	13.5
	Pine Root Pitch	13.5
	Rice Starch	6
Green Fireworks	Potassium Perchlorate	46
	Barium Nitrate	32
	Pine Root Pitch	16
	Rice Starch	6
Purple Flame	Potassium Perchlorate	70
	Polyvinyl Chloride	10
	Red Gum	5
	Copper Oxide	6
	Strontium Carbonate	9
	Rice Starch	5 (additional %)
Blue Flame	Ammonium Perchlorate	70
	Red Gum	10
	Copper Carbonate	10
	Charcol	10
	Dextrine	5 (additional %)
Yellow Flame	Potassium Perchlorate	70
	Sodium Oxalate	14
	Red Gum	6
	Shellac	6
	Dextrine	4
Black Smoke	Potassium Perchlorate	56
	Sulfur	11
	Anthracene	33
Whistle	Potassium Perchlorate	70
	Potassium Bensoate	30

Reference: J.A. Conkling. 1985 Chemistry of Pyrotechnics. Basic Principles and Theory. Marcel Dekker, Inc. New York.

Most of the fireworks consumed in the U.S. are imported from China, with only approximately 3% of the total mass of fireworks produced in the U.S.[24] In 2003, 87.5 million kilograms (192 million lbs) of the 89.2 million kilograms (196 million lbs) of imported consumer fireworks or 98% and 7.5 million kilograms (16.5 million lbs) of the 8.1 million kilograms (17.8 million lbs) or 93% of imported display fireworks were from China.[25]

Potential to Impact Groundwater

Raw perchlorate from fireworks manufacturing facilities and perchlorate residue from detonated fireworks both have the potential to contaminate surface water and groundwater. Although fireworks contain high percentages of perchlorate, it is not currently known how much of the perchlorate finds its way into the environment. If we assume that most of the perchlorate present in the firework is ultimately decomposed with the burning of the firework, it seems necessary to consider only the perchlorate from blind stars, un-ignited display shells, and residues from the fireworks or lift charges.[27] However, statistics on dud rates (fireworks that are launched but not burned) do not exist.[28] To date, housekeeping (i.e., post-event cleanup) related to fireworks displays has been done for safety purposes with the main aim being removal of unexploded fireworks. Typically, dud display shells are removed, but blind stars (which contain perchlorate) are typically not collected. Blind stars are often released at high altitudes and can therefore travel great distances from the launch site. Blind stars can also be released as a result of the breakage of dud shells.

As previously indicated, many fireworks displays occur at the water's edge or on barges, presumably for safety reasons and/or to enhance visual impact. Post-display clean-up becomes more difficult as duds and blind stars can be submerged. The advantage is that there is likely to be less dud breakage. However, perchlorate may leach out of the shell either through the fuse or as the result of de-lamination of the shell casing. The latter is more likely to result in perchlorate releases when the shell casing is comprised of paper/cardboard, as is often the case with fireworks produced in China.

Past and Current Environmental Studies

The number of case studies in the literature discussing extent of soil and water contamination at firework discharge sites is limited. More controlled studies are currently being conducted, which should shed more light on the extent of perchlorate contamination associated with fireworks.

Perchlorate contamination linked to fireworks displays was examined by the Massachusetts Department of Environmental Protection (MADEP) at the

University of Massachusetts at Dartmouth. Eight monitoring wells were installed at a site where fireworks were launched/displayed over the Labor Day weekend of 2004. The campus has been the site of summertime fireworks for more than 10 years. Prior to the 2004 display, soil samples had no detectable levels of perchlorate, while groundwater samples had perchlorate concentrations ranging from <1 to 62 ug/L.[29] Soil samples were collected the day following the display, while groundwater samples were collected periodically throughout the fall. Results of soil sampling immediately after the display indicted a maximum perchlorate concentration of 560 ug/kg. Groundwater concentrations were not substantially different than they were before the display.[29]

Perchlorate contamination may also originate from fireworks manufacturing facilities, given that perchlorate is handled on site. For example, perchlorate was detected at a concentration of 270 µg/L in an inactive well near a defunct fireworks site in Rialto, California.[30] Perchlorate has also been detected at a concentration of 24 µg/L in groundwater from a well near a fireworks manufacturing facility in Mead, NE.[31]

SAFETY FLARES

Safety flares (or fusees) are used in emergency situations for road-side accidents and rail and marine emergencies. The following sections describe the main components of commercial safety flares and assess the potential for perchlorate to impact the environment.

Perchlorate Content in Safety Flares

A flare generally consists of a waxed cardboard tube casing filled with a burn mixture and a cap at the end to ignite the flare. Based on Material Safety Data Sheets (MSDS), the burn mixture contains primarily strontium nitrate (75% by weight), potassium perchlorate (<10% by weight), sulfur (<10% by weight) and sawdust/soil (<10% by weight). Other ingredients present in lesser amounts can include: synthetic rubber, aromatic polycarboxylic anhydride fuel, benzene tetracarboxylic acid (dianhydride and metallic dianhydride), sodium nitrate, polyvinyl chloride case binder, dextrin, magnesium, cellulose nitrate, black powder, wax, and red phosphorus.[32]

Through experiments conducted by the Santa Clara Valley Water District in California, Silva[33] analyzed the contents of an unburned road flare and detected 50,000 mg/kg of perchlorate and 450,000 mg/kg nitrate in a single flare. Comparison of perchlorate leaching from unburned flares that had been damaged (i.e., sliced open) to completely burned flares indicated that the unburned damaged flares leached 2000 times more perchlorate than damaged road flares that were completely burned (3,645 mg versus 1.95 mg).

Production/Use Statistics

In 1997, approximately $101.5 million dollars worth of pyrotechnics (NAICS product code of 325998H107) were produced in the U.S.[34] This classification includes road flares, jet fuel igniters, railroad torpedoes, and toy pistol caps, but not fireworks. Production and trade statistics for road flares alone are not available. In 2003, 7.0 million lbs or $10.6 million dollars worth of pyrotechnics were imported,[25] with 92% from China. While numbers are not available for total domestic flare production, assuming an average cost per flare of $0.50 to $1.00 per flare and annual sales of $20 million by the largest U.S. manufacturer, then at least 20 to 40 million flares may be sold annually.

Potential to Impact Groundwater

Preliminary research by Silva[32,33] of the Santa Clara Valley Water District (SCVWD) indicates that 3.6 g of perchlorate can potentially leach from an unburned, damaged (i.e., run over by a motor vehicle) 20-minute road flare. According to Silva,[33] this amount of perchlorate can potentially contaminate 2.2 acre-feet of drinking water above 4 ug/L (the standard EPA Method 314.0 quantitation limit). Interestingly, even fully burned flares leached 1.9 mg perchlorate/flare.[33] More than 40 metric tons of flares were reported to be used/burned in 2002 in Santa Clara County, California alone.[33] Given this estimate, the potential for perchlorate leaching from road flares and subsequent surface runoff from highways and roads represents a potentially significant and largely uninvestigated impact to surface water and groundwater quality.

Road flare manufacturing has also been implicated in perchlorate contamination at a site in Morgan Hill, California.[35] From 1956 to 1996, highway flares were manufactured at this location.[35] Perchlorate was detected at one on-site monitoring well in 2001 and was detected in a municipal well in March 2002. The perchlorate plume is estimated to be 9 miles long.[36] It is important to note that this site is located in an area that was historically used for fruit and nut production, and perchlorate impacts to soil and groundwater in some areas may also be the result of past fertilizer practices, as discussed earlier.

BLASTING AGENTS

Blasting agents are non-cap sensitive explosives. Generally, they are intimate mixtures of inorganic oxidizers and fuels, rather than the organic explosives commonly used in military applications (e.g., RDX, TNT, HMX). While the main oxidizer employed is usually ammonium nitrate (AN), ammonium perchlorate and other perchlorates (sodium or potassium perchlorate) are compatible with the AN mixtures and can be employed for special applications and to take advantage of perchlorate available from DOD demilitarization activities. Furthermore, sodium nitrate (Chilean origin) historically used in commercial explosives may contain perchlorate as an impurity. Review of MSDS information identifies perchlorate as a common component of many slurry gel explosives (Table 2). The following sections discuss the composition of various commercial blasting agents based on review of MSDS information and examine the potential for perchlorate impacts to soil and groundwater from blasting operations.

Common Blasting Agents and Their Perchlorate Content

Blasting agents, as opposed to explosives, require a booster, in addition to a detonator, to initiate. This is a significant advantage in terms of less stringent and more economical storage and transport considerations. The most common and simplest blasting agent is ammonium nitrate fuel oil (ANFO), which consists of ammonium nitrate prills soaked with fuel oil (about 5 to 6 wt%). ANFO accounts for a large share of the domestic commercial explosives market (about 80% in 1998)[37] and is available in bulk form for on-site mixing or in premixed bags.

As shown in Table 2, some water gels, emulsions, and non-electric detonators can contain substantial amounts of perchlorate (e.g., up to 30%). Furthermore, MSDS sheets for some ANFO products list other "inorganic oxidizers", which may include perchlorate, in their contents. The inclusion of sodium nitrate of Chilean origin, which is known to contain perchlorate, may also have potential to impact groundwater. Further testing is required to determine if these products contain perchlorate.

Consumption/Market

In 2003, the U.S. production of explosives, reported by 23 commercial explosive manufacturers, was 2,520,000 tons.[38] This amount of explosives is typical of the annual U.S. production in the last decade. Of the total U.S commercial production, 2,475,000 tons were classed as blasting agents. Sales of blasting agents were reported in all states with West Virginia, Kentucky,

Wyoming and Indiana consuming the highest quantities[38]. Sixty seven percent of the blasting agents were used in coal mining. Quarrying and nonmetal mining, the second-largest consuming industry, accounted for 14% of total explosives sales. Construction, metal mining and miscellaneous uses accounted for 8%, 8%, and 3% of explosives sales, respectively.[38]

Table 2: Blasting Agents and Explosives Containing Perchlorate
(% Composition)

Type Product	Blasting Agent (1.5) or Explosive (1.1)	NH_4NO_3	$NaNO_3$	$NaClO_4$	other energetic fuel
gel bulk or packaged	blasting agent	55-85	--	0-4	--
packaged gel	blasting agent	33-40	10-15	--	25 - 51
package emulsion	explosive	60-70	0-5	0-15	--
package emulsion	explosive	60-80	0-12	--	--
packaged gel	explosive	<65	<20	<7	--
ANFO	blasting agent	94.5	--	--	--
water gel	blasting agent	<80	--	<5	
water gel	blasting agent	< 75	<5	<5	
water gel	explosive	<65	<20	<7	
water gel	explosive	<65	<20	<7	
water gel, presplit	explosive	<65	<20	<7	
water gel	blasting agent	10-20	10-20	20-30*	

* ammonium perchlorate

Potential to Impact Groundwater

Although most perchlorate should be consumed during detonation of blasting agents, there are instances where groundwater contamination related to perchlorate in blasting agents may occur. The following are examples of practices that could lead to perchlorate contamination:

- Poor housekeeping of perchlorate-containing explosives (i.e., spillage on-site);

- Exceeding the sleep time of the explosive. Sleep time is the length of time that an explosive can remain in the ground after charging and still detonate with full energy. Blast hole conditions have a large impact on the sleep time of explosives in wet conditions;

- Poorly designed initiation of the charge, permitting small pockets of un-detonated material after the blast; and

- Blasting misfires, where a loaded hole(s) fails to detonate or partially explodes. If the blaster follows proper methods of priming, loading, stemming, hooking up the shot and firing it, the likelihood of a misfire is small.[37]

To our knowledge, no detailed studies are publicly available that quantify the amount of perchlorate originating from blasting agents and explosives. There have been several newspaper and internet reports that attempt to link blasting operations to perchlorate in groundwater and surface water, particularly in Massachusetts.[39,40,41] Perchlorate concentrations as high as several hundred parts per billion have been measured in close proximity to blasting sites.

ELECTROCHEMICALLY-PRODUCED CHLORINE PRODUCTS

During the electrochemical manufacture of chlorine products, such as chlorate, from chloride brine feedstocks, small amounts of perchlorate may be formed as an impurity.[42,43] Because perchlorate was not known to be a chemical of environmental concern until quite recently (1997), and because the impurity level was considered small relative to the primary chemical being produced (e.g., chlorate), little attention has been paid to its presence. Therefore, little publicly-available information regarding perchlorate contamination in ECP chlorine products exists. Recent analysis of several sodium chlorate feedstocks being used for large-scale commercial perchlorate manufacturing suggest that perchlorate is present in the chlorate products at concentrations ranging from 50 to 230 mg/kg chlorate, and therefore, potential exists for release of perchlorate to the environment through chlorate manufacture, storage, handling, and use. The following sections provide information related to chlorate manufacturing and use and discuss the potential for impacts to soil and groundwater.

Manufacture of Chlorate

Sodium chlorate is produced electrochemically by the electrolysis of aqueous sodium chloride, and its production is governed by the following equation:[43]

$$NaCl + 3H_2O \rightarrow NaClO_3 + 3H_2$$

During the production of sodium chlorate, sodium perchlorate is often produced as an impurity in the electrolytic cell. Concentrations of up to 500 mg of sodium perchlorate per kg sodium chlorate are not uncommon.[42] Accumulation of sodium perchlorate decreases the solubility of sodium chlorate and is actually undesirable to the manufacturer of the chlorate product. As such, several processes have been developed and patented to improve the efficiency of the electrolytic cell, prevent perchlorate formation, and/or remove the perchlorate from the chlorate.[42,43] The formation of perchlorate stems from anodic oxidation of chlorate during the electrochemical reaction in accordance with the following reaction:[43]

$$ClO_3^- + H_2O \rightarrow ClO_4^- + 2H^+ + 2e^-$$

Significant amounts of ECP chlorine chemicals such as sodium chlorate are produced in the U.S. on an annual basis. The majority of sodium chlorate produced in the U.S. is used domestically, with only 3% of the annual domestic production exported. To satisfy demand for use, it is estimated that an additional 40% is imported for domestic consumption. Table 3 lists the total domestic production and consumption rates of sodium chlorate. The total annual consumption of sodium chlorate is approximately 1.2 million tons.[44]

Table 3: U.S. Production and Consumption of Sodium Chlorate

	Production (tons)	Exports (tons)	Imports for Consumption	Apparent Consumption (tons)
1991[45]	448,908	n/a	n/a	n/a
1992[45]	554,564	n/a	n/a	n/a
1993[45]	539,259	n/a	n/a	n/a
1994[45]	559,015	n/a	n/a	n/a
1995[45]	614,536	n/a	n/a	n/a
1996[46]	600,890	54,375	395,199	941,714
1997[46]	567,797	65,680	411,687	913,804
1998[47]	706,909	49,425	430,384	1,087,868
1999[47]	742,476	57,543	439,567	1,124,500
2000[48]	852,756	48,983	440,461	1,244,234
2001[44]	792,167	32,834	495,379	1,254,712
2002[44]	721,086	39,828	528,239	1,209,497

Chlorate Use

Historic and current uses for chlorate include pulp and paper bleaching, non-selective contact herbicide application, and plant defoliation.[49] Sodium chlorate is also used in limited capacities for water treatment, mining, and in the production of other chemicals such as sodium perchlorate and other metallic perchlorates.

The pulp and paper industry uses approximately 94% of all sodium chlorate consumed in the U.S.[49] In this industry, it is primarily used for the on-site production of chlorine dioxide to bleach cellulose fibers. In 1998, the U.S. EPA ruled that, by April 2001, pulp and paper mills in the U.S. would have to use elemental chlorine free (ECF) bleaching instead of the traditional chlorine bleaching, which has the potential to produce organic halides. Chlorine dioxide produced from sodium chlorate meets this requirement.

In addition to pulp and paper bleaching, sodium chlorate is used as a non-selective contact herbicide and a defoliant for cotton, sunflowers, sundangrass, safflower, rice, and chili peppers.[49] As a defoliant, approximately 99% of sodium chlorate application is used on cotton plants.[50] By removing the foliage, a better yield is obtained during harvest and the cotton does not become stained. The application of chlorate defoliants is generally unique to Arizona and California because of their warm climates. Elsewhere, early frost causes foliage to drop from cotton plants naturally. In California and Arizona, the frost typically occurs too late, if at all, and the leaves remain on the plants during harvesting, requiring the use of defoliants. Depending on the yearly weather conditions, other states including Mississippi, Texas, Alabama, Arkansas, Georgia, Louisiana, Tennessee and North Carolina may use sodium chlorate as a defoliant for cotton.

In terms of quantity of use, California used more than 24 million pounds of sodium chlorate on cotton between 1991 and 2003, with an average application rate of 4.6 lbs/acre. By comparison, Arizona, Mississippi, and Texas had total application rates of 6.3, 4.5, and 1.7 million pounds, respectively, between 1991 and 2003.[51]

Potential to Impact Surface Water and Groundwater

Based on the documented occurrence of perchlorate in sodium chlorate and available use statistics, it appears that chlorate use by the pulp and paper industry and as a defoliant has the potential to introduce perchlorate to the environment. For example, assuming 1.2 million tons of sodium chlorate are consumed annually in the U.S.,[44] and that sodium chlorate may contain perchlorate at concentrations ranging from 50 to 500 mg/kg, this represents the potential handling of 120,000 to 1,200,000 lbs of perchlorate annually, the fate of which is largely unknown.

Chlorine dioxide production for pulp and paper bleaching involves the addition of a sodium chlorate solution and a reducing agent to produce chlorine dioxide. Reducing agents include sulfur dioxide, methanol, chloride ion, and hydrogen peroxide.[52] Chlorine dioxide is produced as a gas and later absorbed into water prior to being used as a bleaching agent. As such, perchlorate originating in the sodium chlorate would not be expected to be present in the gas stream because of its non-volatility. However, perchlorate is likely to end up in the by-product salt-cake from the chlorine dioxide generator, which is generally added back to the kraft liquor cycle, where it may undergo reduction. On occasion, excess salt-cake is sewered. The fate of perchlorate in this process is unknown, but low ppb levels of perchlorate in mill effluents are possible if the perchlorate is not significantly treated by the plant's effluent treatment system. Further study of the fate of perchlorate in pulp and paper mills is warranted.

With respect to sodium chlorate use as a defoliant, the average yearly application of sodium chlorate in California is nearly 2 million pounds, applied directly to agricultural lands. Assuming a perchlorate impurity level of between 0.05 to 0.5% sodium perchlorate, the use of sodium chlorate as a defoliant may result in the application of 1,000 to 10,000 pounds of sodium perchlorate to agricultural lands in California per year. While this annual application appears to be relatively small, repeated application over many years to decades may result in an accumulation of perchlorate in soils because of its recalcitrance in most soil environments. Over time, perchlorate in soils could impact surface waters due to overland flow during rainfall events or groundwater through longer term infiltration.

CONCLUSIONS

The frequency of detection of perchlorate impacts to soil, groundwater and surface water, unrelated to military activities, is likely to increase as water utilities analyze for this constituent as part of their UCMR monitoring programs. Based on emerging product and process information, perchlorate is present (intentionally or not) in many more products and processes than initially understood. Furthermore, evidence exists that perchlorate can be formed naturally in evaporate deposits and through atmospheric mechanisms.

The U.S. DOD, NASA and related defense contractors are likely to be the most significant domestic users of perchlorate in North America, and as such, a significant percentage of identified groundwater perchlorate impacts are attributable to DOD, NASA, and related defense contractor facilities. However, cases exist, and many more are likely to surface, where perchlorate impacts result from combinations of military, non-military, and/or natural inputs.

ACKNOWLEDGMENTS

We gratefully acknowledge the Department of Defense's Strategic Environmental Research and Development Program (SERDP) for funding this work.

REFERENCES

1.   Jackson, W.A., Rainwater, K., Anderson, T.A., Lehman, T.M., Tock, R.W. Rajagopalan, S., Ridley, M. *Distribution and Potential Sources of Perchlorate in the High Plains Region of Texas.* Final Report to the Texas Commission on Environmental Quality. 2004.

2.   Dasgupta, P.K, Kalyani Martinelango, P., Jackson, W.A., Anderson, T.A., Tian, K. ,R. W. Tock, R.W., Rajagopalan, S. The Origin of Naturally Occurring Perchlorate: the Role of Atmospheric Processes. *Environ. Sci. Technol.* 2005; 39:1569-1575.

3.   Urbansky, E.T., Brown, S.K., Magnuson, M.L., Kelly, C.A. Perchlorate levels in samples of sodium nitrate fertilizer derived from Chilean caliche. *Environmental Pollution.* 2001; 112:299-302.

4.   Urbansky, E.T., Collette, T.W. , Robarge, W.P., Hall, W.L., Skillen, J.M., Kane, PF. Environmental Protection Agency. Survey of Fertilizers and Related Materials for Perchlorate. EPA Doc. No. 600-R-01-047. 2001.

5.   Goldenwieser, E.A. United States Department of Agriculture, Bulletin No. 798 A Survey of the Fertilizer Industry, Washington, D.C. October 20, 1919.

6.   Howard, P.E. United States Department of Agriculture, Circular No. 129, Survey of the Fertilizer Industry, Washington, D.C. January, 1931.

7.   Mehring, A.L. United States Department of Agriculture, Circular No. 689, Fertilizer Consumption in 1941 and Trends in Usage, Washington, D.C. October, 1943.

8.   United States Department of Agriculture.   The 1957 Yearbook of Agriculture. Government Printing Office. Washington, D.C. 1957.

9.   Brand, C.J. 1930.   Recent Developments in the Fertilizer Industry. A Memorandum Prepared for the Consideration of the Committee on Military Affairs, House of Representatives, Washington, D.C. Prepared by The National Fertilizer Association. April 10, 1930.

10. Renner, R. Study finding perchlorate in fertilizer rattles industry. Environ. Sci. Technol. 1999; 33: 394A-395B.

11. United States Department of Agriculture.   The 1938 Yearbook of Agriculture. Government Printing Office. Washington, D.C. 1938.

12. Skinner, J.J. United States Department of Agriculture, Bulletin No. 136, Fertilizer for Cotton Soils. 1932.

13. Nelson, M., Burleson, D.J. Extension Service, College of Agriculture, University of Arkansas, Extension Circular No. 214. Fertilizer Recommendations, December, 1925.

14. Moss, E.G. United States Department of Agriculture, Bulletin No. 12, Fertilizer Tests with Flue-Cured Tobacco, Washington, D.C. October, 1927.

15. National Agriculture Statistics Service (www.nass.usda.gov:81/ipedb/tobacco.htm)

16. Bennett, R.R, Hawks, Jr., S.N. Nau, H.H.. Fertilizing Flue-Cured Tobacco for High Quality and Yield. The North Carolina Agricultural Extension Service. Extension Circular No. 376, September, 1953.

17. North Carolina Department of Agriculture and Consumer Services (www.ncagr.com/agronomi/stnote2htm)

18. Collings, G.H. Commercial Fertilizers: Their Sources and Use. Fourth Edition. Blakiston Company, Philadelphia. 1949.

19. Hudson, GB, , Moran, J.E., Eaton, G.F. Interpretation of Tritium-3Helium Groundwater Ages and Associated Dissolved Noble Gas Results from Public Water Supply Wells in the Los Angeles Physiographic Basin Report to the California State Water Resources Control Board. UCRL-AR-151447. 2002.

20. Bohlke, J-K. Groundwater recharge and agricultural contamination. *Hydrogeology Journal* 2001; 10:153-179.

21. Fogg, G.E., LaBolle, E.M., Weissmann, G.S. 1998. Groundwater vulnerability assessment: hydrogeologic perspective and example from Salinas Valley, CA, in Application of GIS, Remote Sensing, Geostatistical and Solute Transport Modeling to the Assessment of Nonpoint Source Pollution in the Vadose Zone, AGU Monograph Series 108, 45-61. 1998.

22. Weissmann, GS, Zhang, Y., LaBolle, EM, Fogg, G.E. Dispersion of groundwater age in an alluvial aquifer system. *Water Resources Research* 2002; 38: 1198-2002.

23. Conkling, J.A. *Chemistry of Pyrotechnics. Basic Principles and Theory.* Marcel Dekker, Inc., New York. 1985.

24. APA. July Fourth Sales to Surpass Last Year's Record for Backyard Fireworks Celebrations. Press Release. Bethesda, MC. June 24, 2004.

25. International Trade Administration. (www.ita.doc/td/industry/otea/trade-detail/latest-december/imports/36/360410)

26. American Pyrotechnics Association. (www.americanpyro.com)

27. Schneider, R.L., Thorne, P.G., J.W. Haas, J.W. Estimating the Firework Industry's Contribution to Environmental Contamination with Perchlorate. Proceedings of the 6th International Symposium on Fireworks. Orlando, FL, December 3-7, 2001.

28. Schneider, R.L. Personal Communication. September 22, 2004.

29. MADEP. Evaluation of Perchlorate Contamination at a Fireworks Display. Dartmouth, MA. Draft Report. August 2005.

30. California Department of Health Services (http://www.dhs.ca.gov/ps/ddwem/chemicals/perchl/earlyfindings.htm)

31. Defense Environmental Network and Information Exchange (https://www.denix.osd.mil/denix/Public/Library/Water/Percholrate/releases.html)

32. Silva, M.A. Perchlorate from Safety Flares. At Threat to Water Quality. Santa Clara Valley Water District Publication. www.valleywater.org. 2003.

33. Silva, M.A., Safety Flares Threaten Water Quality with Perchlorate. Santa Clara Valley Water District Publication. www.valleywater.org. 2003.

34. U.S. Census Bureau. 1997 Economic Census, Manufacturing - Subject Series, Oct. 10, 2001.

35. Santa Clara Valley Water District. (www.valleywater.org)

36. The Mercury News. Olin proposes plan to stop well water contamination. Sept. 10, 2003.

37. ISEE. 1998. Blasters' Handbook. 17th edition. Cleveland, Ohio. USA.

38. Kramer, D.A. Explosives. United States Geological Service. 2003. (http://minerals.usgs.gov/minerals/pubs/commodity/explosives)

39. Ward, P. "My House is Worth Nothing" Fouled Well Leaves Westford Family High and Dry. Lowell Sun (MA). September 2, 2004

40. Wims, M. "Westford Hires Firm to Investigate Water Contamination". Lowell Sun (MA), December 21, 2004.

41. Town of Tewksbury website: http://www.tewksbury.net/web%20survey/Updates/update17.htm.

42. Wanngard, C.J.F. Process for the Reduction of Perchlorate in Electrolytes used for the Production of Chlorate. United States Patent # 5,063,041. November 5, 1991.

43. Betts, J.A , Dluzniewski, T.J. Impurity Removal for Sodium Chlorate. October 28, 1997. United States Patent # 5,681,446.

44. U.S. Department of Commerce, Economics and Statistics Administration, U.S. Census Bureau. Inorganic Chemicals: 2002, product code 325188A141. August 2003.

45. U.S. Department of Commerce, Economics and Statistics Administration, U.S. Census Bureau. Inorganic Chemicals: Fourth Quarter, 1996, product code 325188A141. February 27, 1997.

46. U.S. Department of Commerce, Economics and Statistics Administration, U.S. Census Bureau. Inorganic Chemicals: 1997, product code 325188A141. September 29, 1998.

47. U.S. Department of Commerce, Economics and Statistics Administration, U.S. Census Bureau. Inorganic Chemicals: 1999, product code 325188A141. September 28, 2000.

48. U.S. Department of Commerce, Economics and Statistics Administration, U.S. Census Bureau. 2002. Inorganic Chemicals: 2001, product code 325188A141. August , 2002.

49. OMRI. NOSB TAP Review Compiled by OMRI for Sodium Chlorate. 6 November 2000.

50. PAN Pesticides Database – California Pesticide Use. Sodium Chlorate – Pesticide Use Statistics for 2002.

51. Agricultural Statistics Board, NASS, USDA. Agricultural Chemical Usage Field Crop Summary (1991 to 2003).

52. Dence, C.W. and D.W. Reeve. Pulp Bleaching: Principles and Practice. TAPPI Press. 1996.

Chapter 5

Stable Isotopic Composition of Chlorine and Oxygen in Synthetic and Natural Perchlorate

Neil C. Sturchio,[1] J. K. Böhlke,[2] Baohua Gu,[3] Juske Horita,[3] Gilbert M. Brown,[3] Abelardo D. Beloso, Jr.,[1] Leslie J. Patterson,[1] Paul B. Hatzinger,[4] W. Andrew Jackson,[5] and Jacimaria Batista[6]

[1] University of Illinois at Chicago, Chicago, IL 60607
[2] U. S. Geological Survey, Reston, VA 20192
[3] Oak Ridge National Laboratory, Oak Ridge, TN 37831
[4] Shaw Environmental, Inc., Lawrenceville NJ 08648
[5] Texas Tech University, Lubbock, TX 79410
[6] University of Nevada, Las Vegas, NV 89154

INTRODUCTION

Ammonium perchlorate has been used since the 1940s in the United States as a component of the solid propellant fuel for rockets and missiles. Perchlorate salts, which are strong oxidants, also are used in fireworks, munitions, air-bag inflation systems, highway flares, and matches. Past activities of the military and aerospace industries have led to widespread perchlorate contamination of groundwater.[1] Improvement of routine analytical methods used for the measurement of perchlorate in groundwater since 1997 have resulted in widespread detection of perchlorate throughout the United States in areas where perchlorate salts have been manufactured, stored, or used.[2-4] Lake Mead and the downstream Colorado River contain measurable levels of perchlorate[1]. Perchlorate has been detected in commercial food products, including vegetables and milk.[5,6] Although the total scope of perchlorate contamination in the United States remains unclear, recent estimates indicate that the drinking water supplies of 15 million people are affected by the contamination of the Colorado River alone.[7]

The potentially adverse human health effect of perchlorate ingestion makes it a major public health concern. Perchlorate inhibits iodide uptake by the thyroid causing disruption of normal thyroid function, which can lead to a number of serious health problems, especially pertaining to early neurological development.[8,9] In March 2004, California established a Public Health Goal of 6 µg/L for perchlorate concentration in drinking water, and other states have independently adopted advisory levels ranging from 1 to 18 µg/L. A special committee of the National Academies of Science (NAS) recently reviewed the toxicological data concerning perchlorate, and released a controversial report[10] that recommended a safe daily perchlorate dose more than 20 times higher than that endorsed earlier by the U.S. Environmental Protection Agency (USEPA).[11] The USEPA subsequently accepted the NAS recommendation and raised their reference dose from 0.03 to 0.7 µg perchlorate/kg body wt/day.

The perchlorate anion (ClO_4^-) consists of a central chlorine atom in tetrahedral coordination with four oxygen atoms. Perchlorate salts are soluble in water and some organic solvents. When dissolved in water, perchlorate is nonvolatile, stable and kinetically inert with respect to abiotic reduction and oxygen exchange. For example, the exchange of ^{18}O-labeled water with $HClO_4$ at a concentration of 9 mol/L was insignificant after 63 days at 100 °C, and the half-life for exchange at room temperature is estimated at greater than 100 years.[12] Perchlorate does not adsorb strongly to activated carbon and most mineral surfaces, and it is sufficiently unreactive that it cannot be effectively removed from water by conventional water treatment methods.[13] Perchlorate can be reduced in aqueous solution by some transition metal ions, such as Ti(III), Re(V), V(III), V(II), Mo(III), and Ru(II).[13]

A variety of different microorganisms have been isolated that can reduce perchlorate under anoxic conditions.[14] These organisms are currently being utilized in several full-scale bioreactor systems to remove perchlorate from groundwater at flow rates as high as 5,000 gallons per minute.[15,16] The presence of persistent plumes of perchlorate in groundwater aquifers indicates that biotic and abiotic reduction of perchlorate under typical oxic groundwater conditions does not lead to significant attenuation. However, the addition of a variety of different substrates is being used to stimulate perchlorate biodegradation in subsurface environments where the anion is otherwise long-lived.[16]

The perchlorate used for military and aerospace applications is synthesized by electrolytic oxidation of aqueous chloride brine. Although much of the known perchlorate contamination in the U.S. can be related to such industrial sources, there are some natural sources of perchlorate as well. The origin and abundance of perchlorate from natural sources is poorly known. The best known natural source of perchlorate is within the nitrate-bearing salt deposits of the Atacama Desert (Chile) that have been mined extensively for use as a nitrate source in agricultural fertilizers.[17] Perchlorate of apparent natural origin also has been detected in a number of ancient evaporite salt deposits,[18] and in modern rain and snow samples.[19] At several locations, such as in the West Texas panhandle area, perchlorate of possible natural origin has been detected in ground water.[20]

The presence of natural sources of perchlorate in groundwater, whether through application of imported perchlorate-bearing fertilizers or from other more local but undefined sources, complicates the issue of liability for groundwater remediation near industrial sources of perchlorate. A practical forensic tool is thus needed that would enable the identification of the source(s) of perchlorate in contaminated aquifers and waterways. Successful applications of stable isotope ratio measurements of N and O in studies of nitrate, and C and Cl isotope ratios in studies of chlorinated solvents, have encouraged us to pursue the development of Cl and O isotopic analyses for tracing the sources and understanding the behavior of perchlorate in the environment. In this chapter, we review earlier isotopic studies of perchlorate and present new data to show that synthetic and natural perchlorates have distinct chlorine and oxygen isotopic compositions. We also present data for perchlorate extracted from groundwater samples that demonstrate the potential application of stable isotope ratio measurements for environmental forensics.

PREVIOUS ISOTOPIC STUDIES OF PERCHLORATE

The use of stable isotope ratio measurements for understanding the geochemical behavior of natural and anthropogenic compounds in the environment is a major field of research within the earth, atmospheric, oceanic, and environmental sciences. Recent developments in the measurement and application of the stable isotope ratios of carbon and chlorine in chlorinated aliphatic hydrocarbons led to important advances in characterizing the sources and behavior of these compounds in contaminated groundwater aquifers. Particularly important is the discovery that microbial degradation of chlorinated aliphatic hydrocarbons under aerobic and anaerobic conditions is accompanied by substantial kinetic isotope effects for both carbon and chlorine, allowing evaluation of the extent of biodegradation through measurement of isotope ratios. To date, however, few analogous isotopic studies of perchlorate have been published. The main reason for the paucity of published perchlorate isotopic data is that methods for precise isotope ratio measurements of chlorine and oxygen in perchlorate were not developed until about the past five years.[21-24]

Measurements of the kinetic isotope effect of microbial reduction of perchlorate in laboratory microcosms have been made for chlorine[22,25] and oxygen.[26] Data for the chlorine kinetic isotope effect observed in two microcosms of *Dechlorosoma suillum* JPLRND, which was initially isolated from a groundwater sample collected in southern California, are shown in Fig. 1. The kinetic isotope effect measured for Cl is about 15‰. The preliminary value reported for the oxygen kinetic isotope effect[26] (~30 ‰) is about twice as large as that for chlorine. Similar measurements have not yet been made for abiotic reduction reactions. Such values may prove useful to determine if abiotic perchlorate reduction is a mechanism of loss in some groundwater environments.

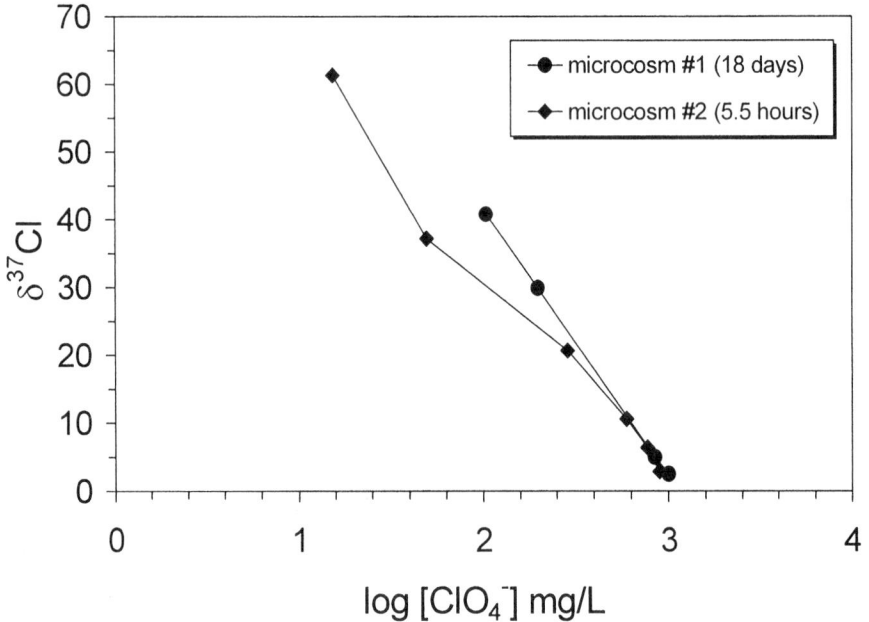

Figure 1. Isotopic composition of residual perchlorate ($\delta 37Cl$) as a function of the perchlorate concentration (mg/L) in two microcosm experiments using liquid cultures of *Dechlorosomas Suillum* JPLNRD[22]. Slopes of lines indicate a chlorine kinetic isotope effect of about 15‰. Analytical errors in $\delta^{37}Cl$ (±0.3 per mil) are smaller than symbols used for data points.

Evidence indicating that synthetic and natural perchlorates have distinct chlorine and oxygen isotopic compositions has been presented recently.[23,24] Oxygen isotopic evidence, especially the presence of ^{17}O in excess of that expected from mass-dependent isotopic fractionation, indicates that both nitrate and perchlorate in the natural Chilean nitrate-bearing salts are dominantly of atmospheric origin.[23,24,27,28] Their anomalous ^{17}O abundances may be attributed to oxidation reactions involving atmospheric ozone.

MATERIALS AND METHODS

Samples

Four types of perchlorate-containing materials were analyzed (Table 1):

(1) *laboratory reagents* from a number of suppliers;

(2) *commercial perchlorate salts and precursor materials*, including samples of NH_4ClO_4, $KClO_4$, and $NaClO_4$ were obtained from the American Pacific Corporation (AMPAC). In addition, AMPAC's perchlorate production facility (the Western Electrochemical Company in Cedar City, UT) supplied water, NaCl feedstock, and $NaClO_3$ and $NaClO_4$ electrochemical synthesis products from a single batch produced during June 2004;

(3) *natural perchlorate* extracted from samples of nitrate-bearing salt deposits from the Atacama Desert (Chile), and from commercial fertilizers derived from this material; and

(4) *groundwater and surface water* from sites in which significant concentrations of perchlorate were detected.

All laboratory reagents and production materials were analyzed as received. Perchlorate from groundwaters and from Chilean nitrate-bearing salts and fertilizers was extracted at Oak Ridge National Laboratory using a highly selective bifunctional anion exchange resin. The perchlorate was eluted in a solution of $FeCl_3$ and HCl, then precipitated by addition of CsCl or KCl to form $CsClO_4$ or $KClO_4$.[23,29] The purities of the purified perchlorate salts were verified by micro-Raman spectroscopy.

Several tests were made to ensure that the extraction procedures did not fractionate the isotopic compositions of Cl and O in perchlorate. First, a laboratory reagent of known isotopic composition (ORNL-6, hydrous Na-perchlorate) was added to a perchlorate-free, nitrate- and chloride-rich salt solution (the residue from extraction of perchlorate from an aqueous solution of a Chilean salt sample), and then the added perchlorate was extracted from this solution. The data for perchlorate recovered from this test (ORNL-15) had $\delta^{37}Cl$ and $\delta^{18}O$ values indistinguishable from those of ORNL-6 (Table 2). In another test, a sample of anhydrous Na-perchlorate reagent (UIC-1) was dissolved; the resulting solution (UIC-4) was later evaporated to dryness at 130°C. The $\delta^{18}O$ and $\Delta^{17}O$ values of this dehydrated perchlorate were indistinguishable from those of UIC-1. In a third test, an aliquot of 60 % perchloric acid (RSIL-2) was mixed with an equal volume of ^{18}O-enriched water ($\delta^{18}O$ = +254 ‰), and perchlorate from this solution was recovered after 38 days as Na-perchlorate salt (RSIL-3) following quick neutralization

Table 1. Identification of perchlorate samples

Sample Identification	Compound	Description
Laboratory Reagents		
UIC-1	$NaClO_4$	Mallinckrodt lot # 1190KHJJ
UIC-2	$KClO_4$	Baker Analyzed lot # 45155
UIC-3	$NaClO_4 \cdot H_2O$	Aldrich lot # 00722CG
UIC-8	$KClO_4$	Sigma lot # 60K3451
UIC-9	$NaClO_4$	Sigma lot # 111K1334
UIC-10	$KClO_4$	Hummel-Croton
ORNL-6	$NaClO_4 \cdot H_2O$	EM lot # SX0693-2
ORNL-8	$KClO_4$	General Chem. Co. lot # 13
ORNL-9	$RbClO_4$	Aldrich lot # AN00625LZ
ORNL-10	$CsClO_4$	Aldrich lot # LI09119JI
ORNL-11	$CsClO_4$	Aldrich lot # 02407AS
RSIL-2	$HClO_4$	Baker 9656-1, lot # 146358
RSIL-4	$KClO_4$	Aldrich, lot # 11921HO
Production Materials		
UIC-5	$NaClO_4$	American Pacific Co. (N3300401)
UIC-6	$KClO_4$	American Pacific Co. (P0900402)
UIC-7	NH_4ClO_4	American Pacific Co. (A2000433)
ORNL-18	$NaClO_3$	Western Electrochemical Co. (TO403E)
ORNL-19	$NaClO_4$	Western Electrochemical Co. (TO403B)
ORNL-20	H_2O	Water used by WE (6/23/04)
NaCl (TO415D)	NaCl	NaCl feedstock used by WE
Natural Perchlorate		
ORNL-2	$KClO_4$	Commercial Hoffman fertilizer
ORNL-4	$CsClO_4$	Commercial Hoffman fertilizer
ORNL-5	$CsClO_4$	Atacama, Chile nitrate salt (AT-74-1)
ORNL-12	$CsClO_4$	SQM fertilizer (RSIL N7791)
Groundwater and Surface Water		
ORNL-14	$CsClO_4$	Groundwater at Edwards AFB, CA
ORNL-16	$CsClO_4$	Surface water, Las Vegas Wash
ORNL-17	$CsClO_4$	Groundwater, Henderson NV
ORNL-21	$CsClO_4$	West Texas groundwater (TTU-G1S)
ORNL-22	$CsClO_4$	West Texas groundwater (TTU-M3L)
CPMW-5*	$NaClO_4$	Indian Head, MD groundwater (2002)
TPMW-5*	$NaClO_4$	Indian Head, MD groundwater (2002)
Method Tests		
ORNL-15	$CsClO_4$	Prepared from ORNL-6
UIC-4	$NaClO_4$	Prepared from UIC-1
RSIL-3	$HClO_4$	Prepared from RSIL-2

* perchlorate extracted from Indian Head samples by method of Ref. 22

of the acid with concentrated NaOH. The $\delta^{18}O$ value of RSIL-3 perchlorate was indistinguishable from that of RSIL-2. A subsequent recovery of perchlorate from this solution, 151 days later, showed no significant change.

Chlorine Isotope Ratio Analyses

Chlorine isotope ratios (Table 2) were measured at the Environmental Isotope Geochemistry Laboratory of the University of Illinois at Chicago. Perchlorate salts were first decomposed at ~650°C in evacuated borosilicate glass tubes to produce alkali chloride salts, which were then analyzed according to well established methods.[30,31] The alkali chloride salts produced by perchlorate decomposition were dissolved in warm 18.2 MΩ deionized water, and Cl was precipitated as AgCl by addition of AgNO$_3$. The resulting AgCl was recovered by centrifugation, washed in dilute HNO$_3$, dried, and reacted in a sealed borosilicate glass tube with excess CH$_3$I at 300°C for two hours to produce CH$_3$Cl. The resulting CH$_3$Cl was purified cryogenically and/or using gas chromatography and then admitted to a Finnigan Delta-Plus XL isotope-ratio mass spectrometer and analyzed in dual-inlet mode by measurements at m/z 50 and 52, with the exception of the West Texas samples which were analyzed in continuous-flow mode. Samples were analyzed in replicate if sufficient amounts were available. Analytical yields of this procedure for reagent salts ranged from 81-97 % with no significant dependence of isotope ratio on yield. Chlorine isotope ratios are reported in units of $\delta^{37}Cl$, in terms of the per mil (‰) difference between the $^{37}Cl/^{35}Cl$ ratio of a sample and that of the isotopic reference material, Standard Mean Ocean Chloride (SMOC):[32,33]

$$\delta^{37}Cl\ (‰) = [(^{37}Cl/^{35}Cl)_{sample}/(^{37}Cl/^{35}Cl)_{SMOC} - 1] \times 1000$$

The analytical precision of $\delta^{37}Cl$ values reported in Table 2 ranges from ±0.1 to 0.3‰, based on replicate analyses of samples and isotopic reference materials.

Oxygen Isotope Ratio Analyses

Oxygen isotope ratios (Table 2) were measured at the Reston Stable Isotope Laboratory of the U.S. Geological Survey. Oxygen isotope ratios are reported as follows:[34]

$$\delta^{18}O\ (‰) = [(^{18}O/^{16}O)_{sample}/(^{18}O/^{16}O)_{VSMOW} - 1] \times 1000$$

$$\Delta^{17}O\ (‰) = [[(1 + \delta^{17}O/1000) / (1 + \delta^{18}O/1000)^{0.525}] - 1] \times 1000$$

For $\delta^{18}O$ determinations, perchlorate salts were reacted with glassy C at 1325°C to produce CO, which was transferred in a He carrier through a

Table 2.   Stable Isotope Ratios of Cl and O*

Sample Identification	$\delta^{37}Cl$	$\delta^{18}O$	$\Delta^{17}O$**
Laboratory Reagents			
UIC-1	+1.2	-16.2	+0.01
UIC-2	+1.1	-24.8	+0.01
UIC-3	+1.3	-16.1	+0.12
UIC-8	+0.5	-16.3	+0.02
UIC-9	+1.0	-17.2	0.00
UIC-10	+0.4	-12.5	-0.01
ORNL-6	-3.1	-17.2	+0.08
ORNL-8	+0.6	-19.1	+0.00
ORNL-9	+1.3	-16.4	+0.11
ORNL-10	+1.6	-16.6	-0.04
ORNL-11	+0.6	-16.9	+0.00
RSIL-2	n.a.	-14.6	+0.01
RSIL-4	+0.6	-17.0	+0.00
Production Materials			
UIC-5	+0.4	-22.3	-0.01
UIC-6	+0.4	-21.5	+0.07
UIC-7	+0.4	-21.3	-0.03
ORNL-18	+1.1	n.a.	+0.06
ORNL-19	+0.9	-20.4	+0.00
ORNL-20	n.a.	-13.5	n.a.
NaCl (TO415D)	+0.2	n.a.	n.a.
Natural Perchlorate			
ORNL-2	-13.7	-8.4	+8.95
ORNL-4	-14.5	-9.3	+8.93
ORNL-5	-11.8	-4.2	+9.57
ORNL-12	-14.2	-7.6	+9.25
Groundwater/ Surface Water			
ORNL-14	-0.9	-15.8	+0.04
ORNL-16	+0.4	-14.5	+0.00
ORNL-17	+0.9	-15.0	+0.02
ORNL-21	+6.2	+4.7	+0.42
ORNL-22	+5.1	+2.5	+0.49
CPMW-5	+0.2	n.a.	n.a.
TPMW-5	+1.1	n.a.	n.a.
Method Tests			
ORNL-15	-3.2	-17.2	n.a.
UIC-4	n.a.	-16.3	-0.13
RSIL-3	n.a.	-14.7	n.a.

* All values reported in per mil

** $\Delta^{17}O = [[(1 + \delta^{17}O/1000) / (1 + \delta^{18}O/1000)^{0.525}] - 1] \times 1000$
   n.a. = not analyzed

molecular-sieve gas chromatograph to a Finnigan Delta Plus XP isotope-ratio mass spectrometer and analyzed in continuous-flow mode by monitoring peaks at m/z 28 and 30. Derived values of $\delta^{18}O$ were calibrated against VSMOW and SLAP by analyzing reference materials NBS-127 sulfate ($\delta^{18}O$ = +8.6 ‰), IAEA-N3 nitrate ($\delta^{18}O$ = +25.6 ‰), and USGS34 nitrate ($\delta^{18}O$ = -27.9 ‰) in a system equipped to prevent the N_2 peak from entering the ion source.[36] Yields of O (as CO) typically were within ± 2 per cent for ClO_4^-, NO_3^-, and SO_4^{2-} reagents and samples. The analytical precision of $\delta^{18}O$ values reported in Table 2 ranges from ±0.1 to 0.3 ‰, based on replicate analyses of samples and isotopic reference materials.

For $\Delta^{17}O$ determinations, perchlorate salts were decomposed for 12 minutes at 650°C in evacuated quartz glass tubes to produce O_2, then quenched in air. The O_2 gas was expanded into a liquid N_2 trap and then admitted to a Finnigan Delta Plus XP isotope-ratio mass spectrometer and analyzed in dual- inlet mode by measurements at m/z 32, 33, and 34. Yields of O (as O_2) typically were within ±5 percent for ClO_4^- reagents, ClO_4^- samples and measured aliquots of tank O_2. $\Delta^{17}O$ was assumed to be 0.00 ‰ for a representative $KClO_4$ reagent designated USGS37 (equal to the average of all anthropogenic ClO_4^- samples). This is consistent with derivation of the O from H_2O during ClO_4^- synthesis, and it results in a measured $\Delta^{17}O$ value of - 0.4 ‰ for tank O_2 from air. The apparent $\delta^{18}O$ difference between UIC-2 and RSIL-4 when analyzed as O_2 was about 0.92-0.96 times the normalized difference when analyzed as CO. Accordingly, the $\Delta^{17}O$ scale was expanded by a factor of 1.06 to account for O blank or exchange during ClO_4^- decomposition. The reproducibility of the $\Delta^{17}O$ measurements was ±0.1‰ or less after normalization, based on replicate analyses of samples and isotopic reference materials.

RESULTS AND DISCUSSION

Perchlorate Reagents and Production Materials

Samples of common reagent-grade perchlorate salts (Table 1) were obtained from the laboratory stocks of some of the authors. In addition, reactants used in the electrochemical synthesis of perchlorate (water and NaCl) as well as intermediate and final products ($NaClO_3$, $NaClO_4$, NH_4ClO_4, $NaClO_4$, and $KClO_4$) were obtained directly from the Western Electrochemical Company or from their parent company, AMPAC, the sole perchlorate supplier to the U.S. military since 1998. Chlorine and/or oxygen isotope ratios were measured in sixteen synthetic perchlorate salts, one chlorate salt, and one perchloric acid (Figure 2). The total range in measured $\delta^{37}Cl$ values of these reagents is −3.1 to +1.6‰. Ader et al.[21] reported three $\delta^{37}Cl$ values for perchlorate reagents ranging from +0.2 to +2.3‰. Excluding sample ORNL-6, the range of $\delta^{37}Cl$ values in Table 2 is from +0.4 to +1.6 ‰. These values are near that of seawater chloride (0.0‰), reflecting the fact that the perchlorate in these materials is synthesized by electrochemical oxidation of brines derived from marine evaporite salt deposits (e.g., the sample of NaCl feedstock has a $\delta^{37}Cl$ value of +0.2‰), and that this process is relatively efficient with respect to Cl oxidation. This is demonstrated by data obtained for the series of samples obtained from the Western Electrochemical Company, including NaCl feedstock, $NaClO_3$ intermediate product, and $NaClO_4$ final product.

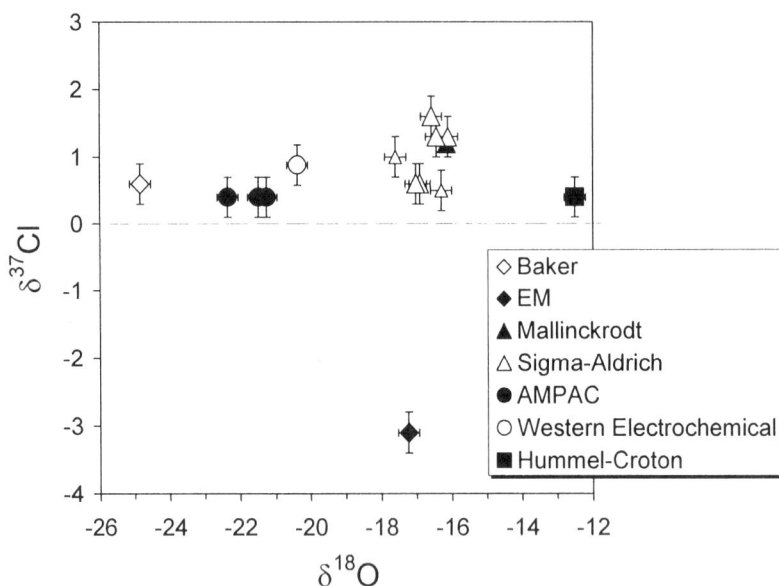

Figure 2. Isotopic composition of synthetic perchlorate salts, grouped by source according to symbols (see legend), displayed in a $\delta^{37}Cl$ vs. $\delta^{18}O$ diagram. Horizontal dashed line shows reference $\delta^{37}Cl$ value of SMOC (Standard Mean Ocean Chloride). Analytical error bars are shown at ±0.3 per mil. Isotopic differences between (and within) sources are significantly larger than the analytical precision.

The range in $\delta^{18}O$ values of synthetic perchlorate (−24.8 to −12.5‰) is significantly larger than that of $\delta^{37}Cl$ values, reflecting both a range in the local meteoric water compositions and the effect of mass-dependent isotopic fractionation involved in the perchlorate synthesis, which is relatively inefficient with respect to H_2O consumption. For example, there was a ~7‰ difference in $\delta^{18}O$ values of $NaClO_4$ (sample ORNL-19) and the local meteoric water used in its synthesis (sample ORNL-20) at the Western

Electrochemical Co. plant during June 2004. All samples of synthetic perchlorate have $\Delta^{17}O$ values of 0.0 ± 0.1‰, indicating that neither the reactants nor the reaction mechanisms involved in perchlorate synthesis exhibit evidence for mass-independent isotopic fractionation.

Natural Perchlorate

Four samples of perchlorate were extracted from nitrate-bearing salts of the Atacama Desert and derivative fertilizer products, and their Cl and O isotopic compositions were measured (Table 2). The resulting $\delta^{37}Cl$ values range from -14.5 to -11.8 ‰. These $\delta^{37}Cl$ values are much lower than the range of synthetic perchlorates reported in Table 2 (-3.1 to $+1.6$ ‰). The $\delta^{18}O$ values of the natural perchlorate samples range from -9.3 to -4.2 ‰, which is, in contrast, higher than that of synthetic perchlorates (-24.8 to -12.5 ‰).

The most diagnostic isotopic characteristic of natural perchlorate is its positive $\Delta^{17}O$ value, which distinguishes natural perchlorate from synthetic perchlorate, which has $\Delta^{17}O = 0.0\pm0.1$ ‰ on the scale adopted here. We measured $\Delta^{17}O$ values in the natural perchlorate ranging from $+8.93$ to $+9.57$ ‰, confirming the earlier results obtained for similar materials.[23]

A graphical comparison of the isotopic compositions of all synthetic and natural perchlorate samples for which Cl and O isotope ratios were measured in this study is shown in Figures 3 and 4. The samples occupy two well-separated fields in the $\delta^{37}Cl$ vs $\delta^{18}O$ diagram (Figure 3). These data indicate that mixtures of synthetic and natural (Atacama) perchlorate should be identifiable, as should perchlorate that exhibits significant kinetic isotope effects caused by partial microbial or abiotic reduction. The $\Delta^{17}O$ vs $\delta^{18}O$ diagram (Figure 4) highlights the ^{17}O excess of natural perchlorate. The initially positive $\Delta^{17}O$ value found in the natural perchlorate should be unaffected by subsequent mass-dependent fractionation.

Groundwater and Surface Water

The distinct isotopic characteristics of synthetic and natural (Atacama) perchlorate may provide a simple tool with which to identify the source(s) of perchlorate in a given sample of groundwater or surface water, as well as the extent to which perchlorate is affected by subsurface chemical processes such as microbial or abiotic reduction. With these potential applications in mind, we extracted and analyzed perchlorate from groundwater at two military sites (Edwards Air Force Base, CA; and the Naval Surface Warfare Center, Indian Head Division, MD); groundwater and surface water from near the former Kerr-McGee perchlorate manufacturing site at Henderson,

NV; and groundwater from the West Texas panhandle area, where there is no known anthropogenic source of perchlorate. The groundwater and surface water samples from all of the military and industrial sites have a narrow range of $\delta^{37}Cl$ values (-0.9 to $+1.1$ ‰) that lies within the range of $\delta^{37}Cl$ values we found for synthetic perchlorate (-3.1 to $+1.6$ ‰) (Figure 3). Their measured $\delta^{18}O$ values (-15.8 to -14.5 ‰) are near the upper end of the range of $\delta^{18}O$ values we found for synthetic perchlorate (-24.8 to -14.6 ‰) (Figure 3), and their measured $\Delta^{17}O$ values were all identical to that of synthetic perchlorate (0.0 ± 0.1‰) (Figure 4). These results indicate that the perchlorate in groundwater preserves the isotopic characteristics of its source, at least on a time scale of decades.

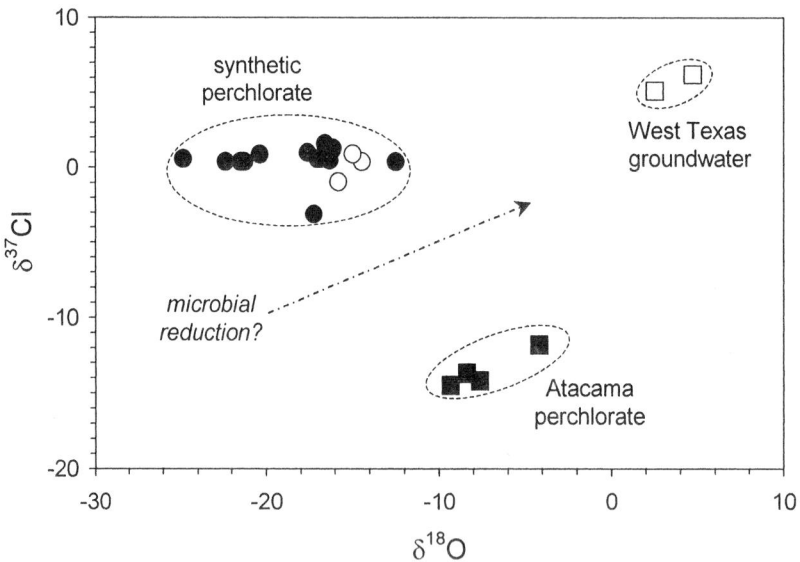

Figure 3.   Comparison of $\delta^{37}Cl$ and $\delta^{18}O$ values of synthetic (filled circles) and natural "Atacama-type" (filled squares) perchlorate. Also shown are groundwater and surface water perchlorate from military and industrial sites (open circles) and groundwater from West Texas (open squares) for which a synthetic source is unknown.[20] Dashed arrow indicates approximately the slope of the trajectory followed by residual perchlorate during partial microbial degradation according to preliminary experimental work.[22, 25, 26]

The results for the West Texas waters are interesting, because they fall outside the ranges that we found for $\delta^{37}Cl$, $\delta^{18}O$, and $\Delta^{17}O$ in all other perchlorate samples (Table 1, Figures 3 and 4). There are at least two plausible explanations for the anomalous isotopic ratios of the perchlorate in these West Texas groundwater samples:

i) The perchlorate is of natural origin, but it represents a type of natural perchlorate that is isotopically distinct from the Atacama perchlorate and has a formation mechanism that is currently unknown.

ii) The perchlorate is a mixture of synthetic perchlorate and "Atacama-type" natural perchlorate with an apparent mixing ratio based on $\Delta^{17}O$ of approximately 19:1 (synthetic:natural), This hypothesis assumes that oxygen exchange between perchlorate and water is negligible and requires that the mixture was significantly affected by mass-dependent isotopic fractionation associated with either microbial or abiotic reduction by about one-third.

Additional studies are underway to better define the origin of the West Texas perchlorate as well as that found at many other sites for which perchlorate sources are currently in question.

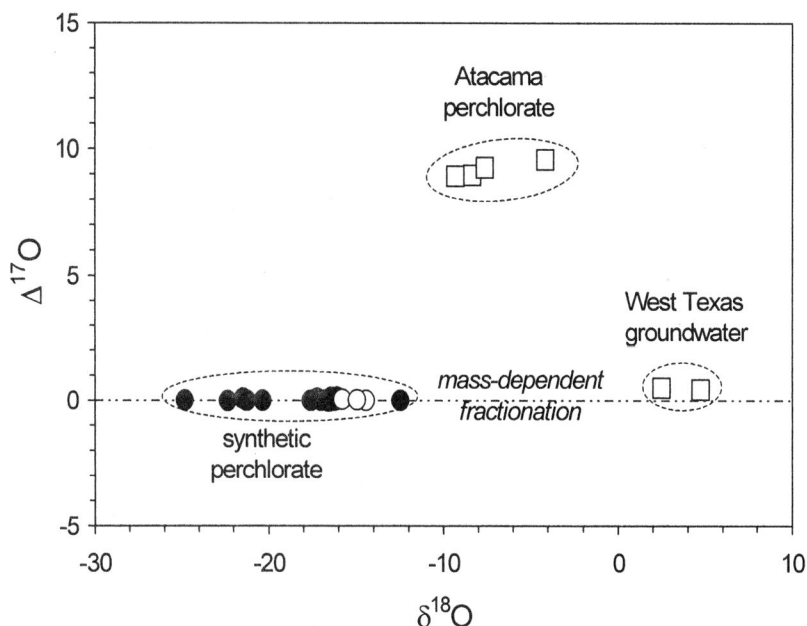

Figure 4.   Comparison of $\Delta^{17}O$ and $\delta^{18}O$ values of synthetic (filled circles) and natural "Atacama-type" (open squares) perchlorate. Also shown are groundwater and surface water perchlorate from military and industrial sites (open circles) and groundwater from West Texas (open squares) where there is no known synthetic source of perchlorate.[20] Dashed horizontal line at $\Delta^{17}O = 0$ indicates the slope of the trajectory followed by residual perchlorate during partial microbial degradation or other mass-dependent isotopic fractionation.[35]

SUMMARY AND CONCLUSION

Representative samples of perchlorate from a variety of synthetic and natural sources, along with trace amounts of perchlorate extracted from groundwater and surface water at several locations, have been analyzed for their Cl and O isotopic compositions. Large differences in Cl and O isotopic compositions were found between synthetic and natural sources of perchlorate. Synthetic perchlorate has $\delta^{37}Cl$ values from 0 to +2‰, $\delta^{18}O$ values from −25 to −12‰, and $\Delta^{17}O$ values of 0.0±0.1‰ (by definition in this study). Natural perchlorate is less well characterized as yet, but samples derived from nitrate-rich Atacama salts have $\delta^{37}Cl$ values from −15 to −11‰, $\delta^{18}O$ values from −10 to −4‰, and positive $\Delta^{17}O$ values around +9 to +10‰. These distinct isotopic characteristics may enable the use of stable isotopic analysis in forensic applications. In some cases, specific batches of synthetic perchlorate may be identified from their isotopic compositions. In addition, it should be possible to determine from measured isotopic composition whether dissolved perchlorate in groundwater is from a synthetic or natural source, or a mixture thereof.

A complicating factor in the application of isotopic measurements to tracing perchlorate in environmental samples is the potential for processes such as natural degradation or isotope exchange to alter the isotopic characteristics of dissolved perchlorate. Although the rates and isotope effects of such processes can be measured in the laboratory, few relevant laboratory studies have been done. Empirical evidence for the occurrence of perchlorate isotope effects in samples of groundwater and surface water also may be gained from additional field studies at well-characterized sites. Additional laboratory data on the kinetic isotope effects of various biotic and abiotic reduction mechanisms should be acquired to enable quantitative estimates of the extent of intrinsic degradation of perchlorate in contaminated aquifers.

Although the data compiled to date are intriguing, the reconnaissance study of perchlorate isotopic compositions reported herein is far from comprehensive. In order for isotope data of Cl and O in perchlorate to be useful and applicable from a forensic standpoint, a much broader database of these values must be obtained from both synthetic and natural sources. Moreover, the origin and behavior of perchlorate in a wide variety of natural settings must be explored in much greater detail.

ACKNOWLEDGEMENTS

We thank Linnea Heraty at UIC, Stan Mroczkowski and Tyler Coplen at the USGS, and Hui Yan and YeeKyoung Ku at ORNL for technical assistance in the laboratory. This study was supported in part by the ORNL Laboratory Directed Research and Development Fund managed by UT-Battelle, LLC for the U. S. Department of Energy under Contract No. DE-AC05-00OR22725, the USGS National Research Program in Water Resources, and the USDOD Environmental Security Technology Certification Program.

REFERENCES

1. Motzer W.E. Perchlorate: problems, detection, and solutions. Environ Forens 2001; 2: 301-311.

2. Urbansky E.T. Perchlorate as an environmental contaminant. Environ Sci Poll Res 2002; 3: 187-192.

3. Jackson P.E., Gokhale S., Streib T., Rohrer J.S., and Pohl C.A. Improved method for the determination of trace perchlorate in ground and drinking waters by ion chromatography. J Chrom A 2000; 888: 151-158.

4. Magnuson M.L., Urbansky E.T., Kelty C.A. Determination of perchlorate at trace levels in drinking water by ion-pair extraction with electrospray ionization mass spectrometry. Anal Chem 2000; 72: 25-29.

5. Jackson W. A., Joseph P., Patil L., Tan K., Smith P. N., Yu L., and Anderson T. A. Perchlorate accumulation in forage and edible vegetation. J Agricult Food Chem 2005; 53: 369-373.

6. Kirk A.B., Martinelango P. K., Tian K., Aniruddha D., Smith E. E., and Das Gupta P. K. Perchlorate and iodide in dairy and breast milk. Environ Sci Technol 2005; 39: 2011-2017.

7. Hogue C. Rocket-fueled river. Chem Eng News 2003; 81(33): 37-46.

8. Porterfield S.P. Vulnerabilities of the developing brain to thyroid abnormalities: environmental insults to the thyroid system. Environ Health Perspect 1994; 102: 125-130.

9. Wolff J. Perchlorate and the thyroid gland. Pharmacol Rev 1998; 50: 89-105.

10. National Research Council, *Health Implications of Perchlorate Ingestion.* Washington, D.C.: National Academies Press, 2005.

11. U. S. Environmental Protection Agency, National Center for Environmental Assessment, *Draft Perchlorate Risk Assessment.* Washington, D.C.: U. S. Govt Printing Press, 2002.

12. Hoering T. C., Ishimori F. T. and McDonald H. O. The oxygen exchange between oxy-anions and water, 11. Chlorite, chlorate and perchlorate ions. J Amer Chem Soc 1958; 80: 3876.

13. Urbansky E.T. Perchlorate chemistry: implications for analysis and remediation. Bioremediation J 1998; 2: 81-95.

14. Coates J.D. and Achenbach L.A. Microbial perchlorate reduction: rocket-fuelled metabolism: Nature Rev Microbiol 2004; 2: 569-580.

15. Hatzinger P. B., Whittier M. C., Arkins M. D., Bryan C. W., and Guarini W. J. In-situ and ex-situ bioremediation options for treating perchlorate in groundwater. Remediation 2002; 12: 69-85.

16. Hatzinger, P. B. Perchlorate biodegradation for water treatment. Environ. Sci. Technol. 2005 (*In press*).

17. Ericksen, G.E. The Chilean nitrate deposits. Amer Scientist 1983; 71: 366-374.

18. Orris G. J., et al. Preliminary analyses for perchlorate in selected natural materials and their derivative products. U.S. Geol Surv Open-File Rept 03-314, 2003.

19. Dasgupta P. K., Martinelango P. K., Jackson W. A., Anderson T. A., Tian K., Tock R. W., and Rajagopalan S. The origin of naturally occurring perchlorate. Environ Sci Technol 2005; 39: 1569-1575.

20. Jackson W. A., Anandam S. K., Anderson T., Lehman T., Rainwater K., Rajagopalan S., Ridley M., and Tock R. Perchlorate occurrence in the Texas Southern High Plains Aquifer System. Ground Water Monitor Remed 2005; 25(1): 137-149.

21. Ader M., Coleman M.L., Doyle S.P., Stroud M., and Wakelin D. Methods for stable isotopic analysis of chlorine in chlorate and perchlorate compounds. Anal Chem 2001; 73: 4946-4950.

22. Sturchio N.C., Hatzinger P.B., Arkins M.D., Suh C., and Heraty L.J. Chlorine isotope fractionation during microbial reduction of perchlorate. Environ Sci Technol 2003; 37: 3859-3863.

23. Bao H. and Gu B. Natural perchlorate has a unique isotopic signature. Environ Sci Technol 2004; 38: 5073-5077.

24. Böhlke J.K., Sturchio N.C., Gu B., Horita J., Brown G.M., Jackson W.A., Batista J., and Hatzinger P.B. Perchlorate isotope forensics. Anal Chem 2005; 77: ##.

25. Coleman M.L., Ader M., Chaudhuri S., and Coates J.D. Microbial isotopic fractionation of perchlorate chlorine. Applied Environ Microbiol 2003; 69: 4997-5000.

26. Coleman M. L., Mielke R., and Coates J. D. Changes in Cl and O isotope compositions of perchlorate to measure extent of microbial reduction, but still identify the fingerprint of the original values. Geophys Res Abstr 2005; 7: 05954.

27. Böhlke J.K., Ericksen G.E., and Revesz K. Stable isotope evidence from an atmospheric origin of desert nitrate deposits in northern Chile and southern California, U.S.A. Chem Geol 2000; 136: 135-152.

28. Michalski G., Böhlke J.K., and Thiemens M.. Long term atmospheric deposition as the source of nitrate and other salts in the Atacama Desert, Chile: New evidence from mass-independent oxygen isotopic compositions. Geochim Cosmochim Acta 2004; 68: 4023-4038.

29. Gu B., Brown G. M., Maya L., Lance M. J., and Moyer B. A. Regeneration of perchlorate (ClO_4^-)-loaded anion-exchange resins by novel tetrachloroferrate ($FeCl_4^-$) displacement technique. Environ Sci Technol 2001; 35: 3363-3368.

30. Eggenkamp H.G.M. $\delta^{37}Cl$: *The geochemistry of chlorine isotopes.* Geologica Ultraiectina 1994; 116.

31. Holt B.D., Sturchio N.C., Abrajano T.A., and Heraty L.J. Conversion of chlorinated volatile organic compounds to carbon dioxide and methyl chloride for isotopic analysis of carbon and chlorine. Anal Chem 1997; 69: 2727-2733.

32. Long A., Eastoe C.J., Kaufmann R.S., Martin J.G., Wirt L., and Finley J.B. High-precision measurement of chlorine stable isotope ratios: Geochim Cosmochim Acta 1993; 57: 2907-2912.

33. Godon A., Jendrzejewski N., Eggenkamp H.G.M., Banks D.A., Ader M., Coleman M.L., and Pineau F. An international cross calibration over a large range of chlorine isotope compositions. Chem Geol 2004; 207: 1-12.

34. Miller M. F. Isotopic fractionation and the quantification of ^{17}O anomalies in the oxygen three-isotope system: an appraisal and geochemical significance. Geochim Cosmochim Acta 2002; 66: 1881-1889.

35. Böhlke J.K., Mroczkowski S.J., Coplen T.B. Oxygen isotopes in nitrate: new reference materials for $^{18}O:^{17}O:^{16}O$ measurements and observations on nitrate-water equilibration. Rapid Commun Mass Spec 2003; 17: 1835-1846.

Chapter 6

Recent Developments in Perchlorate Detection

Pamela A. Mosier-Boss

SPAWAR Systems Center San Diego, San Diego, California 92152

INTRODUCTION

Perchlorate is highly soluble and non-reactive with soil sediments. As a result perchlorate is exceedingly mobile in aqueous systems. Because of its resistance to react with other available constituents, perchlorate can persist for many decades under typical ground and surface water conditions. In 2003, the US EPA[1] compiled a list of potential perchlorate release sources. Sites from 25 states were listed. Table 1 summarizes the upper concentration range of perchlorate contamination found at these sites. Besides surface and groundwater, perchlorate has been found in food crops, such as lettuce, in the southwestern states. The uptake of perchlorate by crops is the result of irrigation using perchlorate contaminated water from the Colorado River.[2] There is a concern that perchlorate will be taken up by alfalfa, a major crop in the southwestern states. Some of this alfalfa is fed to dairy cows, which could conceivably result in perchlorate contamination of dairy products such as milk. In 2003, the presence of perchlorate in milk samples from Lubbock, TX was reported.[3]

Considerable efforts are underway to develop perchlorate detection methods for monitoring purposes, compliance, quality control, and delineation of contaminant plumes. These methods need to detect perchlorate in a wide range of sample matrices – including water, soil, fertilizers, milk, *etc.* Detection of trace quantities of perchlorate in these samples remains a challenging task because of the interferences from other anions, whose concentrations are usually several orders of magnitude higher than that of perchlorate. A number of technologies have been developed to detect perchlorate and include chromatographic and spectroscopic methods, gravimetry, *etc.* In a number of instances, two of these methods have been coupled to improve detection of perchlorate. In this communication, these methods are reviewed with regard to specificity, sensitivity, and reliability. How these methods are implemented is also discussed.

Table 1. Occurrence and potential sources of perchlorate releases in the environment as of April, 2003 (reprinted with permission from the U.S. EPA).[1]

State	Number of sites in state	Maximum ClO_4^- concentration range in state (ppb)
Alabama	2	8.9-19,000
Arizona	6	18-670
Arkansas	1	12,500-640,000
California	18	12-682,000
Colorado	1	180
Georgia	2	5.2-38
Iowa	4	9-30
Kansas	1	9
Maryland	4	4- >1,000
Massachusetts	1	100
Minnesota	2	4.5-6
Missouri	2	70-107,000
Nebraska	2	5-24
Nevada	3	24-3,700,000
New Jersey	2	7-627
New Mexico	10	3-21,000
New York	4	5-3,370
North Carolina	1	5.8
Oregon	2	10-1,000
Pennsylvania	4	4-33
Texas	15	3-169,000
Utah	2	16-42
Virginia	1	4.3
Washington	3	6-200
West Virginia	1	400

SOURCE OF PERCHLORATE CONTAMINATION – MAN-MADE OR NATURAL

Perchlorate contamination has been reported in 25 states.[1] The main sources of this contamination are man-made and are due to run-off from military bases, aerospace installations, and defense contractors involved in the production of rockets. However, there are also natural sources of perchlorate. Natural perchlorate is a minor component in salt deposits in the Atacama desert and in Chilean nitrates.[4] Perchlorate has been found in low levels at >80% of wells in a ~60,000 square mile area in the Texas panhandle.[5] The location of these wells were not near any munitions site suggesting that the source of the contamination was natural in origin. For remediation and forensic reasons, it would be desirable to be able differentiate between natural and man-made sources of perchlorate.

Recently, it was shown that natural and man-made sources of perchlorate can be differentiated by looking at the $^{17}O/^{16}O$ and $^{18}O/^{16}O$ isotopic ratios.[4] Once formed, the oxygen atoms of perchlorate will not exchange with those in the ambient environment thereby retaining the isotopic signatures of its sources. Oxygen isotope composition is defined as

$$\delta^{18}O = (^{18}R_{sample}/^{18}R_{standard} - 1) \times 1000\% \tag{1}$$

$$\delta^{17}O = (^{17}R_{sample}/^{17}R_{standard} - 1) \times 1000\% \tag{2}$$

where R is the abundance ratio of $^{18}O/^{16}O$ or $^{17}O/^{16}O$. For most oxygen-bearing compounds on Earth, there is a highly correlated relationship between their $\delta^{18}O$ and $\delta^{17}O$ values, which results because most oxygen isotope fractionation processes are mass dependent, i.e.:

$$\delta^{17}O_{MDR} \cong 0.52 \times \delta^{18}O \tag{3}$$

Reactions that do not follow the mass-dependent fractionation relationship will result in a $\delta^{17}O$ that deviates from the mass dependent relationship (MDR), $\delta^{17}O_{MDR}$. The deviation of the $\delta^{17}O$ value from the mass-dependent relationship, $\Delta^{17}O$, is given by:

$$\Delta^{17}O = \delta^{17}O - 0.52 \times \delta^{18}O \tag{4}$$

For man-made perchlorates, the $\delta^{18}O$ values were -18.4±1.2% and $\Delta^{17}O$ ranged from -0.06 to -0.20. These results are consistent for production of perchlorate by a mass-dependent process. Man-made perchlorate is produced by electrolysis of aqueous chlorate solution. Therefore oxygen is derived from the water used for its production and the electrolysis process is entirely mass dependent. For natural perchlorates, the $\delta^{18}O$ values ranged from -4.5‰ to -24.8‰ and exhibited a large positive $\Delta^{17}O$ ranging from +4.2‰ to +9.6‰. Atmospheric ozone also exhibits a highly positive $\Delta^{17}O$. In the atmosphere, ozone forms as the result of photochemical reactions that often do not follow the mass-dependent isotope fractionation relationship. These results suggest that natural perchlorate has an atmospheric origin involving ozone.

Isotopic analysis requires approximately 1 mg of perchlorate crystals, which are then thermally decomposed using the method developed by Hoering *et al.*[6] In this procedure,[4] ~1 mg of pure perchlorate crystals are sealed in a silver capsule, which is then placed inside a cone of Pt wire (0.5 mm in diameter) in a vacuum flask. A current is applied until the Pt wire is dark red in color. The resistive heat thermally degrades perchlorate to form O_2. The generated gases are passed through liquid nitrogen traps. The O_2 is collected in a sample tube submerged in liquid nitrogen. The isotopic composition of the sample is then determined by mass spectrometry.

The quantity (~1 mg) of perchlorate crystals, as well as its purity, necessary for isotopic analysis requires extraction, concentration, and purification of perchlorate from the sample matrix. This is accomplished using a bifunctional anion exchange resin that exhibits great selectivity for perchlorate.[4,7] The perchlorate-selective, bifunctional anion exchange resin, which is commercially available as Purolite A-530E, is comprised of tri-*n*-hexylamine and triethylamine groups grafted onto 5% divinylbenzene-cross-linked poly(vinylbenzyl chloride) macroporous copolymer beads.[7,8] Use of these bifunctional anion exchange resins are particularly useful in extracting and purifying perchlorate found in soil samples that often contain concentrations of anions such as sulfate, nitrate, and chloride that are several orders of magnitude higher than that of perchlorate. To extract perchlorate from soils, the samples are crushed and then soaked in doubly deionized water.[4] The extract containing perchlorate, and other water soluble anions present in the soil sample, is then pumped through a packed column filled with the bifunctional anion-exchange resin which preferentially adsorbs perchlorate. The perchlorate sorbed on the resin column is then recovered by a displacement procedure involving tetrachloroferrate ions, $FeCl_4^-$, formed in a solution of $FeCl_3$ and HCl.[9] The $FeCl_4^-$ ion is a large, poorly dehydrated anion which effectively displaces ClO_4^- from the resin. The solution containing the recovered perchlorate is then neutralized and the water is driven off using a rotary evaporator. The perchlorate crystals are then subjected to isotopic analysis as described *vide supra*. To regenerate the bifunctional anion exchange resin column, the $FeCl_4^-$ sorbed on the resin is converted into positively charged Fe^{3+} species (such as Fe^{3+}, $FeCl^{2+}$, and

$FeCl_2^+$) which desorb from the resin by charge repulsion. This is accomplished by decreasing the chloride ion concentration in solution.

GRAVIMETRY AND OTHER ION PAIRING METHODS

Prior to the 1990s, gravimetric methods were used for quantitative determinations of perchlorate. In a gravimetric analysis, an insoluble precipitate is formed from the desired analyte. This method of analysis is generally not very sensitive. However, some reagents used to precipitate perchlorate from solution have applicability in other detection methods, as will be discussed *vide infra*.

Perchlorate is a poor complexing agent and is used extensively as a counter ion in nonaqueous solution studies of metal cation chemistry.[10] Despite this fact, some insoluble or sparingly soluble perchlorate compounds exist that can be used for gravimetry. Reagents used for quantitative gravimetric analysis of perchlorate include 3,5,6-triphenyl-2,3,5,6-tetraazabicyclo-[2.2.1]hex-1-ene (also known as nitron), tetraphenyl-arsonium chloride,[10] methylene blue,[10,11] and cetyltrimethyl-ammonium bromide.[12] The gravimetric determination of perchlorate using nitron suffers from interferences. Nitron causes other large anions to precipitate out including BF_4^-, $WO_4^=$, ReO_4^-, NO_3^-, I^-, and Br^-.[10,13] Using cetyltrimethylammonium bromide, gravimetry has been used to detect perchlorate in the 5-20 mg range.[12] However lower amounts of perchlorate (0.05-1.0 mg) can be detected using cetyltrimethylammonium (CTA) bromide in conjunction with nephelometric and turbidimetric methods of analysis. Measurements using CTA can be made in the pH range from 1.5 to 11.5. However, large anions (*e.g.* MnO_4^-, CrO_4^{2-}, $Cr_2O_7^{2-}$, I^-) were shown to interfere with the determination.

Coupling gravimetry with detection using ion selective electrodes provides a means of improving selectivity and sensitivity. Ion selective electrodes will be discussed in more detail *vide infra*. Two studies were reported in which the potentiometric titration of perchlorate with tetraphenyl-arsonium chloride was monitored using the Orion perchlorate ion selective electrode.[14,15] Titration of perchlorate samples with tetraphenylphosphonium chloride yielded a typical sigmoidal-shaped curve. It was shown that the concentration limit, at room temperature, was 0.25 mmole perchlorate. However titrations done at lower temperatures (2 °C) result in sharper breaks in the titration curves thereby lowering the detection limit for perchlorate (0.05 mmole). Selig[16] compared tetraphenylarsonium chloride, tetraphenylphosphonium chloride, and tetrapentylammonium bromide as potentiometric titrants for perchlorate. The arsonium and phosphonium titrants yielded similar standard deviations and the size of the potentiometic breaks and steepness of the titration curves were almost identical. The ammonium precipitate exhibits greater solubility than the arsonium and phosphonium precipitates. As a result the potentiometric curve using the ammonium titrant exhibited smaller breaks and the precision was lower than that obtained using the arsonium and phosphonium titrants.

A simple, inexpensive colorimetric method to detect perchlorate in water and soil extracts has been developed that relies on the formation of an ion pair between perchlorate and a dye.[17] In this method, either a 500 mL sample of water or 1 mL sample of aqueous soil extract is passed through a styrene-divinylbenzene (SDVB) solid phase extraction (SPE) cartridge that had been conditioned with decyltrimethylammonium bromide (DTAB). DTAB is an ion-pair reagent that retained perchlorate as well as small quantities of chlorate and other anions such as nitrite, nitrate, sulfate, chloride, phosphate, and carbonate. A rinse step with DTAB in 15% v/v acetone/water removes the interferences. The perchlorate is eluted off the cartridge using acetone. The dye, brilliant green (BG), and xylene are added to the sample. The $BG^+ \cdot ClO_4^-$ ion pair partitions into the xylene layer and the absorbance at 630 nm is measured. Using this approach, detection limits of 1 μg L^{-1} and 0.3 μg g^{-1} were achieved for water and soil samples, respectively. It was shown that highly colored, biological materials such as humic and fulvic acids and chlorophyll can interfere with the method. Additional sample treatments can be employed to reduce or eliminate these interferences.

Another ion-pair extraction method to detect trace levels of perchlorate was developed that uses electrospray ionization mass spectrometry (ESI-MS).[18-20] In this method, a cationic surfactant is used to ion-pair aqueous perchlorate. The ion pairs are the extracted into an organic solvent. The ion pair is then detected using ESI-MS. The ESI-MS methodology will be discussed in more detail *vide infra*. Of the cationic surfactants investigated, tributylhepyl ammonium bromide (THAB) was the most efficient in extracting perchlorate. However, THAB had an impurity which interfered with quantitation. Cationic precipitants used in perchlorate gravimetry, tetraphenylarsonium and nitron, surprisingly gave no signal in the mass spectrometry measurements. Based on ion-pairing ability, reagent purity, and sensitivity, DTAB was considered to be the optimum cationic surfactant and dichloromethane the optimum extraction solvent. The method detection limit for perchlorate was 100 ng L^{-1}. This method was used to detect perchlorate in distilled water, tap water, drinking water, and Ohio River water (after filtration through a 0.45 μm filter).

ELECTROCHEMICAL METHODS: ION SELECTIVE ELECTRODES, CHEMICALLY SENSITIVE FIELD-EFFECT TRANSISTORS, AND SOLID PHASE MICROEXTRACTION

Figure 1 shows schematics of an ion selective electrode (ISE) and a chemically sensitive field-effect transistor (CHEMFET). A discussion of the theory of operation of these devices can be found elsewhere.[21] While both of these electrochemical sensors are inexpensive and sensitive, the specificity of response for the analyte is determined by the selectivity of the membrane for the analyte. Membranes typically are comprised of a carrier for the analyte; a polymeric matrix, such as polyvinyl chloride (PVC); and a plasticizer, such as 2-nitrophenyl octyl ether (o-NPOE). The sensitivity, selectivity, and linearity of these sensors not only depend on the nature of the carrier used, but also on the membrane composition and properties of the plasticizer used. It has been shown that the physiochemical properties of the plasticizer influences the dielectric constant of the membrane phase, the mobility and state of the carrier molecules, and the state of the ligands. An ionophore is a carrier that selectively binds to an ionic species. Many synthetic ionophores for inorganic cations have been developed for use in ion selective electrodes (ISEs) as well as voltammetric and optical sensors.[22] Noncovalent anion coordination chemistry has been relatively slow to develop in comparison with hosts for cations and even neutral molecules.[23] While anion hosts obey the same general rules that govern the magnitude of binding constants and host selectivity in cation hosts (primarily preorganization, complementarity, and solvation), their application is made more difficult because of some of the intrinsic properties of anions. For one thing, anions are relatively large and therefore require receptors of considerably greater size than those used to bind cations. Unlike inorganic cations, inorganic anions occur in a range of shapes and geometries, $e.g.$, spherical (halides), linear (SCN^-), planar (NO_3^-), tetrahedral ($SO_4^=$, PO_4^{-3}),

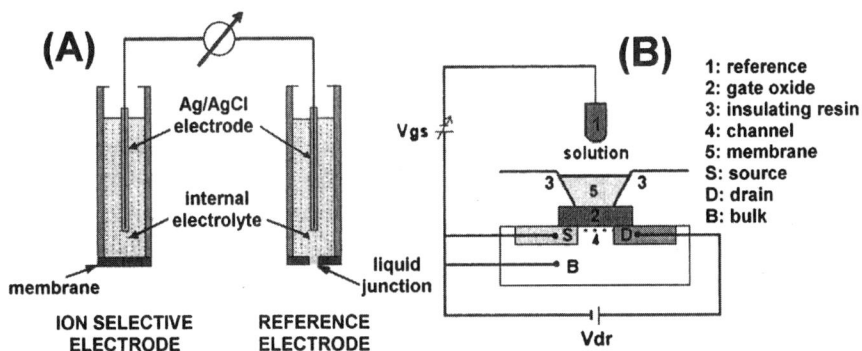

Figure 1. Schematics of (A) an ISE and (B) a CHEMFET.

and octahedral (PF_6^-). In comparison to cations of similar size, anions have high free energies of solvation. Consequently anion hosts must compete more effectively with the surrounding medium. Many anions exist only in a relatively narrow pH window and anions are usually saturated coordinatively and therefore bind only via weak forces such as hydrogen bonding and van der Waals interactions. Despite these challenges, there have been recent efforts to develop ionophores for inorganic and organic anions. These efforts have been summarized in a number of review articles.[24,25] However, the selectivity of the membrane for a given ionic species is not solely dependent upon the ionophore. As discussed, *vide supra*, the polymeric matrix also contributes to the membrane's selectivity. For example, it was shown that a monolayer of a neutral bis-thiourea ionophore on a carbon electrode was selective for the hydrophilic $HPO_4^{2-}/H_2PO_4^-$.[26] The strength of interaction decreased in the order $H_2PO_4^- > F^- \cong SO_4^= > CH_3COO^- > Cl^-$. The stability of the interaction with $H_2PO_4^-$ was attributed to the formation of hydrogen bonds between the oxygen atoms of the anion and the thiourea groups of the receptor. However, incorporating the ionophore in a solvent polymeric membrane resulted in a sensor specific for chloride ion.[27] The selectivity of the solvent polymeric membrane ISE based on the bis(thiourea) ionophore was attributed to the free energies of phase transfer of the anions from the aqueous sample into the organic ISE membrane phase. As a consequence of its lower hydration energy, transfer of chloride from the sample into the membrane phase was energetically more favorable than the phase transfer of phosphate. The results indicated that the selectivity of phosphate binding by the ionophore was not large enough to counterbalance the large difference in the phase transfer energies of the two ions.

In the case of the commercially available Orion ISE from Thermo Electron Corporation, the selectivity of the membrane for perchlorate ion is attributed to synergism between the ionic carrier and the polymer matrix. The perchlorate-sensing membrane incorporates a tris (substituted 1,10-phenanthroline) iron (II) ion exchanger dissolved in an organic solvent, *p*-nitrocymene, to give a non-water soluble ion-exchanging liquid.[20] The tris(substituted 1,10-phenanthroline) iron (II) ion exchanger is a large, inert metal ligand complex that does not complex with small, hard Lewis bases and has no coordination sites. The remaining membrane constituents are hydrophobic. When immersed in an aqueous perchlorate solution, the lipophilic perchlorate ion is extracted into the hydrophobic membrane by the ion exchanger causing a phase-boundary potential to develop at each solution-membrane interface and diffusion potentials arise in the membrane. The resultant potential, E, is given by the following relationship:

$$E = C + \frac{RT}{z_i F} \ln\left[a_i + \sum K_{ij} a_j^{(z_i/z_j)} \right] \qquad (5)$$

where a_i is the activity of the primary ion with charge z_i, a_j is the activity of the interfering ion with charge z_j, K_{ij} is the selectivity coefficient characteristic of a given membrane, and C is a constant that depends upon the choice of the external reference electrode.[28] The effective concentration range of operation for the Orion ISE is 0.7-99,500 ppm (7×10^{-6}-9.95×10^{-1} mol L^{-1}). The electrode can be used over a pH range of 3-10. The selectivity coefficient is defined as the activity ratio of the primary ion i and interfering ion j that gives the same potential change in the reference solution. Selectivity coefficients for the Orion perchlorate ISE are summarized in Table 2. It can be seen that the Orion perchlorate ISE also responds to other lipophilic anions such as iodide and bromide. While ISEs are relatively sensitive, inexpensive, and exhibit rapid response times, they do require recalibration after 10-30 samples in order to compensate for changes in temperature or conditions of the sensing membrane.[21,28]

The use of other ionophores to improve perchlorate selectivity has been explored. Figure 2 shows structures of three ionophores that have been examined. The gold (I) complex, Figure 2a, has triphenylphosphonium groups present that are structurally similar to the reagent used to precipitate perchlorate. The gold (I) complex was dissolved in plasticizer (either o-NPOE or dibutyl phthalate, DBP) and used as a carrier in a PVC membrane of an ISE.[29] The working pH range using this ionophore was 2 - 12, the linear concentration range was 5×10^{-6} to 1×10^{-2} mol L^{-1}, and the detection limit was 1.1×10^{-6} mol L^{-1}. The potentiometric selectivity coefficients for different anions are summarized in Table 2. Of the anions in common, the gold (I) complex exhibited better selectivity for perchlorate than the commercially available Orion ISE, Table 2. The use of an ISE using the ionophore shown in Figure 2a accurately measured perchlorate

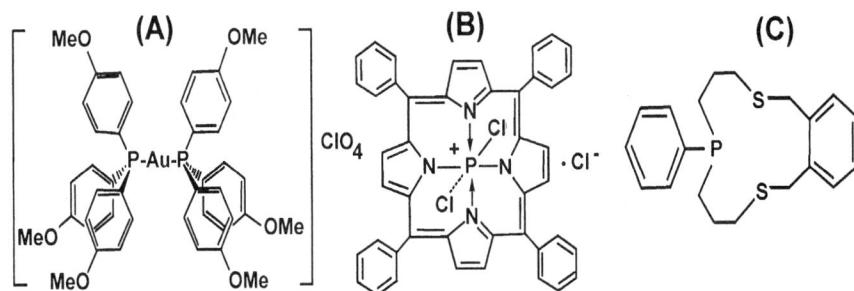

Figure 2. Structures of ionophores used as carriers in perchlorate electrochemical sensors (a reprinted from Anal. Chim. Acta, 2000; 415: 159-164. b reprinted from Sensors and Actuators B, 2003; 89: 9-14, and c from Sensors and Actuators B, 1997; 43: 206-210 with permission from Elsevier Science).[29-31]

concentrations in urine samples without prior separation steps. Polymeric membranes comprised of a phosphorous (V) tetraphenylporphyrin complex, Figure 2b, for use in a perchlorate ISE have been prepared.[30] It has been shown that, for metalloporphyrin compounds, the anion selectivity is determined by specific interactions between the metal and the anion and not by the lipophilicity of the anion or by simple opposite charge interaction with the anion. Optimum results were obtained using polymeric membranes comprised of the phosphorous (V) tetraphenylporphyrin complex shown in Figure 2b, the plasticizer o-NPOE, and the additive cetyltrimethylammonium bromide (CTAB) in a PVC matrix. Recall CTAB was used in gravimetric titrations of perchlorate.[12] The electrodes exhibited a rectilinear range from 8×10^{-6} to 1.6×10^{-1} mol L^{-1} perchlorate and a detection limit of 5.0×10^{-6} mol L^{-1} perchlorate. The operating pH range for these ISEs was between 4 and 10. The selectivity coefficients for ISEs using the phosphorous (V) tetraphenylporphyrin complex as the carrier are summarized in Table 2. The selectivity of this ISE is comparable to that of the commercially available Orion perchlorate ISE. The phosphorous (V) tetraphenyl-porphyrin complex -based ISE worked well under laboratory conditions and gave accurate results when used to measure perchlorate in tap water and human urine.

Figure 2c shows the structure of a phosphadithiamacrocycle that has been used as a neutral carrier in PVC membranes for ISEs and CHEMFETs.[31,32] The plasticizer used in these membranes was o-NPOE. The ISE exhibited a linear range of 1×10^{-6} -1×10^{-2} mol L^{-1} and a detection limit of 8×10^{-7} mol L^{-1} while the CHEMFET exhibited a linear range of 6×10^{-7} -1×10^{-2} mol L^{-1} and a detection limit of 3×10^{-7} mol L^{-1}. The pH range for the ISE and CHEMFET are 1.5-13.5 and 1-11, respectively. Selectivity coefficients for the ISE and CHEMFET are summarized in Table 2. Only SCN⁻ and BF_4^- are significant interfering ions. In general, the selectivity is better than that of the commercially available Orion ISE. In most cases, the selectivity is better for the CHEMFET than for the corresponding ISE. ^{31}P NMR studies indicated that, inside the membrane, the ionophore exists in the oxidized form, suggesting that ion-dipole interactions between perchlorate and the P=O group may be responsible for the perchlorate selectivity.

Another approach to create selective membranes for use in an ISE is molecular imprinting. Molecular imprinting is a technique for constructing tailor-made receptor binding sites in a three-dimensional, cross-linked polymer matrix.[33,34] The result is a polymer having a high affinity for a target molecule. The process involves three key steps: (1) complex formation of the template (print) molecule with the functional monomers (prearrangement step) in a solution containing a high ratio of cross-linker, (2) co-polymerization of the mixture in an inert solvent to form a rigid polymer, and

Table 2. Selectivity coefficients ($\log K_{ClO_4,j}$) measured for perchlorate electrochemical sensors. The ISEs and CHEMFET use polymeric membranes. Structures of ionophores A, B, and C are shown in Figure 2 (reprinted with permission from Elsevier Science and Thermo Electron Corporation).

Anion (j)	Orion ISE[28]	ISE using Ionophore A[29]	ISE using Ionophore B[30]	ISE using Ionophore C[31]	CHEMFET using Ionophore C[31]
I^-	-1.92	-2.0	-1.5	-1.92	-2.54
Br^-	-3.25	-4.0	-3.0	-3.82	-4.53
F^-	-3.60	-	-4.1	-	-
Cl^-	-3.66	-5.0	-3.7	-4.52	-4.74
SCN^-	-	-1.4	-2.0	-1.03	-1.14
CN^-	-	-	-2.9	-	-
OH^-	0	-4.6	-	-	-
NO_3^-	-2.82	-3.2	-3.8	-3.02	-3.49
NO_2^-	-	-4.4	-3.3	-4.04	-4.88
$CH_3CO_2^-$	-3.29	-5.2	-2.8	-	-
ClO_3^-	-	-	-3.9	-	-
HCO_3^-	-3.46	-4.7	-	-4.3	-4.14
HSO_3^-	-	-	-	-4.55	-4.08
BF_4^-	-	-	-	-1.22	-2.10
IO_4^-	-	-	0.11	-	-
$SO_4^=$	-3.80	-5.0	-4.2	-4.42	-4.60
$SO_3^=$	-	-5.0	-	-	-
$CO_3^=$	-	-	-	-4.82	-4.47
$MnO_4^=$	-	-	0.75	-	-
$HPO_4^=$	-	-	-	-4.96	-4.79
PO_4^{3-}	-	-5.1	-	-	-

(3) removal of the template molecule, by hydrolysis or extraction, to afford the imprinted polymer. The resulting molecularly imprinted polymers (MIPs) are macroporous matrices possessing microcavities with a three-dimensional structure complementary in both shape and chemical functionality to that of the template. The high degree of cross-linking enables the microcavities to maintain their shape after removal of the template. As a result, the functional groups are held in an optimal configuration for rebinding the template, allowing the receptor to 'recognize' the original substrate.

An ISE was prepared by polymerizing pyrrole on a glassy carbon surface in the presence of perchlorate.[35] During electropolymerization in the presence of perchlorate ion, the film produced should have cavities selective for perchlorate. Both the size and charge distribution of the cavities within the polymerized film should be complementary to the size and shape of perchlorate ion. The ISE gave Nernstian potentiometric responses over the $10^{-1} - 10^{-4}$ mol L^{-1} perchlorate concentration range and a detection limit of 1.8×10^{-5} mol L^{-1}. Table 3 summaries the selectivity coefficients of the ISE. The results in Table 3 indicate that an ISE prepared by this method exhibits poor selectivity and that NO_3^-, IO_4^-, I^-, CH_3COO^-, and Br^-, are severe interferents. Varying electropolymerization parameters should improve selectivity. The electropolymerization parameters (e.g., fixed potential or constant current method, polymerization time, applied voltage or current, monomer and template concentration, solvent) affect the film morphology (density and porosity). For example, it was shown that constant current techniques produced polypyrrole films that responded better to nitrate than films produced using a fixed potential method.[36]

Table 3. Selectivity coefficients (log $K_{ClO4,j}$) measured for a perchlorate ISE prepared by electropolymerization of pyrrole (reprinted from Electroanalysis, 1989; 1: 271-277 with permission from Wiley Interscience).

Anion (j)	Log $K_{ClO4,j}$	Anion (j)	log $K_{ClO4,j}$
$SO_4^=$	-2.15	F^-	-0.96
$CH_3CO_2^-$	-0.40	Cl^-	-0.96
HCO_2^-	-0.40	Br^-	-0.35
NO_3^-	-0.25	I^-	-0.30
IO_4^-	>1		

Because polypyrrole is a conductive polymer, the applied potential controls the movement of counterions in and out of the polymer film for charge balance. For this reason, the use of polypyrrole films as a coating for solid phase microextraction was explored.[37] The polypyrrole films were prepared by electroreduction on a 200 μm × 1.5 cm long Pt wire. To extract perchlorate, a negative potential is applied to eject counterions. After washing, the electrode is immersed in a 50 mL sample. A positive potential is applied for 15 minutes to extract perchlorate. After extraction, the electrode is washed and immersed in a 1 mL 0.01 mol L^{-1} NaCl solution where the perchlorate is released by applying a negative potential for 2 minutes. The sample is then analyzed for perchlorate using ESI-MS. Results of this method are summarized in Figure 3.

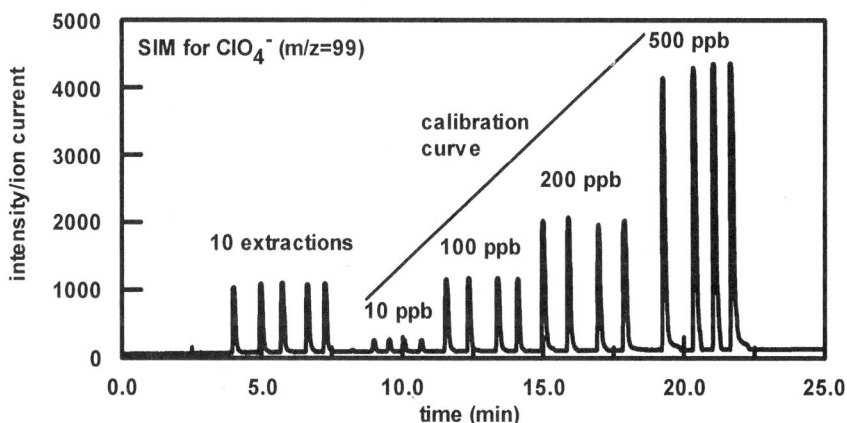

Figure 3. Analysis of perchlorate by SPME with ESI-MS detection (reprinted from Anal. Chem, 2002; 74: 4855-4859 with permission from the American Chemical Society).

SEPARATION METHODS: ION CHROMATOGRAPHY AND CAPILLARY ELECTROPHORESIS

In the 1990s, separation methods such as ion chromatography (IC) and capillary electrophoresis (CE) became preferred methods of perchlorate detection. These methods have been used to detect perchlorate in drinking water, groundwater, soil, plants, fertilizers, and foodstuffs.

In general, in ion chromatography, a column containing an ion-exchange resin is used to separate the anions.[28] The most common detection methods used, after ion separation, is conductivity. For an analysis that requires less that ppm-sensitivity, a second column is used prior to the conductivity detector which suppresses the eluant conductivity, while increasing the analyte conductivity. The result is a quieter baseline and higher sensitivity. This suppressor column is comprised of an ion exchange bed. Although there are a number of mobile phases that have been used in ion chromatography, the two most popular for use with a suppressor are based upon sodium hydroxide and sodium carbonate. In the suppressor column, sodium hydroxide and carbonate are converted into water and carbonic acid, respectively. Anion identification is usually based on retention-time matching.

Perchlorate has a low charge density and is poorly aquated. As a result, on most columns it is strongly retained and often suffers from poor peak shape when it does elute off the column.[13] Methods used to promote elution of perchlorate include addition of an organic solvent, such as methanol, to the mobile phase; use of an anion in the eluant that is more hydrophobic which will displace perchlorate; increase the concentration of eluant anion; and fabrication of more hydrophobic stationary phases.

Separation of perchlorate from other chlorine-containing anions in water in the presence of nitrate and sulfate was demonstrated.[38] The separation column contained a silica-based stationary phase cross-linked with quaternary ammonium groups. Phthalate was used as the eluant anion. Although the retention times for perchlorate were on the order of 30-40 minutes, the peaks due to perchlorate were still well-shaped. Using UV detection, the lower limit of detection (LLOD) was 1.5 μg mL^{-1}.

The first ion chromatographic method to detect trace amounts of perchlorate was developed by the California Department of Health Services.[39] An anion exchange column, Dionex AS5, was used with a mobile phase of NaOH and p-cyanophenol and conductivity suppression. The p-cyanophenol was used to reduce perchlorate's affinity towards the quaternary amine functional groups of the anion exchange resin. Using p-cyanophenol in the eluant, it was shown that, in the presence of chloride, sulfate, and hydrogen carbonate, the shape of the perchlorate peak was satisfactory. The practical quantifiable limit using this method was 6 ppb. However, it was found that the retention time for perchlorate shifted with eluant concentration, making speciation difficult.

Dionex has since developed better columns for separating large polarizable anions like perchlorate. In the AS16 column, the substrate is a 9 μm diameter macroporous bead consisting of a core of ethylvinylbenzene crosslinked with 55% divinylbenzene and an outer anion-exchange layer that is functionalized with very hydrophilic quaternary ammonium groups. Figure 4 illustrates the isocratic separation of anions using this column and a 35 mM NaOH eluant.

The U.S. EPA[40] published Method 314.0 for the detection of perchlorate that uses this next generation anion exchange resin as well as electrolytic conductivity suppression and conductivity detection. The detection limit (DL) of Method 314.0 is 0.53 ppb with a widely achievable minimum reporting limit (MRL) of 4 ppb. Lower DLs and MRLs can be achieved by using large volume injection (2 mL or more), sample preconcentration onto short chromatography columns prior to separation on the analytical column, and sample concentration by evaporation. This ion chromatography method is the predominant technique for determining perchlorate ion concentration in raw and processed drinking water as well as bottled water.[41] Other anion exchange columns designed by Dionex for perchlorate detection are the AS20 and AS21. In the AS20 column, the stationary phase consists of a hyper-branched, anion exchange condensation polymer electrostatically attached to the surface of a wide-pore polymeric substrate. The resultant polymer is extremely hydrophilic and has great selectivity for hydroxide allowing the use of lower hydroxide eluant concentrations. Using suppressed conductivity detection, this results in sub $\mu g\ L^{-1}$ detection of perchlorate. In addition, the AS20 column eliminates interference from 4-chlorobenzene sulfonate, which causes false positives with the AS16 column. The AS21 column is designed for use with mass spectrometric detection. This detection mode will be discussed in more detail *vide infra*. The polymer used in the AS21 column is prepared in a similar manner as the AS20 polymer. However, the AS21 polymer exhibits selectivity for both methylamine and hydroxide. A methylamine eluant allows for better performance in electrospray mass spectrometric detection.

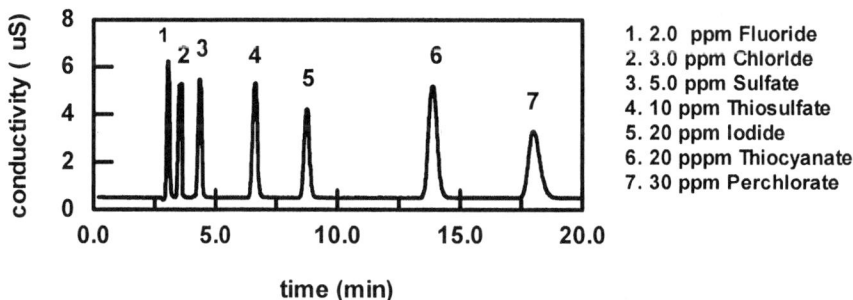

Figure 4. Isocratic separation of polarizable and inorganic anions using the Dionex AS16 column, 35 mM NaOH eluant, 30 °C, flow rate of 1 mL min⁻¹, and 10 μL injection volume (reprinted with permission from Dionex).

EPA Method 314.0 can be subject to matrix interferences, especially in the presence of high amounts of other anions, as shown in Figures 5a and b.[42] High ionic strength samples can result in column overloading, increased detection limits, and overall reduced column performance. SPE can be used to remove some anionic interferences. There are cartridges available that will remove halide ions, sulfate, and carbonate. The use of alumina to remove interferences in the detection of perchlorate in tobacco plants and tobacco products has been demonstrated.[43] Aqueous extracts of green and flue-cured tobacco leaves were prepared from 600 mg quantities of ground leaf extracts that had been freeze dried and 30 mL of 18 MΩ-cm water. Placing the suspension in a boiling water bath resulted in the precipitation of proteins. Solids were removed by filtration through 0.2 μm filters to yield IC extracts. In the chromatograms of these extracts, the perchlorate peak is obscured by the large peak in the 3-15 minute retention time window caused by the elution of weakly retained anions, cations, organic acids, etc. To remove these interferences, a 1 mL sample of the IC extract was added to 500 mg alumina, diluted 1/10, and filtered through a second 0.2 μm filter. Alumina was chosen because it has distinct surface site types for adsorption interactions: acidic or positive field sites, basic or proton acceptor sites, and electron acceptor (charge transfer) sites. The alumina not only removed ionic compounds from the extracts but also neutral material known to foul ion exchange columns. After treatment, the extracts were clear in color – even the black-brown extracts obtained from tobacco products. The resultant chromatograms exhibited a decreased baseline and a corresponding increase in the perchlorate signal. The method detection limit for perchlorate was determined to be 2.0 mg/kg related to dry weight of plant material and 0.4 mk/kg on a wet weight basis. Using this method, the concentration of perchlorate in tobacco products including cigarettes, cigars, and pouch and plug chewing tobacco ranged from no detect (nd) to 60.4±0.8 mg/kg on a wet weight basis.

These extraction methods using SPE cartridges and alumina are cumbersome, they cannot be automated, and they can result in loss of sample. An on-line technique has been developed that concentrates perchlorate and allows   the less strongly held ions to pre-elute before transferring the perchlorate to the principle separation column.[42] A schematic of the system is shown in Figure 5c. The autosampler fills the sample loop L in V1. When the sample loop is filled, V1 switches to the injection position and the loop contents are transferred to the PC by the wash solution pumped by P1. The prewash solution was 10 mM NaOH. The sample that was transferred to the PC was washed to elute the interfering anions while retaining the perchlorate. V2 is switched to the inject position so that the remaining sample is injected into the separation system (GC-SC-ASRS-D) for anion separation and detection. The effectiveness of this method in removing anion interferences is illustrated in Figures 5a and 5b. The samples analyzed were groundwater supplied by the Texas Commission on Environmental Quality (TCEQ) and a solution containing 2000 mg L^{-1} each of Cl$^-$, SO$_4^=$, and CO$_3^=$. Both samples had been spiked with 25 μg L^{-1} of perchlorate. When analyzed using the EPA Method 314.0, the perchlorate peak is a small, barely discernable bump in a sloping baseline due to the interfering anions. Using the on-line preconcentration method, the baseline is significantly decreased and the signal due to the perchlorate is enhanced. The limit of detection (LOD) using this method is 0.77 μg L^{-1} for a 2 mL reagent water sample and decreases proportionately with increasing sample volume, up to 20 mL. This on-line preconcentration method can be automated.

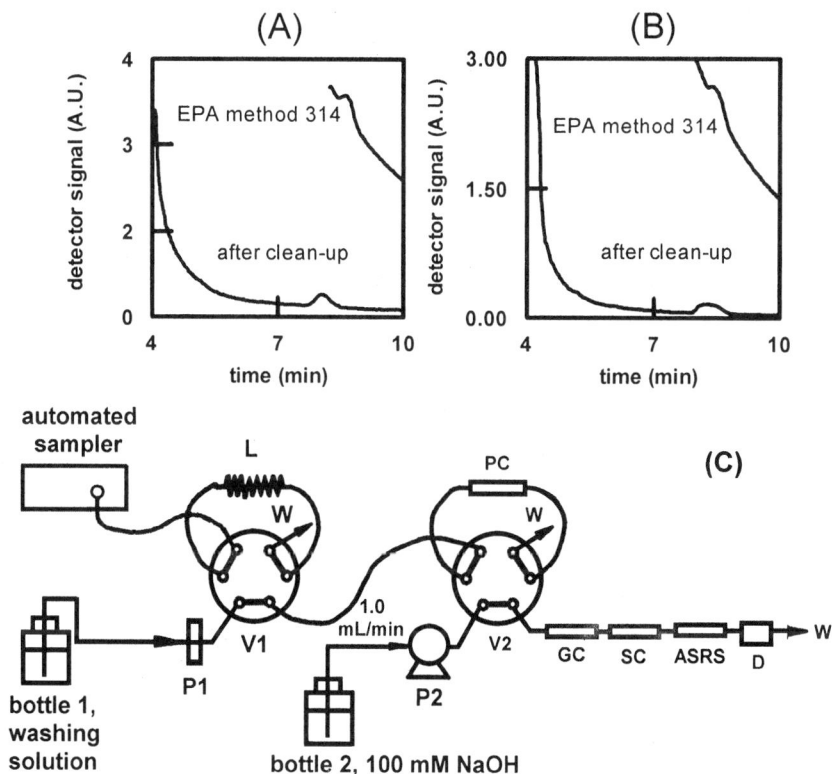

Figure 5. Chromatograms obtained, using the EPA Method 314.0 and on-line preconcentration, for 1 mL samples of (A) 25 μg L^{-1} perchlorate spiked into TCEQ groundwater and (B) 25 μg L^{-1} perchlorate spiked solution containing 2000 mg L^{-1} each of Cl$^-$, SO$_4^=$, and CO$_3^=$. (C) Schematic of the on-line preconcentration system: P1=peristaltic pump; L=sample loop; PC=preconcentration column; GC=guard column; SC=separation column; ASRS=electrodialysis suppressor; D=conductivity detector; V1=low pressure, 6-port valve; V2=6-port chromatographic injector; P2=chromatographic pump; and W=waste (reprinted from Anal. Chem, 2003; 75: 701-706 with permission from the American Chemical Society).

Another approach to detecting perchlorate in high ionic strength matrices is to use a separation column filled with a poly(vinyl alcohol) (PVA) gel resin.[44] When used with 4-cyanophenoxide as an additive in the NaOH eluant, PVA columns satisfactorily separated perchlorate from other anions found in fertilizer products that are high in ionic strength and dissolved matter. The PVA columns were robust in these matrices and allowed the detection and quantification of perchlorate by suppressed conductivity.

The effect of matrix interferences can also be reduced by using mass spectrometry as the detection method in place of conductivity. Mass spectrometry can also be used to improve speciation.[45] When using a conductivity detector, speciation is based solely upon retention time. In complex matrices, the number of peaks in the chromatograms increases, which could lead to false positives. Coupling ion chromatography with mass spectrometry/mass spectrometry (MS/MS) has been used to separate and detect perchlorate in water and soil samples.[46] Soil samples were prepared by leaching 2 g of sample with 20 mL of deionized water. In the MS/MS, the $ClO_4^- \rightarrow ClO_3^-$ reaction transitions were monitored at m/z 99→83 (for the ^{35}Cl isotope) and 101→85 (for the ^{37}Cl isotope). The isotopic ratio of ^{35}Cl to ^{37}Cl was used to improve the specificity of the method and to provide greater confidence that the detected signal was due to perchlorate and not an interfering compound. The theoretical ratio of ^{35}Cl to ^{37}Cl for natural isotopic abundances is 3.08, and the area ratio for the m/z 83 to 85 peaks should be near this value. The method detection limits (MDLs) for perchlorate using this IC-MS/MS procedure were 0.5 μg/kg for soil samples and 0.05 μg/L for aqueous samples. Besides detection of perchlorate in water and soil samples, the IC-MS/MS method has been used to detect perchlorate in foodstuffs such as lettuce,[46] cantaloupe,[46] and milk.[3,46]

Another problem with IC analysis of perchlorate in complex matrices is that retention times migrate with column deterioration.[47] It has also been shown that retention times vary inversely with perchlorate concentration and that large concentrations of anions that elute earlier than perchlorate can alter the retention time of the smaller perchlorate peak. This makes species identification difficult. Spiking the extract can increase the level of certainty, but this significantly increases the analysis time. To avoid carryover between runs and net build-up of anions on the columns, run times need to be sufficiently long to ensure complete elution of strongly retained ions.

Capillary electrophoresis (CE) is another separation method that has been used to detect perchlorate in a wide range of sample matrices. CE developed as a technique in the late 1980s and uses narrow-bore fused-silica capillaries to separate a complex array of large and small molecules. Figure 6 shows the components of a typical CE instrument. In CE, high electric field strengths are used to separate molecules based on differences in charge, size and hydrophobicity. Sample introduction is accomplished by immersing the end of the capillary into a sample vial and applying pressure, vacuum, or voltage. For the separation of ions, CE usually offers better efficiency of separation, shorter analysis time with minimum Joule heating, and lower consumption of reagents.[13,38] Simultaneous detection of 23 ionic species using CE has been demonstrated.[48] Figure 7 shows the electropherogram that was obtained. The concentration for anionic species was 50 nM and the entire time of analysis took 6.5 minutes.

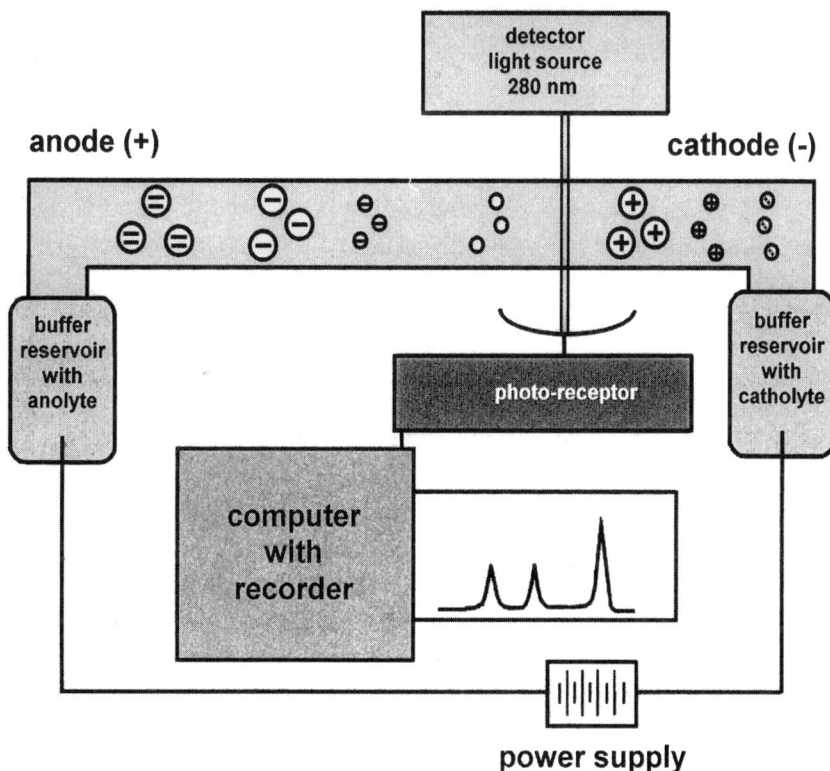

Figure 6. Components of a capillary electrophoresis system.

The low volume characteristics and small diameter of the CE capillary place volume constraints on the method of detection used.[49] The detection device must be sensitive enough to respond to ion concentrations on the order of 20 zmol s^{-1}. Indirect modes of detection are usually used in CE.[13] With indirect detection methods, loss of signal (*e.g.*, absorbance or fluorescence) indicates that the analyte is eluting in place of the background electrolyte. When using indirect methods of detection, speciation is based upon the elution time of the peak. Additives, such as cyclodextrin, can be used to change electrophoretic mobility and thereby improve separations.[13] For example, using a 60 cm long × 0.75 μm diameter fused silica capillary; an electrolyte consisting of 4.6 mK sodium chromate and 0.46 mM CIA-PAK OFM anion BT, an electroosmotic flow modifier, at pH 8.0; an applied voltage of 20 kV; and UV detection at 254 nm, an LLOD for perchlorate of 0.6 μg mL^{-1} was obtained for water in the presence of nitrate and sulfate.[38]

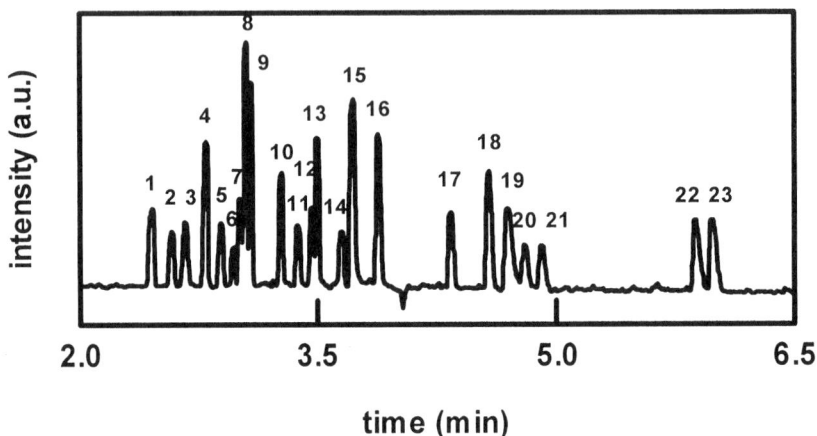

Figure 7. Electropherogram of anions (50 nM concentration) obtained using a Dionex CES 1 with electrokinetic injection and isotachophoretic preconcentration. Peaks: 1=thiosulfate, 2=bromide, 3=chloride, 4=sulfate, 5=nitrite, 6=tetrafluoroborate, 7=nitrate, 8=sulfosuccinate, 9=oxalate, 10=molybdate, 11= perchlorate, 12= thiocyanate, 13=tungstate, 14=chlorate, 15=citrate, 16=malonate, 17=malate, 18=tartrate, 19=fluoride, 20=bromate, 21=formate, 22=phosphate, 23=arsenate (reprinted from Chromatographia, 1997; 45:301-311 with permission from Vieweg-Verlag).

Electrochemical detection by means of ultramicroelectrodes (UMEs) has been used as a direct detection mode for CE. UMEs have diameters on the order of 10 μm or less and exhibit high sensitivity, good selectivity, and are inexpensive to fabricate.[49] The UME response is not limited by a small detection volume. However, the electrochemical detector needs to be completely isolated from the potential used in the separation. Electrochemical techniques used with CE separation and UME detection include amperometry, potentiometry, and conductivity. Amperometric methods have been used primarily in the detection of metal ions and will not be discussed further. To detect anions after separation by CE, ion selective microelectrodes (ISMEs) based on anion-exchanger liquid membranes have been used.[49] The selectivity of these ISMEs depend upon the free enthalpy of hydration of the analyte anions. The more lipophilic the anion, the higher the sensitivity of the ISME. To reduce drift and noise problems, the capillary detector end is etched with hydrofluoric acid to form a conical aperture into which the tip of the ISME is positioned. In this configuration, the field strength at the tip of the ISME is considerably reduced compared with that of the remaining capillary. The use of an ISME to detect 10 nM perchlorate in tap water after CE has been demonstrated.[50] There are a number of advantages of potentiometric electrochemical detectors such as ISMEs. The logarithmic response obtained with potentiometric sensors allow detection over a wide dynamic range.[49] An internal standard with simultaneous analyte addition can be used to compensate for matrix effects and sampling biases. However, under normal CE conditions, ISMEs exhibit a 2-3 day lifetime. In order to achieve accurate results, recalibration every 12 hours is recommended. Conductivity is a suitable alternative direct detection method for ionic species that do not have a chromophore and are not electrochemically active.[49] In this method, the differences in conductivity of the solute zones and the separation electrolyte are measured. Sensitive detection is possible by using separation electrolytes with low conductivity. The conductivity detector can be either in an on-column, off-column, or end-column configuration. Using an on-column configuration, inorganic anions were detected at 50-100 ppb concentration levels in drinking, river, and rain water samples.

Normal Raman spectroscopy has been used as an on-line detector for CE.[51] The advantage of Raman spectroscopy over electrochemical detection and indirect detection methods is that every polyatomic species will exhibit a characteristic Raman scattering spectrum comprised of narrow peaks that can be used for identification purposes. Furthermore water and fused silica are poor Raman scatterers and do not interfere in the detection of polyatomic anions. However, Raman spectroscopy is inherently an insensitive technique with detection limits in the 10^{-2}-10^{-3} M concentration range. To use Raman spectroscopy as a detector for CE requires sample preconcentration. In CE, sample stacking and isotachophoresis are two techniques used to preconcentrate analytes. Isotachophoresis is used for preconcentrating analytes in matrices containing a high-concentration background electrolyte. Sample stacking is the simplest technique but is limited to analytes in dilute solutions with little or no background electrolyte. Field-amplified injection is a form of sample stacking in which the electric field is higher in the lower-conductivity sample plug than in the running buffer. Sample ions move with a high velocity toward the boundary where they slow down and 'stack up' into a smaller zone with a higher concentration. Using the field-amplified injection technique, samples containing nitrate and perchlorate were analyzed using the Raman spectroscopy-CE system shown in Figure 8a. The high voltage end of the capillary was contained inside a Plexiglas box. The running electrolyte was 0.1 M KCl. The Raman excitation source was a Nd:YAG laser operating at 532 nm and 300 mW power. Spectra were taken with 1 second integration times with a 300 ms reset and storage delay between acquisitions. Figure 8b shows the electrophoretic separation of a sample containing 5×10^{-4} M nitrate and perchlorate, which corresponds to 31 ppm nitrate and 50 ppm perchlorate. Using a 15 second injection time at -2 kV into a 75 μm-i.d. capillary, the Raman bands due to nitrate (1047 cm^{-1}) and perchlorate (934 cm^{-1}) were easily detected in the 1-second acquisition spectra. The higher mobility nitrate eluted first. The total analysis time was less than 3 minutes. By increasing the injection time to 40 seconds and using a 50 μm-i.d. capillary, 620 ppb nitrate and 1 ppm perchlorate were detected. With better heat dissipation and narrower-bore capillaries, lower detection limits should be possible.

Recent advances in the field of microfluidics have seen the development of microchip CE systems.[52] In conventional CE, the electrophoretic mobilities of cations and anions prevail over the electroosmotic flow and the cations and anions move in opposite directions. As a result, in conventional CE, only anions or cations can be injected and detected in a given CE run. Microchip CE systems have been fabricated using a single cross one-end injection arrangement. In this configuration, only cations or anions can be injected as only one type of ions migrates to the separation channel depending on the value of the voltage. A bidirectional approach has been developed in which the sample is injected in the middle of the chip and the anions and cations migrate towards the opposite ends where conductivity detectors are positioned. Recently a dual-injection poly(methyl methacrylate) (PMMA) microchip electrophoretic system has been designed and fabricated that also allows the simultaneous detection of anions and cations. The CE microchip system, shown in Figures 9A and B, consists of two sample reservoirs on both sides of a common separation channel. The separation channel, which is between the two injection crosses on both sides of the chip, is 66 mm long. Platinum wires were used to connect the high voltage electrodes in the solution reservoirs. The movable contactlessconductivity detector (MCCD) consists of two 0.8 mm × 24 mm rectangular-shaped electrodes made from 10 μm thick aluminum foil strips. The electrodes were placed in an antiparallel orientation on a PMMA plate equipped with two polyvinyl chloride clips. The position of the detection electrodes can be changed by manually sliding the microchip through the detector holder.

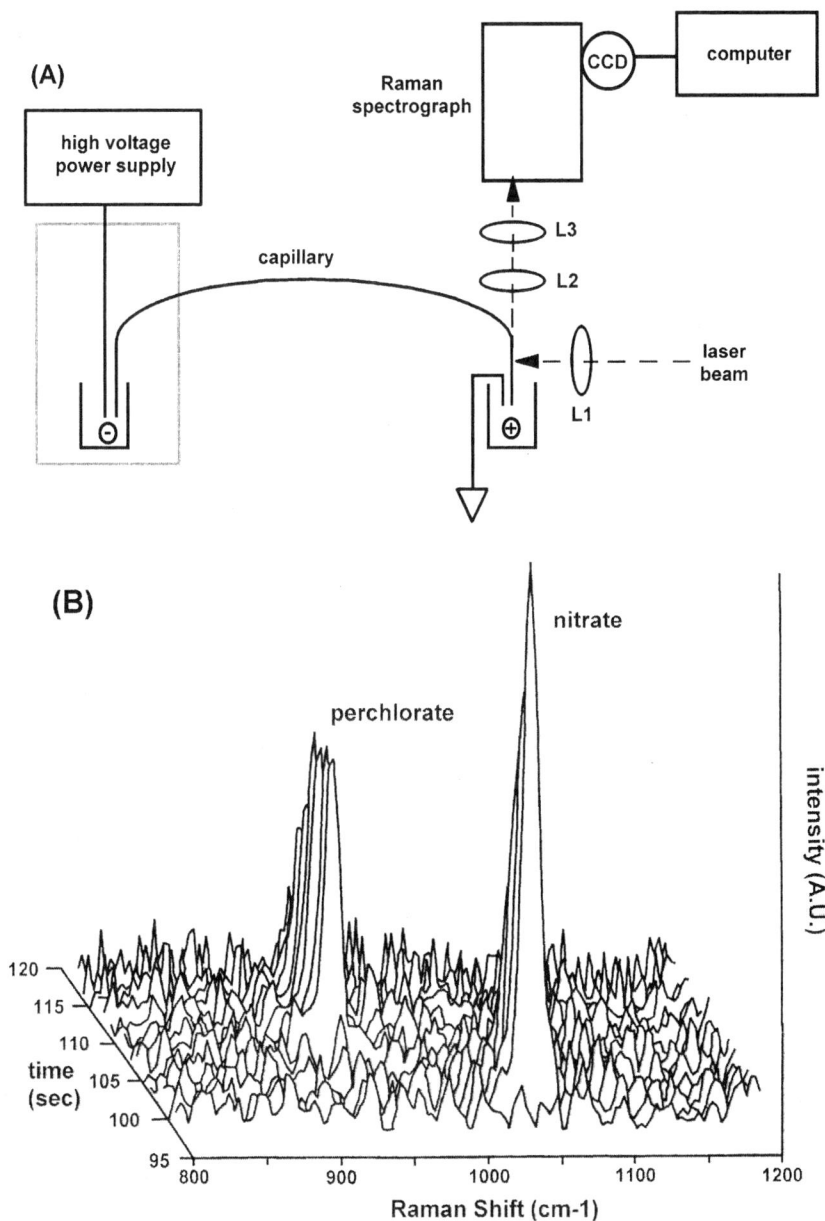

Figure 8. (a) Schematic of the CE-Raman spectroscopy system where L1= 65 mm f1 achromat, L2=20x0.04 NA microscope objective, L3=65 mm f1 achromat. (b) One-second integration Raman spectra of a nitrate/perchlorate separation. Sample concentrations 5×10^{-4} M in each anion; injection time 15 s at -2 kV. The spectra have been flat fielded and ratioed to the background electrolyte spectrum. (reprinted from Appl. Spectrosc., 1995; 49: 1183-1188 with permission from the Society of Applied Spectroscopy).

As shown in Figure 9A, two sample portions (g and h) are electrokinetically injected from two identical reservoirs located at opposite ends of the separation channel. The MCCD, i, is located in the center of the channel; however, it can be moved along the separation channel to a position that offers optimal separation of both cationic and anionic species. Figure 9C shows the effect of applied voltage on the separation on ionic species. Increasing the applied voltage decreases the migration time of the ionic analytes. The shorter migration times are coupled to sharper peaks. Using an applied voltage of 1000 V, a linear concentration response was obtained for prchlorate between 100-1000 μM. Using this microfluidic CE system, the limit of detection for perchlorate, based on a S/N=3, was 130 μM.

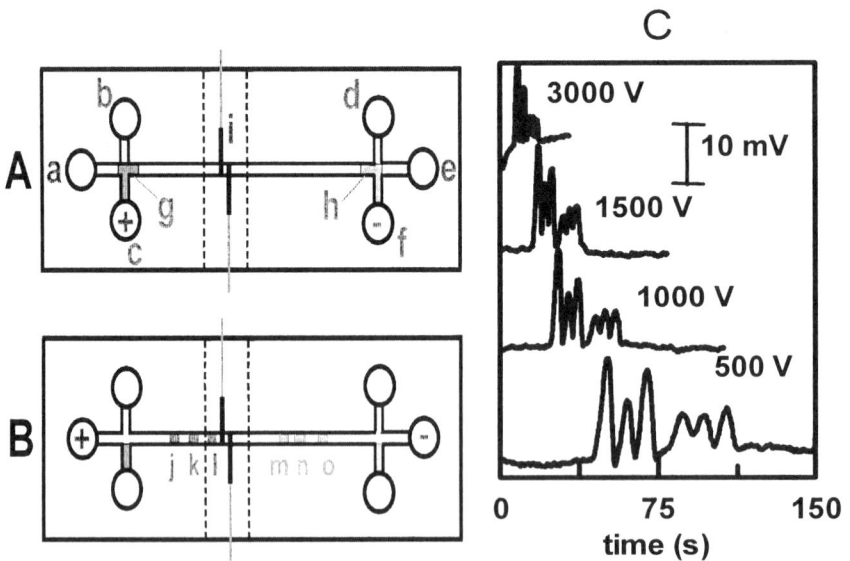

Figure 9. Schematic diagrams of the dual-injection microchip electrophoretic system with an MCCD for simultaneous detection of anions and cations (A) injection mode; (B) separation mode where (a, e) running buffer reservoirs; (b, d) unused reservoirs; (c, f) sample reservoirs; (g) injected cation plug; (h) injected anion plug; (i) MCCD; (j-l) separated cations; (m-o) separated anions. Injection is done by applying a positive voltage to reservoir c and ground to reservoir f. Other reservoirs are floating. Separation is done by applying a positive voltage applied to reservoir a, ground to reservoir e. Other reservoirs are floating. (C) Influence of the separation voltage upon the response of a sample mixture. Separation voltages are indicated. The peaks, from left to right, are due to ammonium, methyl ammonium, sodium, chloride, nitrate, and perchlorate (reprinted from Electrophoresis, 2003; 24: 3728-3734 with permission from Wiley Interscience).

SPECTROSCOPIC METHODS
Nuclear Magnetic Resonance Spectroscopy, Electrospray Mass Spectrometry-Mass Spectrometry, Attenuated Total Reflectance Infrared Spectroscopy, Raman Spectroscopy, and Surface-Enhanced Raman Spectroscopy

A number of spectroscopic methods have been used to both detect and positively identify perchlorate. These include nuclear magnetic resonance (NMR) spectroscopy, electrospray ionization mass spectrometry-mass spectrometry (ESI MS-MS), attenuated total reflectance infrared (ATR-IR) spectroscopy, Raman spectroscopy and surface-enhanced Raman spectroscopy (SERS).

^{35}Cl NMR spectroscopy was used to confirm that perchlorate was present in aqueous extracts of fertilizer and tobacco products.[43] Both ^{35}Cl and ^{37}Cl isotopes have a magnetic moment with spins of 3/2. However, ^{35}Cl has a greater natural abundance and is therefore more sensitive. Both ^{35}Cl and ^{37}Cl have negative quadrupole moments meaning that, unless Cl is in a symmetric environment, the Cl NMR peaks will be broad. In perchlorate, the chlorine atom is surrounded by four oxygen atoms and has tetrahedral (T_d) symmetry. Because chlorine is in the middle of the tetrahedron, it resides in a symmetric environment. Figure 10 shows ^{35}Cl NMR spectra obtained for a perchlorate standard as well as extracts of chewing tobacco and 6-6-18 fertilizer. In all three spectra, only single peaks at 1004.5 ppm versus saturated KCl reference solution were observed. These peaks are attributed to perchlorate. ^{35}Cl NMR spectroscopy also indicated the presence of perchlorate in extracts of tobacco lamina, flue-cured tobacco, and 16-0-0 sodium nitrate fertilizer. The perchlorate peaks in the ^{35}Cl NMR spectra are very narrow. However, any interactions between perchlorate and a complexing agent will disrupt the symmetry resulting in broader lines. In addition, the presence of any paramagnetic materials in the sample will also result in broadening of the lines.

Detection of perchlorate using mass spectrometry has been discussed previously in relation to (1) differentiating man-made versus natural sources of perchlorate, (2) use as a detector for ion-pair extraction methods, and (3) use as a detector for ion chromatography. In the latter two cases, the solvent is removed prior to mass spectrometry using electrospray ionization (ESI). ESI is a liquid introduction technique that uses a high electric field (3-5 kV cm^{-1}) to produce a fine mist of highly charged droplets.[53] The droplets, which carry positive or negative charges depending on the sign of the applied potential, pass into and along a small evaporation chamber. As the droplets pass through the chamber, they evaporate and rapidly become much smaller

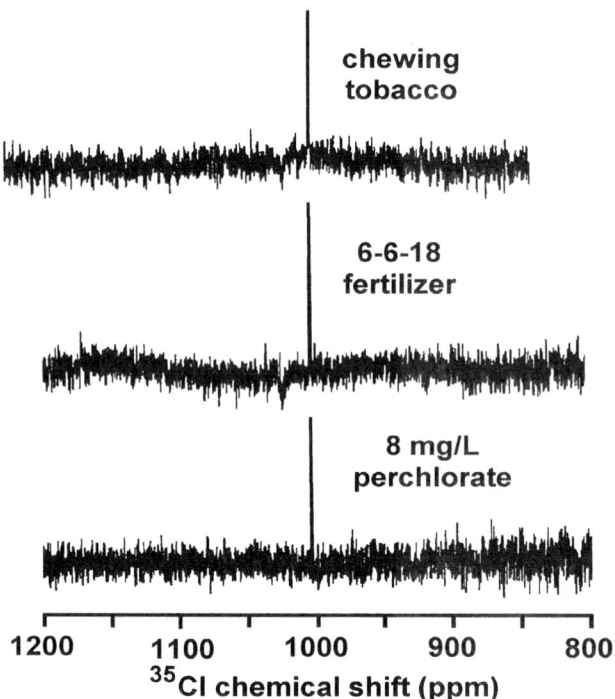

Figure 10. ^{35}Cl NMR spectra of an 8 mg L^{-1} perchlorate standard and aqueous extracts of chewing tobacco and 6-6-18 fertilizer (reprinted from Environ. Sci. Technol., 2001; 35: 3213-3218 with permission from the American Chemical Society).

through vaporization of solvent. As solvent evaporates, the droplets become smaller until the Coulomb repulsion exceeds the surface tension holding them together. At this point, the droplets are effectively 'blown apart' forming smaller charged drops. This process of evaporation and Coulombic explosion is repeated a number of times resulting in gas phase analyte ions. Ions, residual droplets, and vapor formed by electrospray are extracted through a small hole into two more evaporation chambers (evacuated) via a "nozzle" and a "skimmer," which then passes into the analyzer of the mass spectrometer where a mass spectrum of the sample is obtained. ESI is considered to be a soft ionization process so intact molecular ions are observed in the mass spectrum. Because ESI is an atmospheric process, it is easily interfaced to HPLC and CE systems. All molecules that can be charged, and can be dissolved in solvents with low boiling points, are amenable to ESI-MS analysis.

ESI-MS/MS has been used to detect perchlorate in groundwater without prior separation.[54] Samples were filtered through 0.45 µm membrane filters to remove particulates prior to injection. The mass spectrometer was operated in the negative ionization mode and the MS/MS was used to monitor the ClO_4^- to ClO_3^- transition. Calibration curves were linear between 0.8 and 80 µg L^{-1} and the method detection limit was 0.5 µg L^{-1} for groundwater containing relatively high concentrations of bicarbonate (0.8 mM), chloride (2.5 mM), and sulfate (2.3 mM). The sensitivity of ESI-MS is typically lowered by the presence of salts, detergents, and acids.[55] Liquid chromatography, ion exchange, solid phase extraction, and dialysis are methods that have been employed to remove these interferences prior to ESI-MS. In the case of perchlorate detection, these interferences have been overcome by prior separation using either ion chromatography[3,45,46] or ion pair extraction.[20,41] In the later method, a large lipophilic cation is used to ion pair with perchlorate. The ion pair is then extracted into an organic solvent such as methylene chloride or methyl isobutyl ketone. The presence of organic solvents also reduces ion suppression effects observed in ESI-MS.[55]

As noted earlier, perchlorate is a polyatomic anion with tetrahedral symmetry. As such it exhibits four vibrational modes[56] that are summarized in Table 4. As indicated in Table 4, perchlorate has two vibrational modes that are infrared active. However, normal transmission infrared spectroscopy is not a useful method for the detection of trace amounts of perchlorate in aqueous matrices.[57] This is due to the intense bands due to the OH stretching and bending modes of water that dominate large regions of the infrared spectrum. Attenuated total reflection infrared (ATR-IR) spectroscopy provides a means of analyzing samples that absorb too strongly to be analyzed by transmission spectroscopy. In ATR-IR, the infrared radiation is passed through an infrared transmitting crystal with a high refractive index, which allows the radiation to reflect within the ATR element several times. The sample is in intimate optical contact with the top surface of the crystal. Alternatively, the crystal can be coated with a film that is selective for a given analyte in order to

Table 4. Vibrational modes of perchlorate.[56]

Vibrational Mode	Frequency (cm^{-1})	Assignment	Symmetry	Raman (R) or Infrared (IR) Active
ν_1	935	ν_s(Cl-O)	A_1	R
ν_2	460	δ_d(O-Cl-O)	E	R
ν_3	1050-1170	ν_d(Cl-O)	F_2	R, IR
ν_4	630	δ_d(O-Cl-O)	F_2	R, IR

improve sensitivity of ATR-IR. Figure 11a shows a schematic of a selective film that is in contact with an ATR crystal. The infrared radiation from the light source enters the crystal, as shown in Figure 11a and reflects off the sample-crystal interface. However with each reflection along the top surface, a finite amount of radiation penetrates into the sample via the evanescent wave, Figure 11a. The penetration of the evanescent wave into the sample

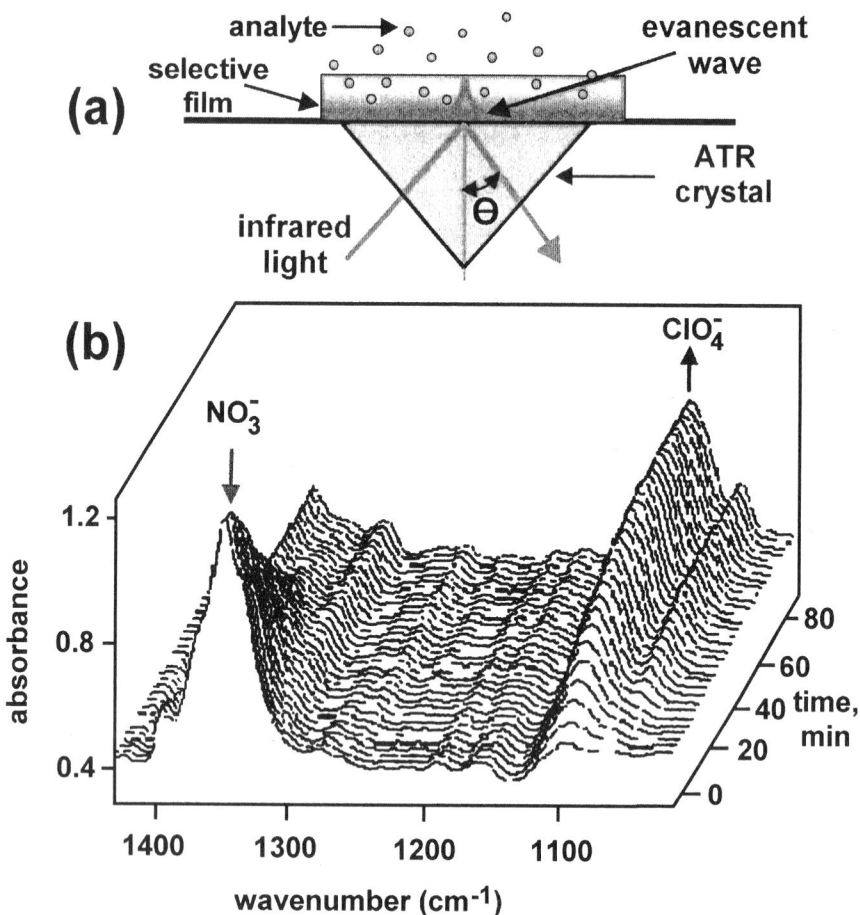

Figure 11. (a) Schematic view of the ATR crystal covered with a selective film and immersed in a solution containing the analyte. (b) Time-dependent, ATR-IR spectra of a thin film film of water-insoluble $DEC^+ NO_3^-$ on a probe, with a silicon ATR crystal, immersed in a pH 5-6 aqueous solution of $LiClO_4$. The perchlorate concentration is 0.5 ppm. The bands at 1092 and 1331 cm are assigned to perchlorate and nitrate, respectively. The appearance of the perchlorate peak and disappearance of the nitrate peak throughout the film are the result of a ClO_4^- /NO_3^- ion exchange reaction (reprinted from the Proceedings of the 222nd National Meeting of the American Chemical Society with permission from the American Chemical Society.

depends on the wavelength of the light, incident angle, and refractive indexes of the sample and the ATR crystal.[58] The output end of the crystal is coupled to the detector. The use of ATR-IR spectroscopy to detect perchlorate has been demonstrated.[57,59] In this method, either a diamond or silicon ATR crystal was used. Measurements were made using both bare and coated ATR probes.[60] In the bare probe measurements, the ATR crystal was in contact with sample under constant stirring. The resultant spectrum was ratioed to that obtained for distilled, deionized water that was recorded by signal averaging continuously for 60 minutes. Using a bare silicon ATR probe and 10 minute acquisition, a detection limit of 288 ppm was obtained for perchlorate. To extract perchlorate from aqueous samples, a highly substituted ferrocenium ion-exchange material, 1,1',3,3'- tetrakis (2-methyl-2-nonyl)ferrocinium nitrate ($DEC^+ \cdot NO_3^-$) was used. The $DEC^+ \cdot NO_3^-$ films on the ATR crystals were prepared by applying dichloromethane solutions of $DEC^+ \cdot NO_3^-$ onto the crystal and allowing the dichloromethane to evaporate. The resultant $DEC^+ \cdot NO_3^-$ film on the ATR crystal had a thickness on the order of 0.2 μm. Each coated ATR probe was used for a single analysis. The coated ATR probe was immersed in water with stirring and allowed to equilibrate for 10 minutes after which time the background spectrum was collected. The probe was then immersed in an aqueous sample containing perchlorate. While the sample is stirred, spectra were obtained every minute over a 10-80 minute time period. Figure 11b shows IR spectra recorded over time when the coated probe was immersed in a 0.5 ppm aqueous perchlorate solution. As can be seen, the band at 1331 cm^{-1} due to nitrate disappears at the same rate as the growth of the perchlorate band at 1092^{-1}. Detection limits of 0.5 ppm and 3 ppb were achieved using analysis times of 2 and 60 minutes, respectively.

Table 4 indicates that all four vibrational modes of perchlorate are Raman active. Earlier the use of Raman spectroscopy as a detector for capillary electrophoresis was discussed.[51] Unlike infrared spectroscopy which is an absorbance technique, Raman spectroscopy is an emission technique that involves inelastic scattering of incident laser energy resulting in spectral peaks that are frequency shifted from the incident energy. Normal Raman spectroscopic measurements of aqueous perchlorate solutions were obtained as a function of perchlorate concentration.[61] The excitation source was a 785 nm CW diode laser operating at 62 mW. Spectra were obtained using 20 second acquisition times. Water did not interfere in the detection of perchlorate. The calibration curve was linear between 100-10^4 ppm perchlorate and the limit of detection was 84 ppm. Lower detection limits can be obtained by either using longer acquisition times, using higher laser powers, or by using an excitation source operating in the visible range instead of the near IR. The disadvantage with visible excitation sources is that they can stimulate fluorescence, which can totally mask the Raman signal.

Normal Raman spectroscopy was used to verify the presence of perchlorate in fertilizers.[62] Prior ion chromatographic and capillary electrophoresis measurements suggested that perchlorate was present in the sample. Because elution times are subject to matrix interferences, the use of an information-rich spectroscopic technique such as Raman spectroscopy was explored to confirm species identification. In this method, a 5 g sample of fertilizer was added to 20 mL of water and centrifuged. After centrifugation, activated charcoal was added to remove trace organic impurities that could potentially cause fluorescence interference. Photobleaching would also be used to quench fluorescence.[47] Photobleaching is a technique that uses the excitation radiation to degrade the photosensitive fluorescent molecules in order to reduce the sample fluorescence. In some instances, both photobleaching and the activated charcoal treatment were not sufficient to quench fluorescence.[47] In those instances, an activated alumina adsorbent was use to remove the fluorescent impurities without loss of perchlorate. After treatment, the sample was then filtered to remove particulates. Figure 12 shows Raman spectra obtained for a fertilizer sample and ammonium perchlorate.[62] In these spectra, the water background has been subtracted and the baseline corrected. In the ammonium perchlorate spectrum, all four peaks due to perchlorate are observed at 463, 629, 934, and 1113 cm^{-1}. The 934 cm^{-1} peak is the most intense while the 1113 cm^{-1} peak is weak and

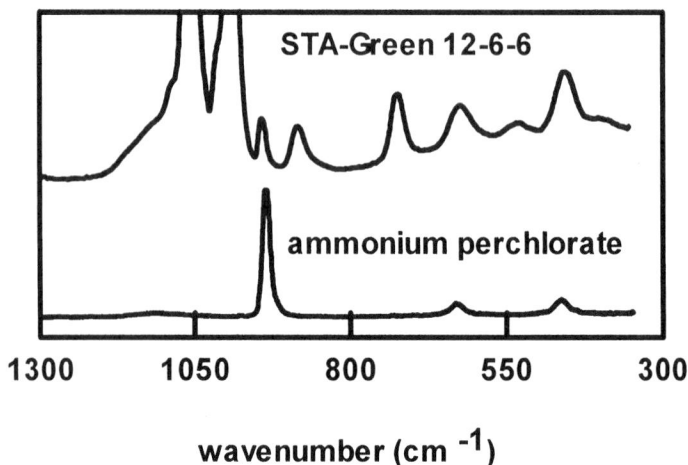

Figure 12. Normal Raman spectra of the STA-Green fertilizer extract (top) and ammonium perchlorate solution (bottom) (reprinted from Environ. Sci. Technol., 1999; 33: 3469-3472 with permission from the American Chemical Society).

broad. In the spectrum obtained for the STA-Green fertilizer extract, the 924 cm^{-1} peak due to perchlorate is clearly visible. Other spectral features are attributed to nitrate (1048 and 719 cm^{-1}), sulfate (981, 618, and 449 cm^{-1}), and dihydrogen phosphate (878 cm^{-1}). The 629 and 463 cm^{-1} peaks of perchlorate are only observed as shoulders on the more intense sulfate peaks. The Raman spectra verify that perchlorate is present in the fertilizer samples investigated. Using 785 nm laser excitation between 100-135 mW power and 5 minute acquisition times, detection limits for the fertilizer extracts were 50 ppm.[47]

Normal Raman spectroscopy was used to detect perchlorate in lettuce extracts.[47] Due to the presence of organics, direct Raman analysis of the lettuce extracts could not be done. A C-18 SPE column was used to remove the nonpolar organics while allowing the polar anions through. The filtrate was then boiled gently to reduce the volume by a factor of 15. In the Raman spectra, the 630 and 463 cm^{-1} perchlorate peaks were obscured by large broad bands at 609 and 505 cm^{-1} that were attributed to some other organic components present in the extract. Only the perchlorate peak at 934 cm^{-1} was observed in the Raman spectra.

In order to detect lower concentrations of perchlorate by Raman spectroscopy will require either preconcentration or enhancement of the signal. To preconcentrate perchlorate, the use of the bifunctional anion exchange described previously in relation to isotopic analysis[4,7,8] was explored.[63] In this investigation, 200 mL solutions of 5mM NaCl and perchlorate at varying concentrations were equilibrated with 15 beads of the bifunctional anion resin for ~24 hours. Afterwards the resin beads were transferred to a glass slide and analyzed using a micro-Raman system. Figure 13 shows Raman spectra obtained using the micro-Raman system. Spectra were obtained using a 785 nm diode laser operating at 60 mW. Multiple measurements of each bead indicated that the perchlorate ions are evenly sorbed and distributed on the resin surface. The ν_1 Raman band of perchlorate occurs at ~930 cm^{-1}. The detection limit was 10 ppb.

In the 1970s,[64,65] it was discovered that Raman scattering from molecules adsorbed on such noble metals as silver, copper, and gold can be enhanced as much as 10^6 to 10^7. The phenomenon is referred to as surface enhanced Raman spectroscopy (SERS). Because SERS provides both spectroscopic information and sensitivity, it can be a useful technique for trace analysis. In order to detect an analyte by SERS, it is necessary for the analyte to adsorb on the SERS-active metal. Detection of perchlorate by SERS has been demonstrated using colloidal silver[63] and cationic-coated ilver substrates.[61] In

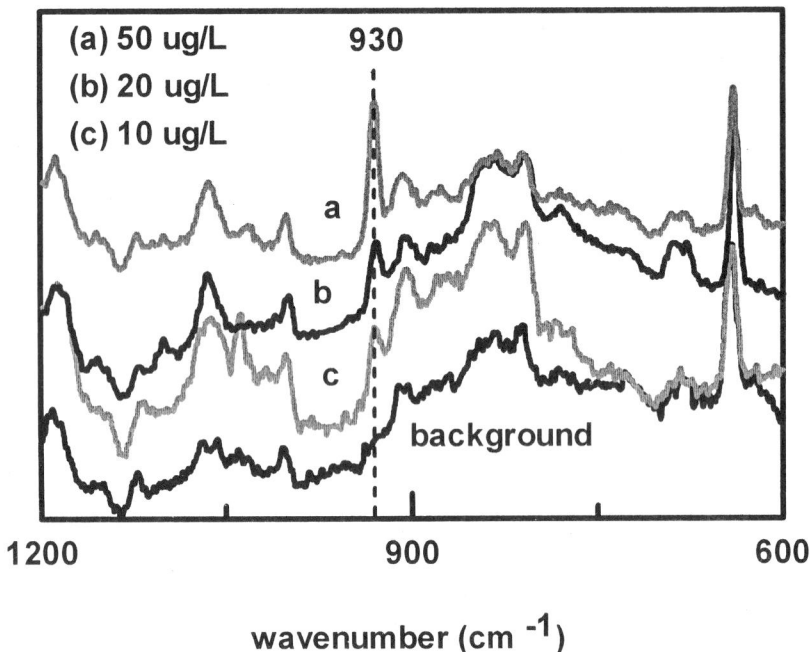

Figure 13. Normal Raman spectra of perchlorate sorbed on bifunctional anion-exchange resins which had been equilibrated with solutions of 5 mM NaCl and varying concentrations of perchlorate (10, 20, and 50 μg L^{-1}). The background spectrum was collected with an untreated resin bead (reprinted from Appl. Spectr., 2004; 58: 741-744 with permission from the Society of Applied Spectroscopy).

the former method,[63] silver colloids were prepared by reduction of silver nitrate using borohydride. Substrates were prepared by placing 0.05 mL of the colloidal suspension and 20 μL of perchlorate solution onto a quartz slide. After drying at ~50 °C for 10 minutes, SERS spectra were obtained using a micro-Raman system and 785 excitation at 60 mW of power. SERS spectra for these substrates are shown in Figure 14. As shown in Figure 14, several Raman bands were observed in the SERS spectra between 500 and 1000 cm^{-1}. The bands at 718 and 877 cm^{-1} were due to residual nitrate and borohydride used in the preparation of the colloidal silver. Perchlorate peaks are observed at 941 and 633 cm^{-1}. The position of the v_1 peak is dependent upon the sample matrix. In aqueous solutions, the v_1 peak occurs at 934 cm^{-1}.[61,62] However, in the solid state and adsorbed on silver, the v_1 peak has frequencies of 952 and 941 cm^{-1}, respectively. Detection limits on the order of 10^{-6}-10^{-7} M were reported.

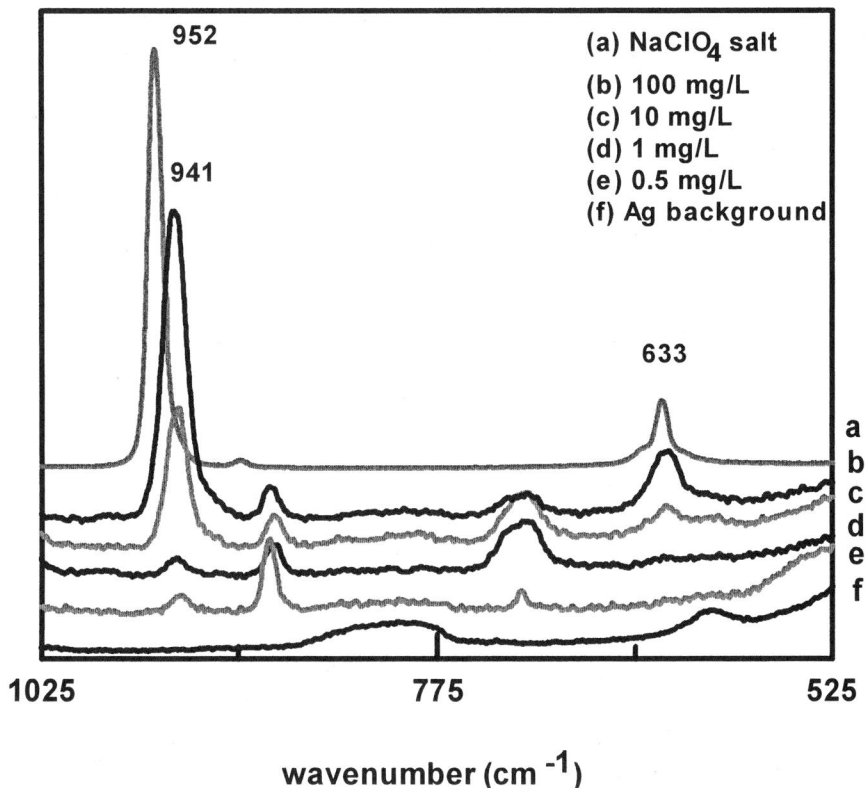

Figure 14. SERS spectra of perchlorate on Ag colloid substrates (b, c, d, e). Concentrations are indicated. (reprinted from Appl. Spectr., 2004; 58: 741-744 with permission from the Society of Applied Spectroscopy).

One problem with using bare silver SERS substrates, as those described above, is poor reproducibility. SERS substrates that are prepared using sol-gel techniques give better reproducibility. Another problem with bare SERS substrates is the lack of selectivity. Organic materials can also chemisorb onto the colloidal particles. To improve selectivity, the SERS substrates can be coated. It has been shown that thiols react on gold and silver surface to form self-assembled monolayers (SAMs). The SAM protects the SERS substrate from surface deterioration thereby extending their lifetimes. Thiols are chosen that have an affinity for the desired analyte species. The use of cationic thiols on the SERS surface to attract perchlorate was explored.[61] There are a number of commercially available cationic thiols. Cysteamine (CY) hydrochloride, dimethylaminoethanethiol (DMA) hydrochloride, diethylaminoethanethiol (DEA) hydrochloride, L-cysteine methyl ester (CYSM) hydrochloride, and L-cysteine ethyl ester (CYSE) hydrochloride are examples of aliphatic, cationic thiols. The thiols 4-(2-

mercaptoethyl)pyridinium (MEP) hydrochloride and 2-mercapto-4-methylpyrimidine (MMP) hydrochloride are both heterocyclic aromatics. The response of SERS substrates prepared with these cationic thiols with perchlorate and other anions was examined. Figure 15a shows SERS spectra of Ag/CY obtained as a function of perchlorate concentration. The spectra were obtained using 6 second acquisition times. It can be seen that the CY SAM exhibits a characteristic SERS spectrum that can be used as an internal standard for calibration purposes. With the addition of perchlorate, a new peak appears at 935 cm^{-1}, which is assigned to the v_1 vibrational mode of perchlorate. The concentration response is shown in Figure 15b. At low anion concentration, the perchlorate peak area increases linearly with concentration. At higher solution concentrations of perchlorate, the response levels off as the adsorption sites on the substrate become fully occupied. The adsorption of perchlorate is described by a Frumkin isotherm:

$$\theta = \frac{cKe^{2g\theta}}{1 + cKe^{2g\theta}} \tag{6}$$

and

$$\theta = \frac{N}{N_T} \tag{7}$$

where θ is the fractional coverage of perchorate on the coating, N is the number of sites on the substrate occupied by perchlorate, N_T is the total number of sites on the substrate, c is the solution concentration of perchlorate in M, K is the ion-pair constant between the anion and coating, and g is the Frumkin parameter. N is directly proportional to the area of the anion peak. The Frumkin parameter takes into account interactions between the adsorbed species. A negative value of g indicates the presence of repulsive forces upon adsorption while a positive value is indicative of attractive forces. Table 5 summarizes the values of K and g obtained for the cationic coatings and anions. Because chloride ion does not exhibit a Raman active vibrational mode, a competitive complexation approach was developed to evaluate the ion-pair constant between chloride and the cationic coating.[66] For the cationic coatings investigated, no interaction was observed to occur between the coatings and dihydrogen phosphate. From the ion-pair constants summarized in Table 5, it can be seen that, with the exception of MEP, perchlorate forms relatively strong ion pairs with the cationic coatings. It can also be seen, from the data summarized in Table 5, that other anions also interact with these cationic coatings. The advantage of this is that, for anions exhibiting a Raman-active vibrational mode, multiple anionic species can be detected simultaneously. However, quantification of these species would require knowing the values of the selectivity coefficients – just as is

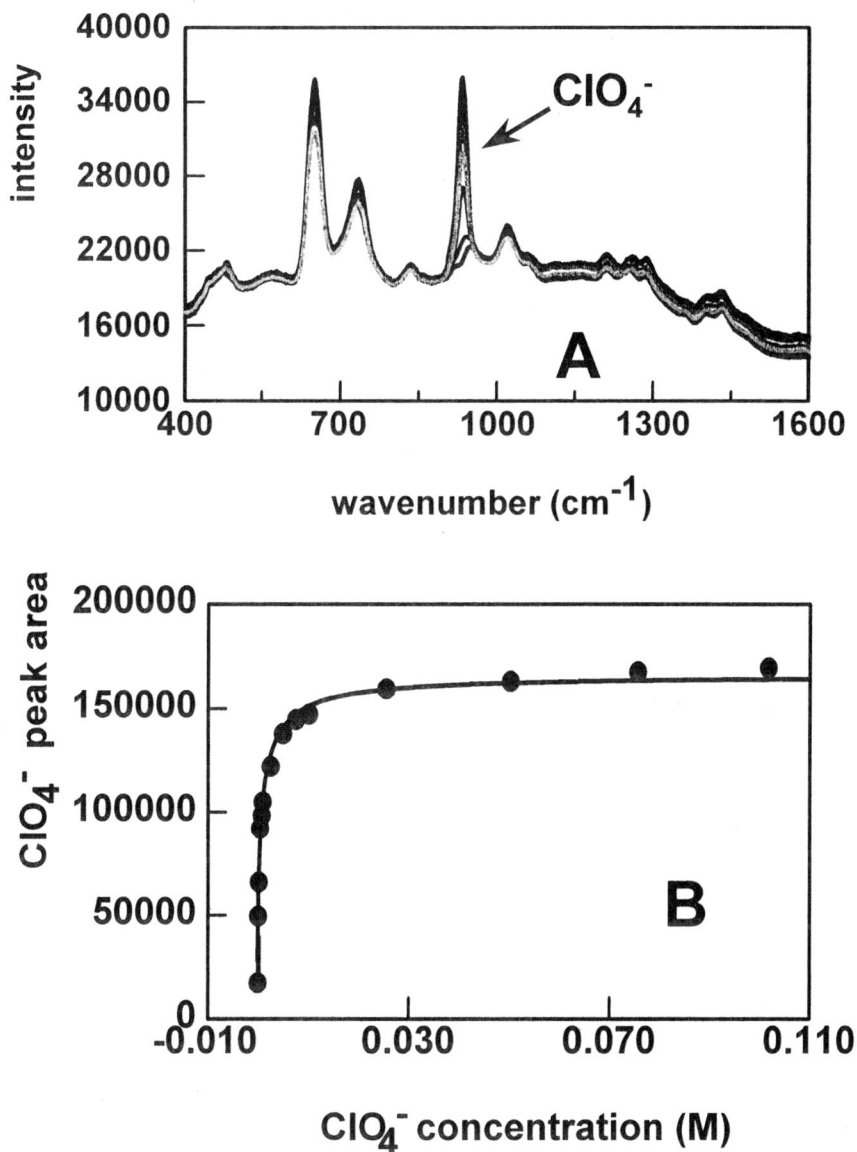

Figure 15. (a) SERS spectra of Ag/CY obtained using 785 nm excitation and 80.2 mW power and averaging five 6 s spectra. Perchlorate concentrations are 5, 100, 500, 1000, and 10^4 ppm. (b) Perchlorate peak area plotted as a function of perchlorate concentration for Ag/CY(reprinted from Appl. Spectr., 2003; 57: 1129-1137 with permission from the Society of Applied Spectroscopy).

Table 5. Summary of ion-pair constants, K (M^{-1}) and Frumkin parameters, g, for perchlorate and other anions (reprinted from Appl. Spectrosc., 2003; 57: 1129-1137).

SAM	ClO_4^-	$SO_4^=$	NO_3^-	Cl^-
CY	K: 6150±830 g: -1.10±0.15	K: 1620±320 g: -.37±0.23	K: 382±60 g: -.30±0.25	K: 146±23 g:-0.30±0.23
DMA	K: 404±59 g: -0.64±0.21	K: 972±85 g: -1.14±0.13	K: 301±78 g: -2.3±1.1	K: 310±180 g: -2.3±1.1
DEA	K: 4950±250 g: -2.42±0.10	K: 770±100 g: -0.07±0.12	K: 228±29 g:-1.22±0.30	K: 107±20 g:-1.22±0.30
CYSM	K: 7380±450 g: -2.71±0.10	No interaction	K: 307±15 g: -1.20±0.11	K:36600± 6800 g: -1.20±0.11
CYSE	K: 4650±500 g: -2.14±0.18	No interaction	K: 513±85 g: -2.06±0.44	K: 1190±260 g: -2.06±0.44
MEP	K: 38±11 g: 1.35±0.24	No interaction	No interaction	K: 92±22 g: 1.35±0.24
MMP	K: 1163±56 g: -2.51±0.21	Spectral interference	K: 2370±200 g: -2.58±0.15	K: 596±77 g: -2.58±0.15

done when compensating for interferents when using ion-selective electrodes. To correct the SERS response of the anion in the presence of a non-Raman active species, such as chloride ion, requires coupling the cationic-coated SERS sensor with a device capable of measuring chloride ion concentrations. One such device is an ion selective electrode. The use of SPE to remove chloride ion from solution prior to analysis by SERS was explored.[67] The commercially available SPE cartridges contain silver ion and chloride is removed from the sample matrix by a precipitation reaction. It was found that for analyte anion concentrations below 100 ppm, the use of the Ag^+-based SPE cartridges removed the chloride ion interference with little or no affect on the magnitude of the analyte anion SERS signal. However for analyte concentrations greater than 100 ppm, enhancements in the analyte anion

SERS signal were observed that were greater than the theoretical limit, N_T. These enhancements were attributed to the formation of colloidal silver which adsorbs analyte anion to form negatively charged particles. These negatively charged particles then accumulate on the surface of the cationic-coated SERS substrate. In these experiments, acquisition times to obtain SERS spectra of the cationic coatings and interacting anions ranged from 6 to 20 seconds.[61,66-68] Limits of detection for perchlorate ranged from 1 to 3 ppm. However, lower detection limits can be achieved by increasing acquisition times, applying a positive overpotential to the SERS electrode, and/or optimizing of the roughening parameters used to prepare the SERS surface.[68] Another means of improving sensitivity is to use SERS substrates comprised of cationic-coated Ag/Au colloidal particles immobilized on magnetic microparticles. The amine groups of amine-terminated magnetic microparticles can be used to immobilize gold and silver colloidal particles.[69] Afterwards the immobilized colloidal particles can be reacted with a cationic thiol to form a SAM on the surface of the colloidal particles to form the SERS substrates shown in Figure 16a. As shown in Figure 16a, the cationic thiol, MMP, attracts perchlorate to the magnetic microparticles. Consequently, an aliquot of these SERS substrates can be added to a large volume of sample. The cationic SAM will extract perchlorate from the sample and concentrate perchlorate near the SERS active surface. A NdFeB magnet is then used to separate the magnetic microparticles from the sample. The effectiveness of this approach are summarized in Figure 16b, which shows SERS spectra of Ag/MMP obtained in the presence and absence of perchlorate.

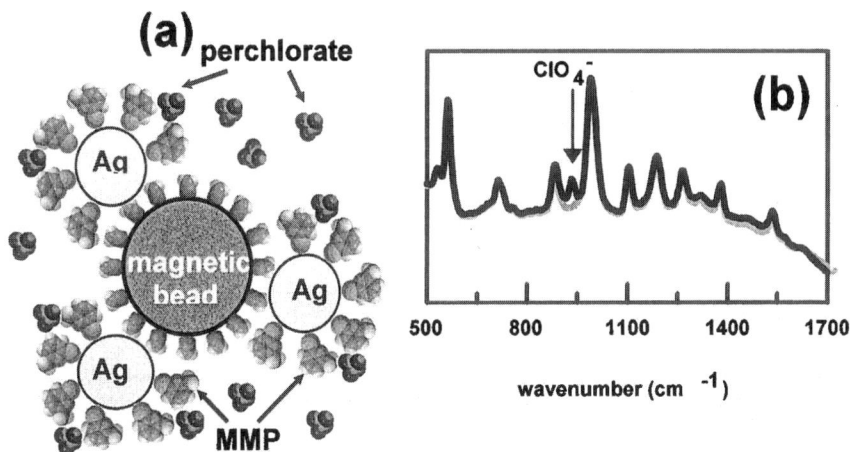

Figure 16. (a) Schematic of a SERS substrate comprised of cationic-coated Ag colloidal particles immobilized on amine-terminated magnetic microparticles. (b) SERS spectra of Ag/MMP colloid immobilized on magnetic microparticles in the presence of absence of 1000 ppm perchlorate.

ACKNOWLEDGMENTS

This work was supported by the Strategic Environmental Research and Development Program (SERDP) and the SPAWAR Systems Center San Diego Independent Research (ILIR) Program.

REFERENCES

1. U.S. EPA (2003). Occurrence and potential sources of perchlorate releases in the environment as of April, 2003, retrieved from http://www.epa.gov/swerffrr/ pdf/perchlorate_occurrence_april03a.pdf

2. Hogue C. Rocket-fueled river. Chem. Eng. News 2003; 81:37-46.

3. Kirk A.B., Smith E.E., Tian K., Anderson T.A., Dasgupta P.K. Perchlorate in milk, Environ. Sci. Technol. 2003; 37:4979-4981.

4. Bao H., Gu, B. Natural perchlorate has a unique oxygen isotope signature, Environ. Sci. Technol. 2004; 38: 5073-5077.

5. Erickson B.E. Tracing the origin of perchlorate, Anal. Chem. 2004; 76: 388A-389A.

6. Hoering T.C., Ishimori F.T., McDonald H.O. Oxygen exchange between oxy-anions and water. II. chlorite, chlorate, and perchlorate, J. Am. Chem. Soc. 1958; 80, 3876-3979.

7. Brown, G.M., Bonnesen P.V., Moyer B.A., Gu B., Alexandratos S.D., Patei V., Ober R. "The design of selective resins for the removal of pertechnetate and perchlorate from water." In *Perchlorate in the Environment*. E.T. Urbansky, ed. New York, NY: Kluwer Academic/Plenum Publishers, 2000.

8. Bonnesen P.V., Brown, G.M., Alexandratos S.D., Bavoux L.B., Presley D.J., Patel V., Ober R., Moyer B.A. Development of bifunctional anion-exchange resins with improved selectivity and sorptive kinetics for pertechnetate: batch-equilibrium experiments, Environ. Sci. Technol. 2000; 34: 3861-3766.

9. Gu B., Ku Y.-K., Brown G.M. Treatment of perchlorate-contaminated groundwater using highly selective, regenerable ion-exchange technology: a pilot-scale demonstation. Remediation. 2002; 12: 51-68.

10. Urbansky E.T. Perchlorate chemistry: implications for analysis and remediation, Bioremediation Journal, 1998; 2: 81-95.

11. Nabar G.M., Ramachandran C.R. Quantitative determination of perchlorate ion in solution, Anal. Chem., 1959; 31:263-265.

12. Cyganski A., Kowalczyk P., Krystek J., Ptaszynski B. New cetyltrimethylammonium methods of determination of perchlorate, Chemical Analysis. 2000; 45: 911-919.

13. Urbansky E.T. Quantitaion of perchlorate ion: practices and advances applied to the analysis of common matrices, Crit. Rev. Anal. Chem., 2000; 30: 311-343.

14. Baczuk R.J., DuBois R.J. Potentiometric titration of perchlorate with tetraphenylarsonium chloride and a perchlorate ion specific electrode, Anal. Chem. 1968; 40: 685-689.

15. Smith M.J., Manahan S.E. Low-temperature precipitation titration of perchlorate and tetrafluoroborate with tetraphenylarsonium chloride and ion-selective electrodes, Anal. Chim. Acta, 1969; 48: 315-319.

16. Selig W. Lower limits of the potentiometric titration of perchlorate using a perchlorate ion-selective electrode, Microchem. Journal, 1977; 22: 1-6.

17. Thorne P.G. *Field Screening Method for Perchlorate in Water and Soil*, U.S. Army Corps of Engineers ERDC/CRREL TR04-8, April 2004.

18. Urbansky E.T., Magnuson M.L., Freeman D., Jelks C. Quantitation of perchlorate ion by electrospray ionization mass spectrometry (ESI-MS) using stable association complexes with organic cations and bases to enhance selectivity, J. Anal. Atomic Spect., 1999; 14: 1861-1866.

19. Magnuson M.L., Urbansky E.T., Kelty C.A. Determination of perchlorate at trace levels in drinking water by ion-pair extraction with electrospray ionization mass spectrometry, Anal. Chem., 2000; 72: 25-29.

20. Magnuson M.L., Urbansky E.T., Kelty C.A. Microscale extraction of perchlorate in drinking water with low level detection by electrospray-mass spectrometry, Talanta, 2000; 52: 285-291.

21. Janata. J. *Principles of Chemical Sensors*, Plenum Press, New York, NY, 1989.

22. Bühlmann, P., Amemiya, S., Nishizawa S., Xiao K.P., Umezawa Y., "Hydrogen-bonding ionophores for inorganic anions and nucleotides and their application in chemical sensors", J. Inclusion Phenom. Mol. Recognit. Chem., 1998; 32: 151-163.

23. Steed J.W., Atwood J.L. *Supramolecular Chemistry*, John Wiley and Sons, Ltd., Chichester, England, 2000.

24. Bühlmann P., Pretsch E., Baker E., "Carrier-based ion-selective electrodes and bulk electrodes. 2. ionophores for potentiometric and optical sensors", Chem. Rev., 1998; 98: 1593-1687.

25. Schmidtchen F.P., Berger M. Artificial organic host molecules for anions, Chem. Rev., 1997; 97: 1609-1646.

26. Xiao K.P., Bühlmann P., Umezawa Y., "Ion-channel-mimetic sensing of hydrophilic anions based on monolayers of a hydrogen bond-forming receptor", Anal. Chem., 1999; 71: 1183-1137.

27. Xiao K.P., Bühlmann P., Nishizawa S., Amemiya S., Umezawa Y., "A chloride ion-selective solvent polymeric membrane electrode based on a hydrogen bond forming ionophore", Anal. Chem., 1997; 69: 1038-1044.

28. Shugar G.J., Dean J.A. *The Chemist's Ready Reference Handbook*, McGraw-Hill Inc., New York, NY 1990.

29. Sánchez-Pedreño C., Ortuño J.A., Hernández J. Perchlorate-selective polymeric membrane electrode based on a gold(I) complex: application to water and urine analysis, Anal. Chim. Acta, 2000; 415: 159-164.

30. Shamsipur M., Soleymanpour A., Akhond M., Sharghi H., Hasaninejad A.R. Perchlorate selective membrane electrodes based on a phosphorous(V)-tetraphenyl porphyrin complex, Sensors and Actuators B, 2003; 89: 9-14.

31. Errachid A., Pérez-Jiménez C., Casabó J., Escriche L., Muñoz J.A., Bratov A., Bausells J. Perchlorate-selective MEMFETS and ISEs based on a new phosphadithiamacrocycle, Sensors and Actuators B, 1997; 43: 206-210.

32. Casabó J., Escriche L., Pérez-Jiménez C., Muñoz J.A., Teixidor F., Bausells J., Escriche L. Application of a new phosphadithiamacrocycle to ClO_4^- selective CHEMFET and ion-selective electrode devices, Anal. Chim. Acta, 1996; 320: 63-68.

33. Jie Z., Xiwen H. Study of the nature of recognition in molecularly imprinted polymer selective for 2-aminopyridine, Anal. Chem. Acta, 1999; 381: 85-91.

34. Al-Kindy S., Badia R., Suárez-Rodríguez J.L., Díaz-García M.E. Molecularly imprinted polymers and optical sensing applications, Crit. Rev. in Anal. Chem., 2000; 30: 291-309.

35. Lu Z., Sun A., Dong S. Study of ClO_4^- – selective electrode based on a conducting polymer polypyrrole, Electroanalysis, 1989; 1: 271-277.

36. Hutchins R.S. and Bachas L.G. Nitrate-selective electrode developed by electrochemically mediated imprinting/doping of polypyrrole, Anal. Chem., 1995; 67: 1654-1660.

37. Wu J., Mullett W.M., Pawliszyn J. Electrochemically controlled solid-phase microextraction based on conductive polypyrrole films, Anal. Chem., 2002; 74: 4855-4859.

38. Biesaga M., Kwiatkowska M., Trojanowicz M. Separation of chlorine-containing anions by ion chromatography and capillary electrophoresis, J. Chromatogr. A, 1997; 777: 375-381.

39. Okamoto H.S., Rishi D.K., Steeber W.R., Baumann F.J., Perera S.K. Using ion chromatography to detect perchlorate", J. Amer. Water Works Assoc., 1999; 91: 73-84.

40. U.S. EPA, *Method 314: Determination of Perchlorate in Drinking Water Using Ion Chromatography*, 1999.

41. Urbansky E.T., Gu B., Magnusin M.L., Brown G.M., Kelty C.A. Survey of bottled waters for perchlorate by electrospray ionization mass spectrometry (ESI-MS) and ion chromatography (IC), J. Sci. Food Agric., 2000; 80: 1798-1804.

42. Tian K., Dasgupta P.K., Anderson T.A. Determination of trace perchlorate in high-salinity water samples by ion chromatography with on-line preconcentration and preelution, Anal. Chem., 2003; 75: 701-706.

43. Ellington J.J., Wolfe N.L., Garrison A.W., Evans J.J., Avants J.K., Teng Q., Determination of perchlorate in tobacco plants and tobacco products, Environ, Sci. Technol. 2001; 35: 3213-3218.

44. De Borba B.M., Urbansky E.T. Performance of poly(vinyl alcohol) gel columns on the ion chromatographic determination of perchlorate in fertilizers, J. Environ, Monit., 2002; 4: 149-155.

45. Winkler P., Mirteer M., Willey J. Analysis of perchlorate in water and soil by electrospray LC/MS/MS, Anal. Chem., 2004; 76: 469-473.

46. Krynitsky A.J., Niemann R.A., Nortrup D.A. Determination of perchlorate anion in foods by ion chromatography-tandem mass spectrometry, Anal. Chem., 2004; 76: 5518-5522.

47. Williams T.L., Martin R.B., Collette T.W. Raman spectroscopic analysis of fertilizers and plant tissue for perchlorate, Appl. Spectrosc., 2001; 55: 967-983.

48. Ehmann T., Bächmann K., Fabry L., Rüfer H., Pahlke S., Kotz L. Optimization of the electrokinetic sample introduction in capillary electrophoresis for the ultra trace determination of anions on silicon wafer surfaces, Chromatographia, 1997; 45: 301-311.

49. Polesello S., Valsecchi S.M. Electrochemical detection in the capillary electrophoresis analysis of inorganic compounds, J. Chromatogr. A, 1999; 834: 103-116.

50. Nann A., Pretsch E. Potentiometric detection of anions separated by capillary electrophoresis using an ion selective electrode, J. Chromatogr. A, 1994; 676: 437-442.

51. Kowalchyk W.K., Walker P.A. III, Morris M.D. Rapid normal Raman spectroscopy of sub-ppm oxy-anion solutions: the role of electrophoretic preconcentration, Appl. Spectrosc., 1995; 49: 1183-1188.

52. Wang J., Chen G., Muck A. Jr., Collins G.E. Electrophoretic microchip with dual-opposite injection for simultaneous measurements of anions and cations, Electrophoresis, 2003; 24: 3728-3734.

53. Cole R.B. Some tenants pertaining to electrospray ionization mass spectrometry, J. Mass Spectrom., 2000; 35: 763-772.

54. Koester C.L., Beller H.R. Halden R.U. Analysis of perchlorate in groundwater by electrospray ionization mass spectrometry/mass spectrometry, Environ. Sci. Technol., 2000; 34: 1862-1864.

55. Iavarone A.T., Udekwu O.A., Williams E.R. Buffer loading for counteracting metal salt-induced signal suppression in electrospray ionization, Anal. Chem., 2004; 76: 3944-3950.

56. Nakamoto K. *Infrared Spectra of Inorganic and Coordination Compounds*, John Wiley and Sons Inc., New York, NY 1963.

57. Hebert G.N., Odom M.A., Bowman S.C., Strauss S.H. Attenuated total reflectance FTIR detection and quantification of low concentrations of aqueous polyatomic anions, Anal. Chem., 2004; 76: 781-787.

58. Kazarian S.G., Flichy N.M.B., Coombs D., Poulter G. Potential of ATR-IR spectroscopy in applications to supercritical fluids and liquefied gases, Am. Lab., 2001; 2001: 44-48.

59. Strauss S. H., Odom M. A., Hebert G. N., Clapsaddle B. J. ATR-FTIR detection of 25 µg/L aqueous cyanide, perchlorate, and perfluorooctylsulfonate, J. Am. Water Works Assoc., 2002; 94: 109–115.

60. Strauss S.H. Monitoring low levels of ionic pollutants in water using attenuated total reflectance infrared spectroscopy. Proceedings of the Themes in Potable Water sponsored by the Division of Environmental Chemistry of the American Chemical Society during the 222nd National Meeting of the American Chemical Society; 2001 August 26-30; Chicago, Illinois; American Chemical Society, 2001.

61. Mosier-Boss P.A., Lieberman S.H. Detection of anions by normal Raman spectroscopy and surface-enhanced Raman spectroscopy of cationic-coated substrates, Appl. Spectrosc., 2003, 57: 1129-1137.

62. Susarla S., Collette T.W., Garrison A.W., Wolfe N.L., McCutcheon S.C., Perchlorate identification in fertilizers, Environ. Sci. Technol., 1999, 33: 3469-3472.

63. Gu B., Tio J., Wang W., Ku Y.-K., Dai S., Raman spectroscopic detection for perchlorate at low concentrations, Appl. Spectrosc., 2004; 58: 741-744.

64. Fleischmann M., Hendra P.J., McQuillan A.J., Raman spectra of pyridine adsorbed at a silver electrode, Chem Phys. Lett., 1974; 26: 163-166.

65. Jeanmaire D.L., Van Duyne R.P., Surface Raman spectroelectrochemistry part i. heterocyclic, aromatic, and aliphatic amines adsorbed on the anodized silver electrode, J. Electroanal. Chem., 1977; 84: 1-20.

66. Mosier-Boss P.A., Boss R.D., Lieberman S.H. Determination of the ion-pair constant between chloride ion and cationic-coated, silver SERS substrates using competitive complexation, Langmuir, 2000; 16: 5441-5448.

67. Mosier-Boss P.A., Lieberman S.H. Detection of anionic nutrients by SERS of cationic-coated silver substrates. Effect of chloride ion, Appl. Spectrosc., 2001; 55: 1327-1336.

68. Mosier-Boss P.A., Lieberman S.H. Detection of nitrate and sulfate anions by normal Raman spectroscopy and SERS of cationic-coated substrates, Appl. Spectrosc., 2000; 54: 1126-1135.

69. Mosier-Boss P.A., Lieberman S.H. SERS substrate composed of chemically modified gold colloid particles immobilized on magnetic microparticles, Anal. Chem., 2005; 77: 1031-1037.

Chapter 7

The Ecotoxicology of Perchlorate in the Environment

Philip N. Smith

The Institute of Environmental & Human Health, Texas Tech University, Lubbock, Texas 79409-1163

INTRODUCTION

Considerable amounts of ecotoxicological data have been generated since the discovery of perchlorate contamination in the environment to assist in evaluating the potential for ecological exposures and subsequent effects. This chapter attempts to provide a synopsis of, and to interpret the available ecological data pertaining to perchlorate, and to identify areas in need of further study.

Perchlorate is an oxidizing anion found as a contaminant in ground and surface waters as a result of the dissolution of perchlorate salts (ammonium, potassium, magnesium, and sodium). Perchlorate salts are highly water-soluble and, in aqueous solution, perchlorate is quite unreactive within the range of temperatures and pHs normally encountered in the environment.[1] Once in water, perchlorate can persist for many years, and water serves as a carrier. Because it is an anion, perchlorate does not adsorb to soils and its adsorption to minerals is weak and reversible.[2]

ECOLOGICAL EXPOSURE

Soon after the unexpected discovery of perchlorate in ground and surface waters, it became evident that identifying possible sources of perchlorate contamination and evaluating biotic tissue concentrations would be critical in determining the extent of exposure among aquatic and terrestrial ecological receptors. However, the unique chemical characteristics of perchlorate and limitations of analytical instrumentation and methodologies made accurate quantitation of perchlorate concentrations in biological matrices difficult. Since the year 2000, instrumentation, extraction and cleanup techniques, and analytical capabilities necessary to measure the presence of perchlorate and its fate and transport through environmental media have improved considerably.

Perchlorate uptake into biota (and bioconcentration) was first documented in vegetation,[3-5] but measurement of perchlorate concentrations in tissues of animals exposed in the environment did not occur until 2001. Other chapters of this book contain more thorough discussions of perchlorate uptake processes in plants. Two of the first studies to address the occurrence and distribution of perchlorate in flora and fauna were conducted by Parsons[6] and Smith and colleagues.[7] Parsons[6] surveyed a wide array of ecological receptors including surface water, groundwater, soil, sediment, aquatic and terrestrial birds, fish, aquatic and terrestrial reptiles, amphibians, terrestrial mammals, aquatic and terrestrial insects, other aquatic and terrestrial invertebrates, and aquatic and terrestrial plants across six different sites in the United States known to be contaminated with perchlorate. Soil and terrestrial vegetation contained the highest concentrations of perchlorate followed by aquatic vegetation, surface water, sediment pore water, and sediments.

Smith et al.[7] also reported perchlorate in aquatic and terrestrial organisms collected at the Longhorn Army Ammunition Plant in Karnack, Texas. Perchlorate concentrations ranged from 0.8 to 2.0 mg/kg in aquatic insects, below detection limits (BDL) to 0.2 mg/kg in fish, BDL to 0.6 mg/kg in frogs, BDL to 2.3 mg/kg in mammals, and 0.6 to 5,557 mg/kg in vegetation. As in the Parsons study, many of the highest perchlorate concentrations were measured in terrestrial plants, followed by water, sediment, and aquatic plants. Concentrations of perchlorate in aquatic fauna were lower than or similar to those found in water, indicating a general lack of bioconcentration.

The Parsons[6] and Smith et al.[7] studies documented the uptake of perchlorate into a wide range of aquatic and terrestrial receptors. Interestingly, perchlorate was detected in terrestrial mammals that were not intimately associated with contaminated surface water sources suggesting that there were alternative exposure pathways. The highest concentrations of perchlorate were observed in vegetation in both studies which provided evidence that vegetation was an important source of exposure for terrestrial organisms.

We now understand that perchlorate readily accumulates in a number of plants including vegetable crops, trees, grasses, and forbs.[4,5,7-12] In general, plants can bioconcentrate perchlorate to concentrations that are several hundred-fold higher than ambient water or soil concentrations.[11] Perchlorate uptake into plants appears to be a function of the perchlorate concentration in soil and water, and the length of exposure.[12] Perchlorate can be detected in the roots, stems and leaves of many vegetation types, but leaves typically contain the highest concentrations of perchlorate.[12,13] Xylem-supplied tissues, such as leaves, appear to accumulate the majority of perchlorate rather than phloem-supplied tissues, such as fruit. Uptake of perchlorate into terrestrial plants is not an effective remedial strategy because Tan *et al.*[11] demonstrated that perchlorate could be released once plant leaves senesce. Vegetation may thus facilitate transport of perchlorate through ecosystems and the environment as well as serve as a source of perchlorate exposure to ecological receptors. Perchlorate has recently been detected in edible portions (fruits and seeds) of plants.[14] Vegetation is therefore a potential source of perchlorate exposure among humans, livestock, and wildlife.

Exposure Pathways

The major route of exposure to perchlorate for wildlife is through ingestion of contaminated water, but food items may also contribute to exposure (particularly vegetation grown in the presence of perchlorate). In addition to water and vegetation, soil can comprise a significant proportion of the diet of many wildlife species,[15] thus soil should also be considered. Although perchlorate does not readily bind to soil particles, it can persist in soils in arid environments or other areas where rapid drying occurs. Hence, water, vegetation, and soil can all be sources of exposure for ecological receptors at perchlorate-contaminated sites, or in areas where perchlorate is associated with surface or groundwater. However, respiratory and dermal exposure should also be considered for aquatic species, and some exposure can occur *in ovo* or transplacentally in mammals. Based on recent eco-related perchlorate research, Smith and coauthors[16] offer a modification of the conceptual site model (CSM) initially developed by the USEPA to illustrate which pathways and taxa are potentially impacted by the environmental occurrence of perchlorate. As noted earlier, freshwater species appear to have the greatest potential for exposure and effects based on the extreme water solubility of perchlorate. However, terrestrial species are also potentially at risk for exposure to perchlorate through water, food, and soil. Given the capacity for uptake and accumulation of perchlorate in plants, exposure via consumption of contaminated vegetation may have previously been underestimated.

Most sites which harbor above-ground perchlorate contamination are small, isolated, and are usually a result of localized releases related to military or industrial activities. However, there are regions in North America that are affected on a very broad scale such as the Lake Mead-Colorado River system. The Colorado River, which provides irrigation and drinking water for Southern California and Arizona, has detectable perchlorate concentrations[1]. Many farms in this arid region are dependent on perchlorate-contaminated irrigation water for production of a wide variety of fruits, vegetables, and grain, and a number of these agricultural crops accumulate perchlorate.[11,12]

Lake Mead receives much of its perchlorate from the waters of the Las Vegas Wash which is located adjacent to Henderson, Nevada and approximately 10 miles southeast of Las Vegas, Nevada. Water flow in the wash is now a result of wastewater and industrial discharge, urban runoff from the cities of Las Vegas and Henderson, and various groundwater seeps located near the Wash.[17] The major source of perchlorate in the groundwater that ultimately enters the Las Vegas Wash is thought to come from a defunct perchlorate production facility in Henderson, NV that was destroyed in 1988.[17] As a result, Salt cedar (*Tamarix ramosissima*) and other terrestrial and aquatic plant species growing in or near the Wash are heavily contaminated with perchlorate.[9,18] Water from the Las Vegas Wash enters Lake Mead, which then empties into the Lower Colorado River.

Water, soil, vegetation, and rodents were collected from different reaches along the Las Vegas Wash to determine the relative contributions of each to rodent perchlorate exposure.[18] Perchlorate was detected at elevated concentrations in water, soil, and vegetation, but was infrequently detected in rodent liver and kidney tissues. Broadleaf weeds contained the highest concentrations of perchlorate among all plant types examined, likely because they have large surface areas for evapotranspiration and movement of large volumes of water. Perchlorate in rodent tissues and vegetation was correlated with perchlorate concentrations in soil as expected, however rodent residues were not highly correlated with plant perchlorate concentrations. This suggested that soil may have been a greater source, or a more constant source of perchlorate exposure among rodents than vegetation. Alternatively, soil may better integrate cumulative exposure potential than vegetation which may have highly variable perchlorate concentrations.[18]

EFFECTS IN ECOLOGICAL RECEPTORS

Perchlorate interferes with normal thyroid function, and can thus be designated an endocrine disrupting chemical. Perchlorate has the capacity to alter endocrine function in exposed ecological receptors as well as humans. Mechanistically, perchlorate competitively inhibits iodide uptake into thyroid gland follicular cells by binding to the sodium (Na^+) - iodide (I^-) symporter (NIS)[19], thereby reducing synthesis of the thyroid hormones thyroxine (T4) and triiodothyronine (T3).[20] Thyroid hormones are important in many species for normal metabolism, growth, and brain and physical development.[21] Thyroid hormones play homologous roles in many vertebrate and invertebrate taxa, so there is potential for disruption of endocrine homeostasis in exposed ecological receptors.[22] It is also conceivable that minor changes in maternal thyroid hormones concentrations can have adverse effects on developing fetuses.[23] Furthermore, low-level, long-term exposure could potentially impact adult organisms that rely on hormonally-regulated processes for survival or reproduction.[24] Following are summaries of eco-related perchlorate effects data for different taxa.

Invertebrates

There is relatively little information on the effects of perchlorate in invertebrates. Available data suggest that perchlorate is not particularly lethal to earthworms at environmentally relevant soil concentrations with possible exceptions including spills or point sources.[25-27] Acute (14 day) and sub-chronic (28 day) LC50s (lethal concentration of perchlorate in water that is expected to kill 50% of animals tested) for sodium and ammonium perchlorate in *Eisenia fetida* are above 4000 mg/kg.[27] Perchlorate does appear to have slight effects on earthworm reproduction at concentrations nearing the upper end of the range of concentrations normally found in the environment.[27] EC50 (effective concentration of perchlorate in soil expected to alter cocoon production in 50% of animals tested) values for cocoon production in 14 day sand tests using sodium perchlorate were approximately 260 µg/g. The sub-chronic (21 day sand tests) EC50 for sodium perchlorate to *E. fetida* reproduction (cocoon production) was 1.3 µg/g. The sub-chronic (28 day artificial soil tests) EC50 for sodium perchlorate to *E. fetida* reproduction (cocoon production) was 350 µg/g.[27] Chronic perchlorate exposure may also result in a reduction in earthworm weight, which could influence survival over the long term.

Fishes

Thyroid follicular hyperplasia, hypertrophy, colloid depletion, and increased angiogenesis have been observed in laboratory studies with mosquitofish (*Gambusia holbrooki*) exposed to 0.1 mg/L sodium perchlorate for 30 days.[28,29] Thyroidal effects (hyperplasia, hypertrophy, colloid depletion) have also been observed in fish collected from perchlorate-contaminated waters (less than 80 µg/L) in the field.[30] The most sensitive indicator of exposure in fish is colloidal T4 ring intensity, which can be indicative of exposure to perchlorate concentrations as low as 0.011 mg/L.[29] Increased occurrence of macrophage aggregates was observed in zebrafish at 18 mg/L and mosquitofish at 92 mg/L.[31] There were no adverse effects on fecundity and growth in mosquitofish exposed to perchlorate for 6 weeks.

Therefore, perchlorate has the potential to induce perturbations in piscine thyroidal homeostasis at concentrations that have been observed in some surface waters ($<$ 1 mg/L).[32] How these thyroid alterations translate into reduced fitness among fishes is unclear. There is equivocal evidence for fitness effects of perchlorate in fish,[33] but growth and pubertal development of sub-adult zebrafish (*Danio rerio*) were not affected by exposure to ammonium perchlorate at 0 to 11.48 mg/L.[29] More data are needed related to the effects of perchlorate on the early life stages of fishes that are typically more sensitive to environmental contaminants. Full life cycle and/or partial life cycle tests would be helpful in this regard. Since thyroid hormones can directly or indirectly influence embryonic neuronal development and behavior, immune function, and expression of thyroid-responsive and fitness-related genes in fish, additional research in these areas is needed for a more complete understanding of the potential effects of perchlorate in fishes.[32]

Amphibians

The process of metamorphosis in amphibians is driven, in large part, by thyroid hormones. Accordingly, perchlorate exposure among amphibians can potentially have significant impacts on the ecological fitness of wild amphibians. In laboratory studies, perchlorate dramatically inhibits amphibian metamorphosis at environmentally relevant concentrations. Goleman et al.[34] reported delays in metamorphosis in *Xenopus laevis* at concentrations ranging from 0.005 to 0.1 mg/L and inhibition of metamorphosis at 0.147 mg/L. McDaniel[35] also noted inhibition of metamorphosis in *X. laevis* at 0.050 mg perchlorate/L, and Tietge and Degitz[36] reported similar findings at 0.250 mg/L. However, it appears that the antimetamorphic effects of perchlorate are reversible to some extent under laboratory conditions.[37] Histological changes in thyroid tissue have been documented in *X. laevis* exposed to ammonium and sodium perchlorate at concentrations of 0.056 mg/L[37] and 0.016 mg/L,[36] respectively. Goleman et al.[37] also reported that perchlorate exposure altered sex ratios in developing *X. laevis* resulting in greater than expected numbers of females. *X. laevis* exposed to perchlorate exhibited increased susceptibility to UV-induced DNA damage and mortality at concentrations as low as 50 μg/L in laboratory studies.[35]

Thyroid histological changes have also been observed in amphibians collected from perchlorate-impacted sites. Follicle cell hypertrophy and colloid depletion have been observed in wild chorus frog (*Pseudacris triseriata*) tadpoles inhabiting a pond containing approximately 10 mg/L perchlorate as compared to stage-matched tadpoles from a reference pond.[38] Similarly, thyroid follicle cell height and follicle cell hypertrophy were altered in adult male cricket frogs (*Acris crepitans*) collected from perchlorate contaminated streams in central Texas.[39]

These data indicate that amphibians can be quite sensitive to the thyroid effects caused by perchlorate. In fact, no lowest observed adverse effect level (LOEL) has been reported for amphibians because effects are consistently observed in the lowest concentration ranges. The effects of perchlorate exposure on gonadal differentiation[37] may have implications for reproductive success and population structure and stability. Multigenerational exposure studies are needed to better understand the potential biological significance of perchlorate-induced changes in sex ratio. Further, most data on the effects of perchlorate on amphibians come from studies on *X. laevis*, a species that is not native to North America. Information on the responses of other amphibian species is lacking.[40]

Like most vertebrates, amphibians obtain iodide through their diet, and aquatic vertebrates can also take up iodide through skin or respiratory surfaces such as gills. Iodide appears to prevent perchlorate-induced disruption of metamorphosis in amphibians.[41,42] The seemingly protective effect of environmental iodide has not been examined in other species, thus potential effects of environmental perchlorate exposure are likely overestimated in areas where ambient iodide concentrations are sufficiently high.

Birds

As noted above, most perchlorate contaminated sites are small in scale which may limit exposure among the highly mobile avian species. However, birds that concentrate foraging effort on contaminated sites, those whose migratory routes transect contaminated regional watersheds like the Colorado River, or those whose movement is restricted by nesting/reproductive activities may be exposed to biologically meaningful concentrations of perchlorate. While adult organisms may be relatively resistant to subtle alterations in thyroid hormone concentrations, the potential effects on developing organisms remains unclear, particularly in wildlife species. Perchlorate is depurated into eggs of Northern bobwhite (*Colinus virginianus*) hens exposed to ammonium perchlorate in drinking water[43] with egg perchlorate concentrations increasing as a function of maternal dose.

In birds and other animals, thyroid function parameters (primarily thyroid hormone concentrations and histology) are altered at lower perchlorate exposures and at shorter exposure times than those that result in hypothyroidism. Thyroid glandular T4 content was significantly decreased in 21-day bobwhite quail embryos from eggs with a concentration of ≥ 0.5 $\mu g/g$ ammonium perchlorate and in hatchlings from eggs with ≥ 50 mg/L ammonium perchlorate.[44] However, egg concentrations of up to 300 $\mu g/g$ did not result in hypothyroidism in embryos or hatchlings. Perchlorate sensitivity also appears to be species specific in birds.[45]

Bobwhite quail chicks experience alterations in thyroid function within two weeks of ammonium perchlorate exposure, but can adapt their thyroid function to compensate for the anti-thyroidal effects of perchlorate, especially at lower concentrations.[46] Hypothyroidism only occurred in those chicks exposed to water that contained very high concentrations (≥ 500 mg/L) of perchlorate. In adult bobwhites, the effects of perchlorate were slower to develop, increased with time, and higher perchlorate exposures were tolerated as compared to chicks. Adult quail do not appear to have the thyroid adaptive capacity that has been observed in chicks.[46]

Site specific occurrence data indicate that the levels of perchlorate required to affect avian thyroid histology may be found in potential avian food items and drinking water at some sites. However, to date, there have been no published field studies documenting perchlorate exposure and subsequent effects on bird development, thyroid function, behavior, survival or reproduction. Since the development of thyroid function and amounts of circulating thyroid hormones are different in precocial compared with altricial bird species, both should be examined in future studies. Thyroid hormones play important roles in the development of the mammalian central nervous system, but data are needed on potential neurodevelopmental effects associated with perchlorate exposure in birds. Thyroid hormones also influence metabolism and thermoregulation which have obvious ecological significance for wild birds with high metabolic demands, thermoregulatory demands, or those with high energetic demands due to parental care of nestlings.[47]

Mammals

There are more perchlorate-related effects data on mammals than any other taxa, due mainly to concerns for human health and the historical use of perchlorate as a treatment for Grave's disease. A complete review of mammalian effects data is outside the scope of this chapter, hence the emphasis herein will be on studies using wild-type mammalian models and mammalian wildlife species. In general, long-term exposure to perchlorate in mammals results in decreased plasma thyroid hormone concentrations, and changes in thyroid histology.

A number of perchlorate-related studies have been conducted using the deer mouse (*Peromyscus maniculatus*), a wild rodent widely distributed throughout North American as the research model. Thuett and co-workers[48,49] evaluated the effects of *in-utero* and lactational perchlorate exposure among juvenile deer mice. Offspring from breeding pairs dosed continuously from the onset of cohabitation with perchlorate (0, 0.117, 117, and 117,000 μg/L ammonium perchlorate salt) exhibited changes in body mass and heart mass, but there were no differences in survival or litter sizes.[48] Deer mouse pups also had significantly decreased colloid area and decreased numbers of active follicles/unit area among treatment groups.[49] In contrast to most other studies, total plasma T4 concentrations were significantly higher in the two lowest treatment groups (0.117 and 117 μg/L) than controls in this study.

Mean plasma thyroxine concentrations decreased in a dose-dependent manner in Prairie voles (*Microtus ochrogaster*), an herbivorous wildlife species, and a marked reduction in T3 (but not plasma T4 concentration) was observed following exposure for 51 days to 0-10 mg/kg/day ammonium perchlorate salt.[24] However, mean thyroid glandular T4 content in the perchlorate-exposed voles decreased in a dose-dependent manner. There were also differences in mean thyroid gland T4 concentrations in 42-day exposure treatment groups with thyroidal T4 content from a 10.0 mg/kg treatment group significantly lower than controls. Glandular T4 content, compared to plasma thyroid hormone concentrations, may be a more sensitive indicator of perchlorate-induced alterations in mammalian and avian thyroid function.[24,47]

A wide range of dosages and exposure periods can affect thyroid-related endpoints in mammals, but secondary or "downstream effects" of thyroid alterations have not been well documented. However, alterations in mammalian brain morphology have been observed in a rodent model following exposure to perchlorate during gestation and lactation.[50] These findings may have implications for the neurobehavioral development of wild mammals which could ultimately impact fitness and survival in exposed individuals and populations. In addition, testicular morphological changes have been observed in perchlorate treated deer mice[49] suggesting the potential for reproductive effects.

It is important to note that the mammalian thyroid gland can store large amounts of thyroid hormone;[51] therefore, physiological effects of thyroid hormone deficiency may not be observed for extended periods depending on reserve capacity. Adult mammals exposed for short periods, or those exposed intermittently may never experience significant reductions in thyroid hormone production. Other than some rodent species, few mammals concentrate foraging efforts at spatial scales which are typical of most perchlorate-contaminated sites. Therefore rodents would appear to be among the mammals most likely to experience sufficient exposure to experience thyroid alterations. As an example, Smith and colleagues[52] were unable to identify perchlorate-related changes in thyroid hormone concentrations among raccoons (*Procyon lotor*) inhabiting a contaminated site in east Texas which may indicate that higher trophic level species, species that eat a variety of food items, or those with large home ranges may never be exposed to sufficient perchlorate concentrations to elicit a noticeable effect. Alternatively, iodide available via food resources may mitigate the effects of episodic or chronic exposure to perchlorate.[53]

Livestock

In contrast to free-ranging mammalian wildlife, livestock are often confined to limited areas and correspondingly few sources of forage and water. In areas where perchlorate-contaminated groundwater is used for irrigation and/or a source of drinking water, livestock may consume considerable amounts of perchlorate on a daily basis. Perchlorate exposure and effects in livestock were evaluated in heifer calves placed on a site with water access restricted to a stream fed by a perchlorate-contaminated groundwater spring (~25 ng/mL) in central Texas near the Naval Weapons Industrial Reserve Plant (NWIRP).[54] Blood was collected from calves on the perchlorate-impacted site (and two control calves from an uncontaminated site) approximately every 2 weeks over 14 weeks for analysis of perchlorate residues and thyroid hormones. Perchlorate was detected in blood plasma twice (15 ng/mL and 22 ng/mL) in one of the heifer calves on consecutive sampling periods. Near constant exposure to perchlorate in drinking water and presumably via the grass growing near the contaminated stream had no effect on circulating thyroid hormones in the heifer calves. No perchlorate was detected in any of the heifer tissues after completion of this study (although the thyroid gland was not tested), but tissue concentrations may have been below the limit of detection in this particular study.[54] Interestingly though, perchlorate has been detected in some commercial milk samples.[55]

Other Ecotoxicological Considerations

Thyroid hormones help regulate heat production, therefore perchlorate-induced alterations in thyroid function may be significant for mammals and other homeotherms inhabiting regions with wide seasonal temperature ranges or daily temperature fluctuations. Perchlorate-induced disturbances in an animal's overall metabolic and thermogenic capacity would presumably reduce fitness. However, Isanhart *et al.*[24] exposed prairie voles to perchlorate at 10 mg/kg/day for 51 or 0.75 mg/kg/day for 180 d, and reported no significant alterations in resting metabolic rates or peak metabolic rates during cold exposures. Thus, alternative mechanisms involved with thermoregulation may compensate for perchlorate-induced reductions in thyroid hormone production.

Additional research is needed to fully understand the primary and secondary effects associated with mammalian exposure to perchlorate at environmentally relevant concentrations. Further, these studies should examine maternal exposure and milk transfer to neonatal and developing offspring.[53] A more definitive understanding of potential neurobehavioral, metabolic and thermoregulatory effects of perchlorate would be most helpful for evaluating effects in wild mammals.

CONCLUSIONS

Given the occurrence, distribution, and concentrations of perchlorate at contaminated sites in North America, ecological receptors can be exposed to levels that cause adverse effects. For example, wild vertebrates can consume perchlorate in doses that alter thyroid function as measured by plasma thyroid concentrations and thyroid histology.[38] However, there are mitigating factors which may prevent the occurrence of adverse effects resulting from perchlorate exposure including thyroid hormone storage capacity in mammals, home range size, and uptake of iodide from natural sources. Further, young birds possess some capacity for adaptation to perchlorate. Amphibians appear to be the most sensitive indicators of environmental perchlorate contamination, with environmentally relevant concentrations of perchlorate having the potential to alter metamorphosis in amphibians.[34,37] Altered amphibian metamorphosis would be most critical in those species inhabiting ephemeral ponds and those having rapid developmental rates. However, there are no empirical data linking perchlorate exposure to population-level effects in amphibians or other taxa. Fish are not expected to experience acute toxic effects of perchlorate, but chronic exposure to perchlorate in the laboratory and the field has resulted in thyroid gland structural perturbations as well as alterations in thyroid hormone concentrations. There is little evidence of perchlorate-induced effects on growth, survival, and reproduction in fish at environmentally relevant concentrations. Perchlorate is not acutely toxic to earthworms at environmentally relevant soil concentrations, but chronic perchlorate exposure may reduce earthworm mass and reproduction.

A significant amount of research effort since 2000 has produced a basic database on exposure and effects in a variety of ecological receptors. Despite these efforts, additional information on the ecological effects of perchlorate in developing organisms, homeotherms inhabiting temperature extremes, neurobehavioral effects that may affect ecological fitness, and population-level effects would be useful to fully understand the ecotoxicology of perchlorate.

ACKNOWLEDGEMENTS

The work of many of the students, staff, and my colleagues at The Institute of Environmental & Human Health at Texas Tech University must be acknowledged, for if not for their efforts, much less would be known about the ecological effects of perchlorate and this chapter could not have been written. Researchers include Drs. Todd Anderson, Stephen Cox, Ken Dixon, Angie Gentles, Mike Hooper, Ron Kendall, Scott McMurry, Ernest Smith, Chris Theodorakis. Dr. Jim Carr from the Department of Biological Sciences, Dr. Andrew Jackson from the Department of Civil Engineering, and Dr. Reynaldo Patino from the TTU USGS Cooperative Unit at Texas Tech were integral collaborators in the perchlorate-related research conducted at Texas Tech. Additionally Dr. Anne McNabb at Virginia Tech University, Dr. Shane Snyder of the Southern Nevada Water Authority, Dr. Erin Snyder of Black and Veatch, Dr. Jody Wireman formerly of the Air Force Institute of Operational Health, Kevin Mayer of the US Environmental Protection Agency Region 9, and Mr. John Isanhart were all major contributors. Finally, I wish to acknowledge the Strategic Environmental Research and Development Program (SERDP) for their long-term support of perchlorate research at Texas Tech (Dr. Ron Kendall, Principal Investigator). Additional support was provided by the US Department of Agriculture, US Corps of Engineers (Ft. Worth District), and The Institute of Environmental & Human Health at Texas Tech University.

REFERENCES

1. Urbansky, E.T. Perchlorate chemistry: Implications for analysis and remediation. Bioremediat J 1998; 2:81-95.

2. Urbansky, E.T., Collette, T.W., Robarge, W.P., Hall, W.L., Skillen, J.M., Kane, P.F., Survey of fertilizers and related materials for perchlorate (ClO$_4^-$). EPA/600/R-01/049. Final Report. U.S. Environmental Protection Agency Office of Research and Development, Cincinnati, OH. 2001.

3. Susarla, S., Bacchus, S.T., Wolfe, N.L., McCutcheon, S.C. Phytotransformation of perchlorate and identification of metabolic products in *Myriophyllum aquaticum*. Intern J Phytoremed 1999; 1(1): 97-107.

4. Nzengung, V.A., Wang, C.H., Harvey, G. Plant-mediated transformation of perchlorate into chloride. Environ Sci Technol 1999; 33:1478-9.

5. Ellington, J.J., Wolfe, N.L., Garrison, A.W., Evans, J.J., Avants, J.K., Teng, Q. Determination of perchlorate in tobacco plants and tobacco products. Environ Sci Technol 2001; 35: 3213-8.

6. Parsons. 2001. Scientific and technical report for perchlorate biotransport investigation: A study of perchlorate occurrence in selected ecosystems. Interim Final Report. Air Force Institute for Environment, Safety and Occupational Health Risk Analysis, Brooks AFB, TX, USA.

7. Smith, P.N., Theodorakis, C.W., Anderson, T.A., Kendall, R.J. Preliminary assessment of perchlorate in ecological receptors at the Longhorn Army Ammunition Plan (LHAAP), Karnack, TX. Ecotoxicology 2001; 10:305-13.

8. Susarla, S., Bacchus, S.T., Harvey, G., McCutcheon, S.C. Phytotransformations of perchlorate contaminated waters. Environ Technol 2000; 21:1055-65.

9. Urbansky, E.T., Magnuson, M.L., Kelty, C.A., Brown, S.K. Perchlorate uptake by salt cedar (*Tamarix ramosissima*) in the Las Vegas Wash riparian ecosystem. Sci Total Environ 2000; 256:227-232.

10. Aken, B.V., Schnoor, J.L. Evidence of perchlorate (ClO4-) reduction in plant tissues (poplar tree) using radio-labeled $^{36}ClO_4^-$. Environ Sci Technol 2002; 36: 2783-88.

11. Tan, K., Anderson, T.A., Jones, M.W., Smith, P.N., Jackson, W.A. Uptake of perchlorate in aquatic and terrestrial plants at field scale. J Environ Qual 2004; 33:1638-46.

12. Yu, L., Canas, J.E., Cobb, G.P., Jackson, W.A., Anderson, T.A. Uptake of perchlorate in terrestrial plants. Ecotoxicol Environ Saf 2004; 58:44-9.

13. Urbansky, E., Schock, M. Issues in Managing the Risks Associated with Perchlorate in Drinking Water. J Environ Manage 1999; 56:79-95.

14. Jackson, W.A., Joseph, P., Laxman, P., Tan, K., Smith, P.N., Yu, L., Anderson, T.A. Perchlorate accumulation in forage and edible vegetation. *Journal of Agric Food Chem*; 53:369-373.

15. Beyer, W.N., Connor, E.E., Gerould, S. Estimates of soil ingestion by wildlife. J Wildl Manage 1994; 582:375-82.

16. Smith, P.N., Wireman, J., Carr, J.A., Smith, E.E., Kendall, R.J. "Ecological Risk Assessment" In *Perchlorate Ecotoxicology*, Kendall, R.J., P.N. Smith, eds. Pensacola, FL: SETAC Press, in press.

17. Hogue, C. Rocket-fueled river. Chem Engin News 2003; 81 (33): 37-46.

18. Smith, P.N., Yu, L., McMurry, S.T., Anderson, T.A. Perchlorate in water, soil, vegetation, and rodents collected from the Las Vegas Wash, Nevada. *Environmental Pollution* 2004; 132:121-7.

19. Wolff, J. Perchlorate and the thyroid gland. Pharmacol Rev 1998; 50:89-105.

20. Capen, C. "Toxic Responses of the Endocrine System" In *Casarett and Doull's Toxicology The Basic Science of Poisons, 6th ed.* Klaassen C ed. New York: McGraw-Hill Companies 2001.

21. Pop, V., Kuijpens, J., van Baar, A., Verkerk, G., van Son, M., Vijlder, J., Vulsma, T., Wiersinga, W., Drexhage, H.,Vader, H. Low maternal free thyroxine concentrations during early pregnancy are associated with impaired psychomotor development in infancy. Clin Endocrinol 1999; 50:149-155.

22. Kendall, R.J., Dickerson, R.L., Giesy, J.P., Suk, W.P. *Principles and Processes for Evaluating Endocrine Disruption in Wildlife.* Pensacola, FL: SETAC, 1998.

23. Escobar, G., Obregon, M., Escobar del Rey, F. Is neuropsychological development related to maternal hypothyroidism or to maternal hypothyroxinemia? J Clin Endocrinol Metab 2000; 85(11):3975-87.

24. Isanhart, J.P., McNabb, F.M.A., Smith, P.N. Effects of perchlorate exposure on resting metabolism, peak metabolism, and thyroid function in the prairie vole, *Microtus ochrogaster.* Environ Toxicol Chem 2005; 24(3):678-84.

25. Parsons. 2002. Sub-chronic toxicity and bioconcentration of perchlorate by the earthworm Eisenia fetida. Contract F41624-01-D-8009, Brooks AFB, TX.

26. EA Engineering, Science, and Technology. 1998. Results of acute and chronic toxicity testing with sodium perchlorate. Report 2900, Brooks AFB, TX.

27. Anderson, T.A., Landrum, M., Snyder, E. "Perchlorate Effects on Invertebrates" In *Perchlorate Ecotoxicology,* Kendall, R.J., P.N. Smith, eds. Pensacola, FL: SETAC Press, in press.

28. Bradford, C.M., Carr, J.A., Rinchard, J., Theodorakis, C. Perchlorate affects thyroid function in eastern mosquitofish (*Gambusia holbrooki*) at environmentally relevant concentrations. Eviron Sci Technol; 39:5190-5195.

29. Mukhi, S., Carr, J.A., Anderson, T.A., Patino, R. Development and validation of new biomarkers of perchlorate exposure in fishes. Environ Toxicol Chem 2005; 24:1107-1115.

30. Theodorakis, C.W., Rinchard, J., Park, J.W., McDaniel, L., Liu, F., Carr, J.A., Wages, M. Thyroid endocrine disruption in stonerollers and cricket frogs from perchlorate-contaminated streams in east-central Texas. Ecotoxicology, in press.

31. Capps, T., Mukhi, S., Rinchard, J., Theodorakis, C.W., Blazer, V.S., Patiño, R. Exposure to perchlorate induces the formation of macrophage aggregates in the trunk kidney of zebrafish and mosquitofish. J Aquat Animal Health 2004; 16:145-51.

32. Theodorakis, C., Patiño, R., Snyder, E., Albers, E. "Perchlorate Effects in Fish" In *Perchlorate Ecotoxicology,* Kendall, R.J., P.N. Smith, eds. Pensacola, FL: SETAC Press, in press.

33. Brown, S.B., Adams, B.A., Cyr, D.G., Eales, J.G. Contaminant effects on the teleost fish thyroid. Environ Toxicol Chem 2004; 23:1680-1701.

34. Goleman, W.L., Urquidi, L.J., Anderson, T.A., Smith, E.E., Kendall, R.J., Carr, J.A. Environmentally relevant concentrations of ammonium perchlorate inhibit development and metamorphosis in *Xenopus laevis*. Environ Toxicol Chem 2002; 21:424–30.

35. McDaniel, L.N. Ultraviolet Radiation-Induced Genotoxicity in Xenopus laevis Exposed to Sodium Perchlorate. MS. Thesis Texas Tech University, Lubbock, TX, 2004.

36. Tietge, J.E., Degitz, S.J. The effects of perchlorate on *Xenopus laevis*. Technical Report, Mid-Continent Ecology Division, National Health and Environmental Effects Research Laboratory, Office of Research and Development, U.S. EPA, Duluth, MN. 2003.

37. Goleman, W.L., Carr, J.A., Anderson, T.A. Environmentally relevant concentrations of ammonium perchlorate inhibit thyroid function and alter sex ratios in developing *Xenopus laevis*. Environ Toxicol Chem 2002; 21:590-597.

38. Carr, J.A., Urquidi, L.J., Goleman, W. L., Hu, F., Smith, P. N., Theodorakis, C.W., "Ammonium perchlorate disruption of thyroid function in natural amphibian populations: Assessment and potential impact" In *"Multiple Stressor Effects in Relation to Declining Amphibian Populations"*, ASTM STP 1443, G. Linder, ed. West Conshohocken, PA: ASTM International, 2003. pps. 131-142.

39. U.S. Army Corps of Engineers (USACE). Final Report: Bosque and Leon River Watersheds Study.http://www.swf.usace.army.mil/ppmd/Perchlorate/index.html 2004.

40. Carr, J.A., Theodorakis, C. "Effects of Perchlorate in Amphibians" In *Perchlorate Ecotoxicology,* Kendall, R.J., P.N. Smith, eds. Pensacola, FL: SETAC Press, in press.

41. Sparling, D.W., Harvey, G., Nzengung, V. "Interaction between perchlorate and iodine in the metamorphosis of *Hyla versicolor*" In *Multiple Stressor Effects in Relation to Declining Amphibian Populations"*, ASTM STP 1443, G. Linder, ed., West Conshohocken, PA: ASTM International, 2003. pps. 131-42.

42. Hu, F., Gentles, A., Goleman, W.L., Carr, J.A. Developmental stage and environmental iodide influence the antimetamorphic effects of perchlorate. Proc Soc Environ Toxicol Chem 24th Annual Meeting, 2003.p. 293.

43. Gentles, A., Surles, J., Smith, E.E. Evaluation of Adult Quail and Egg Production Following Exposure to Perchlorate-Treated Water. Environ Toxicol Chem; 24:1930-1934.

44. McNabb, F.M.A. Biomarkers for the assessment of avian thyroid disruption by chemical contaminants. Avian Poult Biol Rev; 16:3-10.

45. McNabb F.M.A., Larsen, C.T., Pooler, P.S. Ammonium perchlorate effects on thyroid function and growth in bobwhite quail chicks. Environ Toxicol Chem 2004; 23:997-1003.

46. McNabb F.M.A., Jang, D.A., Larsen, C.T. Does thyroid function in developing birds adapt to sustained ammonium perchlorate exposure? Toxicol Sci 2004; 82:106-13.

47. McNabb, F.M.A., Hooper, M.J., Smith, E.E., McMurry, S.T., Gentles, B.A. "Perchlorate Effects in Birds" In *Perchlorate Ecotoxicology,* Kendall, R.J., P.N. Smith, eds. Pensacola, FL: SETAC Press, in press.

48. Thuett, K.A., Roots, E.H., Mitchell, L.P., Gentles, B.A., Anderson, T.A., Smith, E. E. In utero and lactational exposure to ammonium perchlorate in drinking water: effects on

developing deer mice at postnatal day 21. J Toxicol Environ Health A. 2002; 65(15):1061-76.

49. Thuett, K.A., Roots, E.H., Mitchell, L.P., Gentles, B.A., Anderson, T., Kendall, R.J., Smith, E.E. Effects of in utero and lactational ammonium perchlorate exposure on thyroid gland histology and thyroid and sex hormones in developing deer mice (*Peromyscus maniculatus*) through postnatal day 21. J Toxicol Environ Health A, 2002; 65:2119-30.

50. York, R.G., Barnett, J. Jr, Brown, W.R., Garman, R.H., Mattie, D.R., Dodd, D. A rat neurodevelopmental evaluation of offspring, including evaluation of adult and neonatal thyroid, from mothers treated with ammonium perchlorate in drinking water. Int J Toxicol 2004; 23:191-214.

51. Guyton, A.C., Hall, J.E. *Textbook of Medical Physiology,* 10th ed. Philadelphia, PA: W.B. Saunders, 2001.

52. Smith, P.N., Utley, S.J., Cox, S.B., Anderson, T.A., McMurry, S.T. Monitoring perchlorate exposure and thyroid hormone status among raccoons inhabiting a perchlorate-contaminated site. Environ Monit Assess 2005; 102:337-347.

53. Smith E.E., Isanhart, J., Smith, P.N., Dixon, K., McNabb, F.M.A. "Perchlorate Effects on Mammals" In *Perchlorate Ecotoxicology,* Kendall, R.J., P.N. Smith, eds. Pensacola, FL: SETAC Press, in press.

54. Cheng, Q., Perlmutter, L., Smith, P.N., McMurry, S.T., Jackson, W.A., Anderson, T.A. A study on perchlorate exposure and absorption in beef cattle. J Agric Food Chem 2004; 52:3456-61.

55. Kirk, A.B., Smith, E.E., Tian, K., Anderson, T. A., Dasgupta, P.K. Perchlorate in milk. Environ Sci Technol 2003; 37(21):4979-81.

Chapter 8

Perchlorate Toxicity and Risk Assessment

David R. Mattie,[1] Joan Strawson,[2] and Jay Zhao[2]

[1] *Air Force Research Laboratory, WPAFB, OH 45433-5707*
[2] *Toxicology Excellence for Risk Assessment, Cincinnati, OH 45211*

PERCHLORATE OCCURRENCE AND EXPOSURE

Sources of Perchlorate

Ammonium perchlorate is the oxidizer ingredient in solid propellant mixtures for rockets, missiles and munitions such as Titan, Minuteman, Peacekeeper, Hawk, Polaris and the Space Shuttle. Perchlorate salts may also be used in medicine, matches, munitions and pyrotechnics (illuminating and signaling flares, colored and white smoke generators, tracers, incendiary delays, fuses, photo-flash compounds and fireworks). Perchlorate is also found in lubricating oils, finished leather, fabric fixer, dyes, electroplating, aluminum refining, manufacture of rubber, paint and enamel production, as an additive in cattle feed, in magnesium batteries and as a component of automobile air bag inflators.[1]

Exposure Routes/Pathways

Perchlorate salts are highly water-soluble and fully ionize in water.[2] The resulting perchlorate ion is identical whether it is from the ammonium salt or another salt such as potassium and its toxicity is due to the perchlorate ion (ClO_4^-). Chemicals may enter the human body in several ways, known as routes of exposure or pathways. Routes of exposure can include ingestion, dermal (skin) absorption and inhalation. The primary route of exposure for perchlorate is through ingestion of water from contaminated drinking water supplies, although ingestion of contaminated food and milk are other potential sources of exposure.

Absorption through skin and inhalation of perchlorate are minor exposure pathways. Compounds most readily absorbed through the skin are primarily organic chemicals. Because perchlorate is an inorganic compound and completely ionized in water, the potential for dermal absorption of perchlorate through intact skin while bathing and washing is unlikely.

Occupational exposure of workers to perchlorate is primarily through inhalation of ammonium perchlorate dust during the commercial production or use of perchlorate salts. Occupational exposure in ammonium perchlorate production facilities was shown to be higher than potential exposures from drinking water or food sources.[3,4] Exposure by inhalation results in absorption into the body through the mucous membranes in the respiratory and gastrointestinal tracts. Some direct ingestion through the oral route is possible, as is dermal contact, but both of these pathways are minor in occupational settings.

PERCHLORATE HEALTH EFFECTS

Mechanism of Toxicity

Thyroid hormones are essential to the regulation of oxygen consumption and metabolism throughout the body. Thyroid iodine metabolism and the levels of thyroid hormone in serum and tissues are regulated by a number of fairly well understood homeostatic mechanisms.[5] Thyrotropin (TSH), a hormone synthesized and secreted by the anterior pituitary gland, is the primary regulator of thyroidal iodide uptake and other aspects of thyroid function.[6] The secretion of TSH is regulated by a negative feed-back pathway through serum level of T_3 and T_4. Formation of thyroid hormones is depressed if sufficient inhibition of iodide uptake occurs. As a result, increased TSH secretion from the anterior pituitary gland will stimulate the thyroid to produce more T_3 and T_4. Inhibition of iodine uptake is the basis for the current and former pharmacological uses of perchlorate.

Perchlorate, like many chemicals and drugs, disrupts one or more steps in the synthesis and secretion of thyroid hormones, resulting in subnormal levels of T_4 and T_3 and an associated compensatory increase in secretion of TSH.[7] The perchlorate ion, because of its similarity to iodide in ionic size and charge, competes with iodide for uptake into the thyroid gland by the sodium-iodide symporter, a transport mechanism in the membranes of thyroid cells. At high doses, this competitive inhibition results in reduced production of the thyroid hormones [triiodothyronine (T_3) and thyroxin (T_4)] and a consequent increase in thyroid stimulating hormone (TSH) via a negative feedback loop involving the thyroid, pituitary and hypothalamus.[8] Subsequent events include decreases in serum T_4 (and T_3), leading to the potential for altered neurodevelopment if observed in either mothers or fetuses/neonates, and increases in serum TSH, leading to the potential for thyroid hyperplasia and tumors. The repeated observation of thyroid effects such as alterations of hormones, increased thyroid weight, and alterations of thyroid histopathology (including tumors) from a large number of rat studies on perchlorate provide supporting evidence for the proposed mode-of-action, and confirms that the perturbation of thyroid hormone economy as the primary biological effect of perchlorate.

Toxicokinetics

Studies of absorption, distribution, metabolism and elimination (ADME) to measure perchlorate kinetics revealed that there was no metabolism of perchlorate in either adult rats or humans. Perchlorate is rapidly excreted, with urinary half-lives on the order of 4 hours in the rat and 6 hours in humans.[9,10] Kinetics studies were also conducted for fetal and lactational time points in rats in parallel with the "Effects" Study. Kinetic studies were designed to aid quantitative interspecies extrapolation and form the basis for physiologically based pharmacokinetic (PBPK) models for adult rats and humans, as well as pregnant and lactating rats.[11,12,13,14] The kinetic studies were also used to evaluate the mechanism of iodide inhibition by perchlorate and the subsequent hormone response. By determining the relative sensitivity of rat versus human, the results helped aid interspecies extrapolation. The results of PBPK model simulations confirmed that the fetus is the most sensitive developmental time point based on extent of iodide inhibition.[13]

Animal Toxicity Studies

In order to determine the health effects of perchlorate in humans, a fairly extensive database of animal toxicity studies was developed in keeping with risk assessment guidelines and the mechanism of action of perchlorate. Several studies of perchlorate in rodents have been conducted in which hormone measurements and thyroid histopathology have been evaluated. Data are available in male and female rats following 14 and 90 days of exposure,[15,16] female mice following 90 days of exposure,[17,18] rat dams on gestation day 20, post natal day 5, postnatal day 10,[9,19,20] and male and female pups on post natal days 5, 10, and 22.[9,19,20]

A 90-day subchronic bioassay determined that the thyroid was the only target organ in male and female rats exposed to perchlorate in drinking water (0, 0.01, 0.05, 0.2, 1.0, and 10 mg/kg-day) for 90 days. The no observable adverse effect level (NOAEL) based on thyroid changes was 1 mg/kg-day but hormone changes, decreased T_4 and increased TSH, were still seen at the lowest doses.[16]

Genotoxicity assays showed that perchlorate has no toxic effects on genes or chromosomes in cells.[21] Perchlorate is not a teratogen as no birth defects were found at doses as high as 100 mg/kg-day in the rabbit[22] or as high as 30 mg/kg-day in the rat.[23] Immunotoxicity studies were motivated by case reports of aplastic anemia and leukopenia in humans when perchlorate was used in high doses as an anti-thyroid drug. Studies using female mice did not demonstrate any adverse effects to the immune system. Evaluation of thyroid responses identified no alterations in T_3 and TSH, while T_4 was decreased after exposure to 1.0, 3.0 or 30 mg/kg-day. Thyroid changes detected histologically were not seen in all animals until the 30.0 mg/kg-day dose.[17]

Developmental neurotoxicity studies exposed pregnant rats to perchlorate in drinking water (0, 0.1, 1.0, 3.0 and 10 mg/kg-day) during pregnancy through day 10 of lactation.[24] No pup behavioral effects were seen except marginal motor activity results at one time point. An additional motor activity study with the same doses found no statistically significant effects in motor activity.[25,26] However, a Bayesian statistical analysis conducted using data from the two different motor activity studies combined resulted in a NOAEL of 1.0 mg/kg-day. Hormone changes, decreased T_4 and increased TSH, were again seen at lower doses. Brain histology and morphometry changes seen in this developmental study were not dose dependent and resulted in larger measured values. Iodine deficiency should produce decreases in brain structure size.[24] Another study was planned as a result of the brain data.

The additional major study performed was referred to as the "Effects" Study. The objectives were to refine understanding of the effects of perchlorate in the thyroid gland and to evaluate brain development at critical time points.[19,20] The brain analysis attempted to provide better brain morphometry data than the developmental study described above. The study was designed to provide critical information for the perchlorate risk assessment, so data were collected at one fetal and three postnatal (during lactation) time points. The fetal time point also provided rat teratogenic results to compare with the rabbit teratogenic study.[22,23] Hormone changes, decreased T_4 and increased TSH, were again seen at lower doses. However, the brain morphometry was equivocal because the changes seen were once again not dose dependent or resulted in larger measured values. Subsequent evaluation of the methodology used in this and the previous developmental neurotoxicity raised questions about preparation of the brain tissue for microscopic analysis, including orientation of the slices through the brain. The interpretation of the brain morphometry data has been debated as a result.

A two-generation reproductive toxicity study was used to evaluate fertility in adult rats and viability/toxicity in their offspring.[27] Reproductive parameters were tested over two generations of drinking water exposure to perchlorate. The NOAEL for reproduction effects is greater than the highest dose tested, 30 mg/kg-day. Thyroid histology changes were seen starting at 3 mg/kg-day. There were three rare benign thyroid tumors identified by a Pathology Working Group in two first generation (F1) pups at the 30 mg/kg-day dose. However, the occurrence of this tumor was not statistically significant as it was also found in control animals.

Perchlorate Use as a Drug in Humans

Potassium or sodium perchlorate was used clinically to treat Graves' disease (hyperthyroidism, thyrotoxicosis) in the 1950s and 1960s. Anti-thyroid drugs act by either blocking iodide uptake into the thyroid or by blocking thyroid hormone (T_3 and T_4) synthesis in the thyroid. The recommended dose was 800 to 1000 mg/day, although doses varied from 200 to 2000 mg/day. Perchlorate was given as either a single large dose or as a daily dose for weeks, months or even years in some cases. The use of perchlorate as a drug became limited in the mid-1960s due to reports of agranulocytosis and fatal aplastic anemia.[28]

More recently, amiodarone has been used to treat patients with ischemic heart disease or with ventricular tachycardia; this drug has a potential side effect of hyperthyroidism. A dose of 1000 mg/day potassium perchlorate has been used to treat amiodarone-induced hyperthyroidism. Potassium perchlorate may also be used if a patient is sensitive to standard antithyroid drugs. The perchlorate discharge test was used clinically to diagnose thyroid function. In this test, a single large dose of perchlorate caused iodide discharge from the thyroid if there was a problem with the organification process, which is the incorporation of iodide into thyroid hormones.[28]

Human Health Effects

Epidemiological studies have examined the associations of thyroid effects and environmental exposure to perchlorate in public drinking water at levels ranging from 4 to 120 ppb (4 to 120 μg/L). A number of the studies have small sample sizes and cannot detect differences in frequency of outcomes between exposure groups. Adjustments for potentially confounding factors were limited. Nearly all the studies were ecologic (epidemiological studies without individual exposure characterization), including those in newborns and children, the groups potentially most vulnerable to the effects of perchlorate exposure. Ecologic studies can provide supporting evidence of a possible association but cannot provide definitive evidence regarding cause.

Perchlorate exposure of individual subjects is difficult to measure and was not assessed directly in any of the published studies conducted outside the occupational setting.

Perchlorate was found in Las Vegas and the surrounding county at levels averaging 14 ppb (μg/L). Congenital hypothyroidism data from the 1996 and 1997 neonatal screening program were examined; no increase in congenital hypothyroidism was observed.[29] From April 1998 to June 1999, the monthly mean T_4 levels of neonates from Las Vegas (an area with perchlorate-contaminated drinking water at 9 to 15 ppb (μg/L) for eight months and non-detectable ($<$ 4 ppb) contamination for 7 months) were compared with those of neonates from Reno (an area with no detectable perchlorate in its drinking water). There were no differences in neonatal T_4 levels between the cities.[30] An analysis of newborn TSH levels from Las Vegas and Reno was conducted for babies born December 1998 through October 1999 at normal birth weight and having their blood sampled within the first month of life. The mean blood TSH levels were not different between the cities.[31] Another Las Vegas vs. Reno comparison examined thyroid disease prevalence rates among Medicaid-eligible residents. Thyroid disease included simple goiter, nodular goiter, thyrotoxicosis, congenital hypothyroidism, acquired hypothyroidism, thyroid cancer or other thyroid diseases. Again there were no differences between the cities.[32] One study examined a possible relation between perchlorate exposure and adverse neurodevelopmental outcomes in children and did not see any association between perchlorate and attention deficit hyperactivity disorder (ADHD) or autism.[33]

Data on thyroid hormone and TSH production in newborns and congenital hypothyroidism were also examined by Brechner et al.,[34] Schwartz,[35] Kelsh et al.,[36] Lamm[37] and Buffler et al.[38] Although hormone changes were reported in areas of California and Arizona, the prevalence of congenital hypothyroidism was not associated with perchlorate exposure. Morgan and Cassady[39] addressed the incidence of cancer after dual exposure to perchlorate and trichloroethylene. Incidence of cancers was not observed more often than expected in children.

A number of studies collected clinical data from adult human volunteers in a controlled setting after exposure to known amounts of perchlorate in drinking water ranging from 0.007 to 12 mg/kg per day. These clinical studies measured inhibition of iodide uptake into the thyroid glands as well as TSH and thyroid hormones. Serum TSH levels did not increase and thyroid hormones did not decrease in any group. Brabant et al.[40] conducted a 4 week study with subjects given 12 mg/kg per day. Thyroid iodide and TSH decreased and thyroid hormones did not change. Although only preliminary data are available, a study in which volunteers were given 0.007 and 0.04

mg/kg-day perchlorate for 6 months resulted in no effects on thyroid function.[41]

Two 14-day studies were conducted in which 10 mg/day was provided in water to 10 male subjects and 3 mg/day was provided in drinking water to 8 male subjects.[42,43] In both studies, each subject served as their own control by having measurements taken before and after perchlorate consumption. Iodide[123] was measured in the thyroid to obtain inhibition data, and iodide and perchlorate were determined in blood and urine. Perchlorate, at both 3 and 10 mg/day, caused inhibition of iodide uptake into the thyroid (38% and 10%, respectively). There were no changes seen in TSH or thyroid hormone levels in the blood. The extrapolated no observable effect level (NOEL) for iodide inhibition was 2 mg/day based on these two exposures.

Another 14-day study employed 10 subjects (5 male/5 female) for each dose (0.5, 0.1, 0.02 and 0.007 mg/kg-day) who also served as their own control.[10] The parameters measured were iodide[123] uptake in the thyroid for inhibition data and iodide and perchlorate in blood and urine for kinetic data. There were no changes seen in TSH or thyroid hormone levels in the blood. The NOEL for iodide inhibition, measured as decreased uptake of iodide[123] in the thyroid, was 0.007 mg/kg-day. Data from these studies were used to develop the human PBPK model for perchlorate.

Employees were examined at an ammonium perchlorate production facility in Nevada and their findings compared to those of a large unexposed control population from the same industrial complex.[3] The average working-lifetime cumulative dose in the higher exposure group was estimated to be 38 mg/kg. Based on both cumulative and single-shift perchlorate exposures, there were no adverse effects on thyroid, kidney, liver or bone marrow function. A cross-sectional health study of two similar worker populations, a group of ammonium perchlorate workers divided into three exposure groups and a comparison group of other workers from the same industrial complex, was conducted at a perchlorate manufacturing plant in Utah.[44] More than 40% of the exposed employees had been working with perchlorate for more than five years. There were no effects on blood and clinical chemical parameters at any level of exposure up to 34 mg per day.

A second occupational study of perchlorate exposure was conducted at the same site in Utah. Serum perchlorate levels averaged 838 ppb (μg/L) during exposure with a calculated daily dose of 0.167mg/kg-day perchlorate. Although iodide inhibition averaged 38%, TSH did not change and thyroid hormones actually increased slightly but significantly.[4]

School-age children were examined in three cities in northern Chile where levels of naturally occurring perchlorate in the water supply were undetectable, 5-7 and 100-120 ppb (ug/L). In the school children (mean age 7.3 years), 127 of whom had lifelong residence in their respective cities, mean TSH, T_4, and T_3, were similar among the three cities. Incidence of goiter in the lifelong residents was similar in all three cities; although the residents in Taltal self-reported a higher incidence of family history of thyroid disease. No changes were found in congenital hypothyroidism, and clinical differences between children from the three different cities. This was the only ecologic study with measures of perchlorate made directly from drinking-water samples taken from homes in the area where subjects lived.[45] Free T_4 was significantly increased in children living in Taltal and Chanaral, compared with Antofagasta (control city), a change in the opposite direction than expected. A variable introduction of iodized salt started in 1982 and may have affected these observations. Crump et al.[45] also studied newborns screened for hypothyroidism by a heel-stick blood sample between February 1996 and January 1999 in the same three Chilean cities. TSH levels were significantly lower in Taltal than in the other two cities, a trend opposite to that hypothesized. The authors concluded that the differences did not appear clinically significant.

A second epidemiological study was conducted in the same Chilean cities. Water was analyzed from the tap in the home of every subject. Perchlorate, at daily levels as high as 114 ppb (μg/L) throughout pregnancy, did not affect maternal thyroid status early in gestation, alter fetal thyroid status at birth, or reduce breast milk iodine concentrations.[46]

Summary of Toxicity

The competitive inhibition of iodide uptake is the only direct perchlorate effect on the thyroid, leading to a reversible chemical-induced iodine deficiency. Alteration of hormones (T_4, T_3, TSH) would be the first observed biological effect of perchlorate exposure. Following a prolonged increase in TSH, thyroid hyperplasia progressing to thyroid tumors would be expected to occur in rodents. However, the relevance of these tumors to humans has been questioned, since this progression has not been observed in humans.[47] In contrast, human data show that decreased T_4 levels, both in pregnant women and in neonates, can lead to neuro-developmental deficit; although this has not been confirmed in animals following perchlorate exposure. Therefore, of the two pathways to altered structure and function proposed by a mode-of-action analysis for perchlorate, decreased T_4 leading to potential neurodevelopmental effects is more relevant to an assessment of human health.

This conclusion is also supported by National Research Council.[48] The NRC, on behalf of the DoD, EPA, DOE and NASA, formed a committee in 2003 to conduct an assessment of key scientific issues associated with the health effects of perchlorate. This committee in its report[48] judged that the development of thyroid tumors as an ultimate result of perchlorate exposure is an unlikely outcome in human based on two considerations: (1) rats are sensitive to the development of thyroid tumors because their thyroid function is easily disrupted; (2) humans are much less susceptible than rats to disruption of thyroid function and therefore are not likely to develop thyroid tumors as a result of perchlorate exposure. Therefore, NRC committee concluded that the most reasonable pathway of events after changes in thyroid hormone and TSH secretion would be thyroid hypertrophy or hyperplasia, possibly leading to hypothyroidism.[48] The hypothyroidism would lead to two potential outcomes (as summarized in Figure 1): (1) metabolic sequelae, such as decreased metabolic rate and slowing of the function of many organ systems, occurring at any age, and (2) abnormal growth and development in fetuses and children.

Following oral exposure in drinking water, serum perchlorate levels increase and provide a measure of the perchlorate internal dose. The mechanism of action also identifies that perchlorate has a threshold for effects and that the degree of effects are dependent on the dosage. Severe or sustained iodine deficiency can lead to hypothyroidism. However, a temporary disruption of iodine leads to temporary effects that are reversed with the proper levels of iodine ingestion in the diet. In rodents, sustained exposure to a chemical that causes hypothyroidism can lead to thyroid tumors, a mechanism that has not been demonstrated in humans. There remains disagreement on what is the first adverse or critical effect after perchlorate exposure.

PERCHLORATE RISK ASSESSMENT

Identification of critical effect

One risk assessment goal is to determine what exposure might be considered "safe." "Safe" or subthreshold doses are defined by a number of health agencies worldwide. Although many of the underlying assumptions, judgments of critical effect, and choices of uncertainty factors are similar among health agencies in estimating these subthreshold doses, this section will follow U.S. EPA's Reference Dose (RfD) methods.[49,50,51]

The first step in defining an RfD is to identify the critical effect(s). U.S. EPA [52] and Haber et al.[53] define critical effect(s) as the first adverse

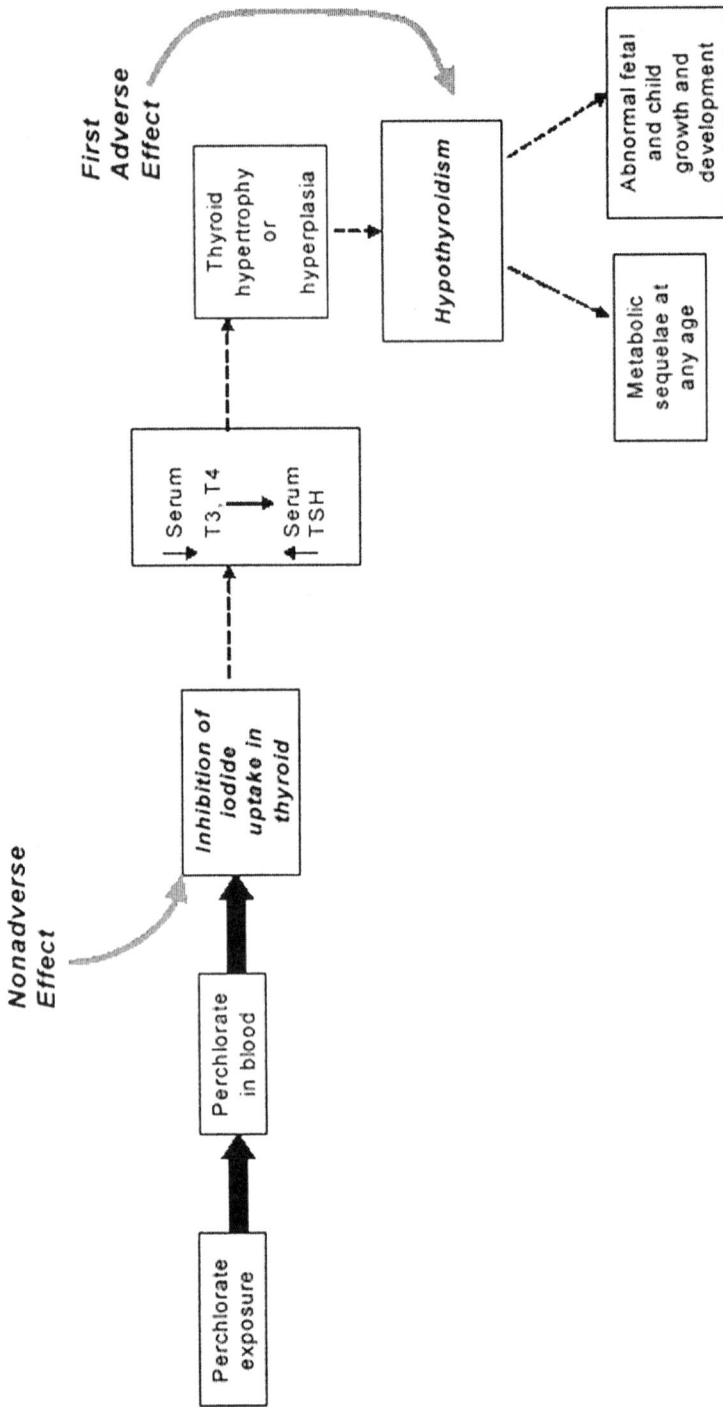

Figure 1. NRC Committee's suggested mode-of-action model for perchlorate toxicity in humans indicating first adverse effect in the continuum. Adopted from NRC.[48]

effect(s), or its known precursor, that occurs as dose rate or exposure level increases. In the determination of critical effect, it is crucial that distinctions be drawn between adverse effects and adaptive effects. An adaptive effect enhances an organism's performance as a whole and/or its ability to withstand a challenge; an adverse effect is a biochemical change, functional impairment, or pathological lesion that impairs performance and reduces the ability of an organism to respond to additional challenge.[49,52] Available animal studies as described above clearly suggest that the thyroid is the primary target organ for perchlorate. Thus, distinguishing adaptive from adverse effects in the thyroid and determining the most appropriate adverse effect on which to base an RfD was the first step in the perchlorate risk assessment.

Inhibition of iodine uptake in thyroid, the key event in the ultimate disruption of thyroid function, can be considered as a marker of the biologically effective dose for perchlorate. However, inhibition of iodine uptake, itself, cannot be considered an adverse effect because in humans we do not yet know what levels of iodine uptake inhibition would decrease T_4 levels. Alteration of hormone levels, including decrease of serum T_4 and T_3 with a corresponding increase of TSH, is considered to be the early biological effect of exposure to perchlorate. The human body has a large reserve capacity of circulating thyroid hormone; serum levels of T_4 and T_3 are highly variable. No clear-cut information is available on how much decrement of circulating serum T_4 can be tolerated without resulting in permanent alteration of thyroid function. However, subclinical hypothyroidism is generally considered to be present when circulating TSH levels are elevated by 2-fold, with, or without decreased levels of T_4.[54]

Based on a mode-of-action analysis, it is clear that altered hormone levels are an early biological effect of perchlorate exposure. If allowed to persist, increased TSH levels, at least in rodents, will eventually lead to thyroid hyperplasia and possible thyroid tumors. Even if this pathway is not relevant to humans, persistent decreases in T_4 levels increase the potential for neurodevelopmental deficits in children. In this case, decreased T_4 can be considered to be a precursor to an adverse effect, rather than an adverse effect in itself, however, because changes in T_4 are routinely compensated by normal, and well understood, homeostatic processes. Therefore, decreases in serum T_4 in the pregnant population should be considered to be the critical effect most relevant to human health. The NRC committee[48] also reached the same conclusion and considered that the first adverse effect in the model-of-action model is hypothyroidism.

Choice of appropriate species and study

US EPA's policy when developing RfDs in many of its program, regional offices, and Office of Research and Development (ORD) has been to use human data first and foremost in the determination of critical effect and choice of uncertainty factors. The preference for use of human data is found in many EPA publications, risk positions, risk methods documents, and practice.[49,51] In general, using human data as the basis for developing an RfD will reduce the uncertainty inherent in extrapolating from rat data.

The available data on the effects of perchlorate in experimental animals consistently points to thyroid disturbance as the sentinel effect. This disturbance may lead to subsequent thyroid and neurological damage. This information in experimental animals is consistent with the available, but more limited, human data. However, these data also demonstrate that rats may respond to perchlorate exposures in a very different manner than humans.[55] The reason that such comparisons are not definitive is that the human data do not include information on pregnant individuals.

Although the rat data set includes the sensitive subgroup (the pregnant animal and its fetus), whereas the human studies only include measurements of TSH and T_4 in adults, infants and children (and not pregnant individuals), rats are known to be more sensitive than humans to thyroid hormone replacement therapy, needing 10 times more T_4 than humans to achieve a euthyroid condition.[56] Because the overall uncertainty in determining an RfD is greater from the rat data, when compared with the human data, the human data are the appropriate choice for determining an RfD. The NRC[48] also concluded that the human data provided a more reliable point of departure for the risk assessment than the animal data.

Of the available human studies, one clinical study[10] and one epidemiology study[45] were considered to yield sufficient information to determine an RfD. Greer study was a well-conducted study underwent a rigorous quality assurance audit and conforms to the "Common Rule" the Federal Agency Guidelines on the ethical conduct of human studies.[57] This study observed no statistically significant effects in serum T_4, T_3, or TSH even at the highest dose tested, and defines a NOAEL of 0.5 mg/kg-day for the healthy adult human population for short-term exposure. The Crump et al.[45] study found no evidence that perchlorate in drinking water at concentrations as high as 120 ug/L suppressed thyroid function in newborns or school-aged children. Crump et al.[45] also studied newborns screened for hypothyroidism by a heel-stick blood sample between February 1996 and January 1999 in the same three Chilean cities. TSH levels were significantly lower in Taltal (high exposure city) than in the other two cities, a trend opposite to that hypothesized. The authors concluded that the differences did not appear clinically significant.

One issue to address in the use of Crump et al.[45] study as a basis of an RfD is the apparent iodine excess when compared with other populations, such as the U.S. A 1 to 2.5-fold excess in urinary iodine seen in the Chilean school children may serve to protect this population from perchlorate exposure. However, a second study with pregnant women in the same cities of Chile with lower iodine supplementation had the same results as the Crump et al.[45] study.[46]

Point-of-departure analysis

Following accepted risk assessment approaches, a point-of-departure analysis establishes the threshold dose that serves as the starting point for developing the RfD. Traditionally, the point of departure for an RfD has been the No Observed Adverse Effect Level (NOAEL), which is the highest experimental dose that is without adverse effect. More recently, risk assessors have attempted to incorporate more of the data about the dose response curve by using benchmark dose (BMD) modeling. BMD modeling uses quantitative dose response models to estimate the dose that results in a specified change (such as 10%) in the critical effect, or its precursor.

No human study involved exposures high enough to cause a decrease in T_4; therefore, all of the human studies can be said to have identified "freestanding NOAELs" for the critical effect. The highest NOAEL identified in the body of human studies is approximately 0.5 mg/kg-day. This dose was achieved in workers exposed for an average of 8 years[3,44] and in healthy adults exposed for 14 days in a clinical study.[10] The lowest NOAEL observed in a human study[45] is an estimated NOAEL of 0.006 mg/kg-day (actual exposure is an average of 0.112 mg/L) measured in school-age children who had been exposed in utero and for their entire lifetime (about 7 years). Because, these children were exposed in utero and as neonates, the NOAEL from this study is a freestanding NOAEL in a sensitive population. Therefore, a NOAEL of 0.5 mg/kg-day could be considered as a point-of-departure for the general human population, while 0.006 mg/kg-day could be a point-of-departure for a sensitive human population.

However, use of a freestanding NOAEL does incorporate some uncertainty into the risk assessment because the true threshold for the critical effect has not been identified. In other words, the true threshold, or true NOAEL, is likely to be higher than the NOAEL used as the point-of-departure.

Therefore, using this freestanding NOAEL would result in a more conservative risk value.

The Greer et al.[10] study adequately characterizes the dose-response curve for inhibition of iodine uptake in humans. This effect of perchlorate is a key event of the mode of action because it is the essential step in the cascade leading to adverse effects. Without inhibition of iodine uptake, there will be no alteration of T_4 or TSH or subsequent adverse effects on neurological development and thyroid hyperplasia. Therefore, a point-of-departure based on inhibition of iodine uptake is a health-protective surrogate that can be used to replace a freestanding NOAEL for decreased T_4. The lowest dose evaluated by Greer et al.,[10] 0.007 mg/kg-day, did not cause a statistically significant inhibition of iodine uptake. Based on a regression analysis taking into account the variability of the experimental population, the authors predicted that the dose that would result in 0% inhibition of iodine uptake is 0.0064 mg/kg-day; the 95% upper confidence limit on iodine uptake inhibition at this dose is 8.3%. Greer et al.[10] concluded that an iodine uptake inhibition less than 10% would not be biologically significant. This threshold of 0.006 mg/kg-day, as well as the lowest dose evaluated by Greer et al.,[10] 0.007 mg/kg-day, are reasonable point-of-departures for estimating an RfD.[48,55]

Currently, insufficient data exist to adequately define the level of iodine uptake inhibition in humans that can be tolerated for a lifetime without altering serum T_4 and TSH levels. Greer et al.[10] demonstrated that for 14-day exposure, inhibition of iodine uptake up to about 70%, has no effect on serum T_4 or TSH. Occupational studies[3,44] demonstrated that workers exposed to perchlorate for several years demonstrated no altered T_4 or TSH serum levels. When the serum hormone levels from these studies are plotted against serum perchlorate AUC predicted by the human PBPK model, it can be seen that chronic exposure in workers had no effect on serum T_4 or TSH at serum AUC values that resulted in approximately 50% I uptake inhibition.[55] Thus, it might be reasonable to conclude that an appropriate benchmark response would be the perchlorate dose that resulted in a 50% inhibition of iodine uptake. Nonetheless, benchmark response levels of 10% inhibition of iodine uptake was modeled in order to be public health protective and take into account the uncertainties involved in extrapolating data from healthy adults to potential sensitive populations such as iodine deficient people, pregnant women, and neonates.

Based on BMD modeling results from Strawson et al.,[55] the perchlorate dose that caused a 10% inhibition of iodine uptake is 0.01 mg/kg-day (BMD10); the its 95% lower limit (BMDL10) estimate ranges from 0.004 mg/kg-day (Hill model) to 0.008 mg/kg-day (Power model). These results are consistent with the conclusions of Greer *et al.*,[10] which indicated that the no effect level for iodine inhibition ranges from 0.006 (predicted) to 0.007 (measured) mg/kg-day.

Therefore, for the purpose of developing a perchlorate RfD, four different points-of-departure could have been used: a freestanding NOAEL of 0.5 mg/kg-day for the general, healthy population, a freestanding NOAEL of 0.006 mg/kg-day for a sensitive subpopulation; the lowest dose evaluated by Greer *et al.*,[10] 0.007 mg/kg-day; and a threshold for iodine uptake inhibition of 0.006 mg/kg-day used as a health-protective surrogate for the freestanding NOAELs.

Choice of uncertainty factors

The judgment of the appropriate uncertainty factor is based on a review of the information supporting the choice of critical effect, and issues associated with extrapolation from experimental animals to humans and to sensitive humans. The noncancer risk assessment by U.S. EPA[51] incorporates five different uncertainty factors to address issues of variability and uncertainty. Interspecies and intraspecies factors are used to address the uncertainty between experimental animals and humans, and the variability within different human populations. Three other factors (Subchronic, LOAEL, Database) are used to address lack of information. Typically, the maximum total uncertainty factor that U.S. EPA will apply is 3000. If all five areas of uncertainty/variability are present warranting a total UF of 10,000, then U.S. EPA[51] generally concludes that the uncertainty is too great to develop an RfD.

1. Interspecies Variability (UF$_A$)

This factor accounts for the differences that occur between animals and humans and is also thought to be composed of subfactors for toxicokinetics (how the body distributes and metabolizes the chemical) and toxicodynamics (how the body responds to the chemical). If no information is available on the quantitative differences between animals and humans, then a default value of 10 is used. If information is available on one of the two subcomponents, then this information (chemical specific adjustment factor, CSAF) is used along with a default value of 3 for the remaining subfactor.[58] If data are available to adequately describe variability in both subfactors, then actual data may be used to replace default values. If an RfD is based on human data, then a value of 1 is appropriate for this factor.

The body of data in experimental animals demonstrates that the rodent response to perchlorate is dramatically different than the human response. In rats, doses that cause only about 10% iodine uptake inhibition cause variable, but statistically significant changes in hormone levels. While in humans, doses that cause 70% iodine uptake inhibition have no effect on hormone levels. Basing the RfD on animal data will introduce greater uncertainty to the RfD than use of human data. Therefore, human data is the best basis for the RfD. If the proposed points-of-departure which are obtained from human studies are used, a factor of 1 is appropriate for this area of uncertainty. Otherwise, an uncertainty factor as high as 10 should be applied when animal data are used as the point-of-departure.

2. Intraspecies Variability (UF$_H$)

This factor accounts for the natural differences that occur between human subpopulations and for the fact that some individuals may be more sensitive than the average population. Similar to the interspecies uncertainty factor, this factor is also composed of two subfactors – one to account for toxicokinetic differences and one to account for toxicodynamic differences. If no information is available on human variability, then a default value of 10 is used. However, if adequate information is available on one or both of the two subcomponents, then this information is used along with a default value of 3 for the remaining subfactor.[58] In addition, if an RfD is based on human data gathered in the known sensitive subpopulation, a value of less than 10, perhaps even 1, may be chosen for this factor.

The judgment of appropriate intraspecies uncertainty factor depends in part on the choice of study as the basis of the RfD. For perchlorate exposure, the available data are insufficient to develop a CSAF for human variability at this time. When the RfD is based on the freestanding NOAEL of 0.5 mg/kg-day identified in the healthy adult population,[10] a full factor of 10 is appropriate to use because this NOAEL does not account for the fact that a NOAEL in sensitive subgroups (i.e., children or pregnant mothers with their fetuses) could be lower. In contrast, a lower factor is appropriate for the freestanding NOAEL of 0.006 mg/kg-day identified in children[45] because this study covers at least one of the sensitive subpopulation, children. In the Crump et al.[45] study, the presence of perchlorate in the water has been a long-term problem. The mothers of the children evaluated were exposed before pregnancy, so that if perchlorate were affecting thyroid function in these women, they would already be hypothyroid at the start of pregnancy. The children themselves were exposed as fetuses in the uteri, as neonates, and throughout their lifetimes. Therefore several of the life stages that are considered sensitive have been studied in the Crump et al.[10] study. Therefore, the observation of a freestanding NOAEL in this study gives greater confidence that fetuses, neonates, and children will be protected by an RfD based on this point-of-departure. However, an uncertainty factor of 3, rather than 1, is appropriate to use with this point-of-departure because there are no data to suggest how the other sensitive subpopulation, e.g., pregnant women, may respond.

When iodine uptake inhibition is used as a point-of-departure, the use of an uncertainty factor should be based on data addressing the variability of iodine uptake inhibition in human population. Mattie *et al.*[59] have used physiologically based pharmacokinetic models for both rats and humans to predict perchlorate doses that will result in a 5% iodine uptake inhibition in different life stages. In rats, the predicted doses that result in a 5% inhibition are 0.03 mg/kg-day, 0.05 mg/kg-day, and 0.13 mg/kg-day for male rats, pregnant rats, and lactating rats, respectively. In humans, the predicted doses that result in a 5% inhibition are 0.01 mg/kg-day, 0.025 mg/kg-day, and 0.061 mg/kg-day for healthy adult males and females, pregnant women, and lactating women, respectively. This analysis suggests that pregnant women are not more sensitive to iodine uptake inhibition than healthy adults. In addition, it confirms that the physiology of pregnancy serves to conserve iodine uptake, making pregnant women less sensitive to iodine uptake inhibition than non-pregnant adults. Thus, Strawson *et al.*[55] concluded that no extra uncertainty factor is necessary when iodine uptake inhibition data from the human study is used as a point-of-departure to derive an RfD. However, the NRC[10] recommended use of a full factor of 10 to protect the most sensitive population, the fetuses of pregnant women who might have hypothyroidism or iodide deficiency. The NRC committee viewed its recommendation as conservative and health-protective because the point of departure is based on a non-adverse effect that precedes the adverse effect in the continuum of possible effects of perchlorate exposure (see Figure 1).

Therefore, the appropriate choice for intraspecies uncertainty factor is either 10-fold with the use of the Greer *et al.*[10] NOAEL for T_4 decrease in adults, 3-fold with the use of the Crump *et al.*[45] NOAEL for T_4 decrease in children, or 10 fold with the use of the Greer *et al.*[10] threshold for iodine uptake inhibition.

3. Subchronic to Chronic Extrapolation (UF_S)

Because the RfD protects for a lifetime exposure, this factor is applied when the database lacks information on the health effects of the chemical following a chronic exposure. If the database contains no information on chronic exposure, a default value of 10 is often applied, unless other data suggest a lack of progression with exposure duration. If the database contains adequate chronic bioassays, then a value of 1 is generally appropriate. If there are data addressing only one of the two issues, then a default of 3 may be applied. Thus, the need for a duration UF for perchlorate depends on whether more sensitive effects are expected after increasing duration of exposure, or whether longer durations of exposure increase the severity or decrease the point of departure for perchlorate's critical effect.

While there are no studies that cover a full lifetime in either animals or humans for the thyroid effects of concern, there are studies that evaluate longer exposures in humans and studies that demonstrate no increase in the severity of effects with increasing duration in animals. In Gibbs *et al.* (1998) study, workers' tenure ranged from 1 to 27 years, with an average of 8 years. In Lamm *et al.*[44] study, 40% of the workers had a tenure greater than 5 years. In Crump *et al.*[45] study, children age 6-8 years who had been exposed their entire lives were evaluated.   In all three of these studies parameters investigated include general physical exam, tests of kidney and liver function, and blood counts, as well as tests of thyroid function. No effects on any of these parameters were observed in the exposed populations in these studies.   When compared to the results of the 14-day clinical studies in humans,[10,42,43] these longer-term studies show that increasing duration of exposure in humans does not increase the incidence or severity of thyroid effects, nor does it induce effects in other target organs that were not identified by the short-term studies.   Although only preliminary data are available, a study in which volunteers were given 0.007 and 0.04 mg/kg-day perchlorate for 6 months resulted in no effects on thyroid function and further supports these conclusions.[41]

The available animal studies also support the conclusion that increasing exposure duration does not result in an increase in incidence or severity of thyroid effects nor does it reveal non-thyroid effects that are not detected by shorter-term studies. Several studies have evaluated perchlorate after either 14 days[15,16,17,60] or 90 days.[16,17,60] These studies have evaluated systemic and immunotoxic effects in addition to thyroid effects. None of these studies observed any non-thyroid effects after either 14 or 90 days of exposure, suggesting that increased exposure duration will not result in systemic effects that occur at lower doses than thyroid effects. Although the thyroid response is variable, particularly the hormone changes, these studies also show that animals exposed for 90 days do not show a clear pattern of more severe hormone changes nor an accelerated progression of thyroid pathology to hyperplasia compared with animals exposed for 14 days.

Strawson *et al.*[55] also reviewed duration affect on inhibition of iodine uptake by perchlorate in animals and humans. Rats up-regulate iodine uptake very quickly during 14-day treatment and that inhibition actually decreases with time.[9] In contrast, humans do not up-regulate iodine uptake within the same time period.[10] However, these data do show that iodine uptake inhibition does not increase with increasing duration in either rats or humans.

Animals or humans exposed to perchlorate for a prolong duration may develop other effects which didn't occur in short-tem studies, e.g., thyroid adenomas in F1 generation male rats of a two-generation study.[27] However, it is known that thyroid tumors in rats are ultimately caused by constant stimulation of the thyroid by TSH. It is also known that perchlorate at 30 mg/kg-day caused dramatic increases in TSH in these animals. Thus, it is not necessarily surprising that tumors were evoked at such a high level of exposure. Nevertheless, the development of thyroid tumors in rats is not a duration effect per se, but rather a threshold phenomenon. If perchlorate doses stay below a level that induce increased TSH levels, then the production of thyroid tumors is not possible according to the proposed mode of action (Figure 2).[21,47] Increased duration of perchlorate at doses that are below this threshold will not increase the risk of thyroid tumor formation. In addition, while the development of thyroid tumors in rats can be considered to be qualitatively relevant to humans, there are questions about whether humans do, in fact, develop thyroid tumors by the same mechanism.

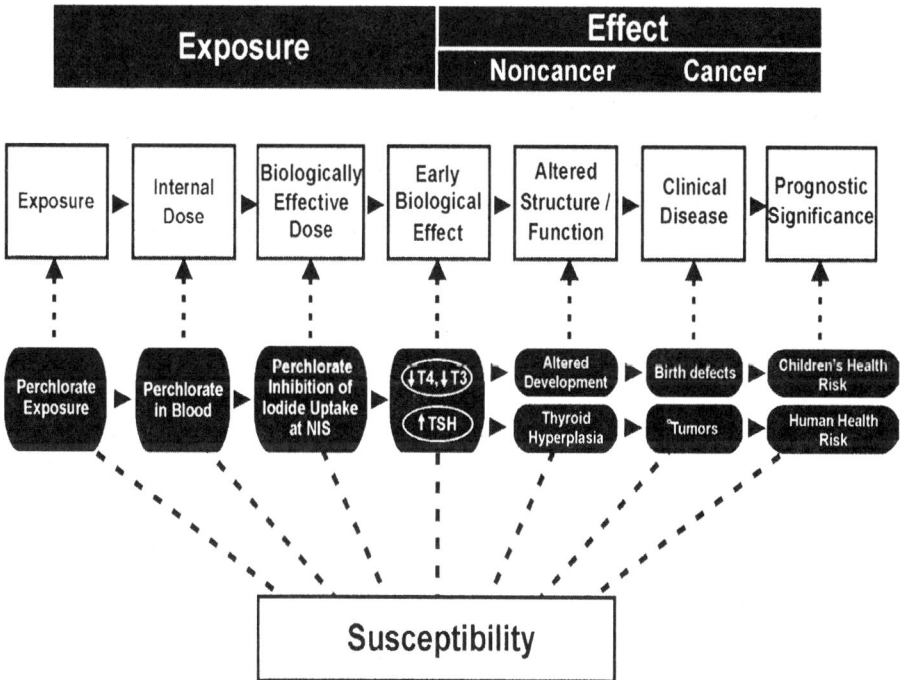

Figure 2. Mode of action model for perchlorate toxicity proposed by U.S. EPA (2003b). Perchlorate interferes with the sodium (NA+)-iodide (I-) symporter (NIS) present in various tissues, particularly thyroid. The model shows the exposure-dose response continuum considered in the context of biomarkers (classified as measures of exposure, effect, and susceptibility) and level of organization at which toxicity is observed (adapted directly from U.S. EPA.[21]

Based these considerations, a value of 1 appears to be appropriate to address this area of uncertainty. Both the human and animal studies demonstrate that increasing exposure duration does not result in the appearance of non-thyroid effects at doses lower than the thyroid hormone effects. Thyroid effects in humans and rodents do not increase in incidence or severity with increasing exposure duration. Inhibition of iodine uptake does not increase in humans or rats with increasing exposure duration. The NRC[48] agreed that this uncertainty factor should be 1.

4. LOAEL to NOAEL Extrapolation (UF$_L$)

Because the RfD is considered to be a subthreshold value that protects against any adverse health effects, this factor is applied when the database lacks information to identify a NOAEL. If the database does not identify a NOAEL, then a default of 10 is used for this factor. If a NOAEL is used, a value of 1 is appropriate. Often, if the database does not identify a NOAEL, but the adverse effects observed are of minimal severity, then a default of 3 will be considered appropriate for use of a "minimal LOAEL".

Both the Greer et al.[10] and the Crump et al.[45] studies identified freestanding NOAELs for the critical effect of decreased T$_4$. When either of these NOAELs is used as the point-of-departure for the development of an RfD, an uncertainty factor of 1 for this area would be appropriate. A point-of-departure at the threshold for iodine uptake inhibition (the lowest dose in Greer et al.)[10] is, likewise, not considered to be a LOAEL, because inhibition of iodine uptake is a key event in the mode of action rather than an adverse effect.[48] In addition, this point of departure represents a dose at which no inhibition of iodine uptake occurs, so that adverse effects cannot occur following exposure to this dose. This conclusion is confirmed by the body of human data, which demonstrate that no effect on serum hormone levels has been observed at doses equal to or higher than this point of departure. Therefore, this point-of-departure should be considered as a NOAEL surrogate, rather than a LOAEL surrogate, and the appropriate value for this factor is 1. The NRC[10] agreed that this uncertainty factor should be 1.

5. Database (UF$_D$)

Based on US EPA's risk assessment methodology, the database for deriving a high confidence RfD includes at a minimum two chronic bioassays by the appropriate route of exposure in different species, one two-generation reproductive toxicity study, and two developmental toxicity studies in different species. The minimal database required for deriving an RfD is a single subchronic bioassay, that includes a full histopathology examination. The database factor is used to account for the fact that a potential health effect may not be identified if the database is missing a particular type of study. This factor may also be used if the existing data indicate the potential for a heath effect that is not fully characterized by the standard bioassays, for example neurotoxicity or immunotoxicity. If the database is complete, a value of 1 is appropriate. If only the minimal database is available, then a default of 10 is used. A value of 3 may be used if the database is missing one or two key studies.

The database for perchlorate includes an large number of experimental animal studies, including chronic (but older) studies that show tumors at high doses,[61] numerous shorter-term bioassays that unequivocally demonstrate that thyroid disturbance occurs at lower doses than other systemic, immunotoxic, genotoxic, or other effects, developmental toxicity studies in two species, a 2 generation reproduction study that also monitored systemic effects in young rats, a developmental neurotoxicity study, a specialized developmental toxicity study to monitor hormone changes in early life and during late pregnancy and lactation, and a specialized neurobehavioral study to confirm earlier findings. The database also includes human clinical, experimental, epidemiology, and occupational studies.

All of this information demonstrates that the thyroid is the most sensitive organ system. In humans, the threshold for iodine uptake inhibition is well characterized and additional studies are not likely to provide different information that would change the risk assessment. In humans, the perchlorate dose that causes a decrease in T_4, the critical effect, is not well characterized since no human population has been exposed to a dose high enough to alter hormone levels. However, if these studies could be performed, their effect would likely be to raise the NOAEL. The mode of action analysis suggests a potential for adverse effects as a result of serum T_4 levels that are consistently depressed by at least 60%. The doses that cause this degree of T_4 decrease are not well characterized in either humans or animals. However, by selecting a point-of-departure that is below the threshold for any T_4 change, subsequent effects will not develop. Therefore, the overall perchlorate database is complete, and any new studies that are conducted to fine tune our knowledge of the perchlorate mode of action will not identify lower points-of-departure than can be estimated from the existing database. Based on these considerations, the appropriate value for this factor is 1.

In summary, when human data are used as the point-of-departure, the only area of uncertainty for a perchlorate RfD that needs to be addressed by the use of uncertainty factors is human variability and the difference in response between pregnant women and the groups for which data are available. A factor of 1 is appropriate to address all other areas of uncertainty. The NRC committee[48] recommended the use of a composite uncertainty factor of 10 to protect the fetuses of pregnant women who might have hypothyroidism or iodide deficiency.

Developing the RfD

As shown by extensive animal studies, the critical effect of perchlorate is T_4 serum decrease. Pregnant rats are demonstrated to be the most sensitive subgroup, likely followed by the young rat. Several human studies exist that monitored for this critical effect. These human studies do not include pregnant women, but they do include children. In addition, comparative data between the experimental animal and human indicate that humans are not more sensitive than the experimental animal species tested to T_4 serum decrease by perchlorate; in fact based on toxicodynamics parameters they are much less sensitive.[56] The most relevant data for developing the RfD for perchlorate exposures comes from human epidemiology and clinical studies, supplemented with available and extensive information on experimental animals. All these data support the use of the human data for development of an RfD.[48]

The NRC Committee report,[48] "Health Implications of Perchlorate Ingestion," recommended using the data from the Greer study for deriving a reference dose (RfD) for perchlorate. Animal and clinical studies conducted since 1997 were all designed based on the mechanism of action for perchlorate, inhibition of iodide uptake. This mechanism of action has been a point of agreement between toxicologists and risk assessors throughout the process to develop the perchlorate risk assessment. There remains disagreement on what is the first adverse or critical effect after perchlorate exposure. The NRC committee avoided all disagreements by using the NOEL for inhibition of iodide uptake from the Greer study as the point of departure when they recommended an RfD for perchlorate. According to the NRC report,[48] there is an added level of conservatism or protection, since inhibition of iodide uptake is the first step in a multi-step process prior to causing an adverse effect. The NRC committee[48] recommended use of a composite uncertainty factor of 10 to protect the fetuses of pregnant women who might have hypothyroidism or iodide deficiency. The US Environmental Protection Agency (USEPA) agreed and developed their RfD based on the NRC recommendation (0.007 mg/kg-day point-of-departure x UF of 10). The USEPA listed the oral reference dose (RfD) for perchlorate as 0.0007 mg/kg-day (24.5 ppb drinking water equivalent) in their Integrated Risk Information System (IRIS).

SUMMARY AND IMPLICATIONS

There are currently many unanswered questions surrounding the prevalence of perchlorate in the environment, the possible exposures and effects, and the risk management strategies available to manage perchlorate contamination. Further information must be obtained in each of these areas in order to develop successful and appropriate strategies for handing perchlorate contamination now and in the future. Because of the uncertainty involved in evaluating toxicological data, the states and federal agencies using this information may use and interpret it somewhat differently to promulgate different standards and advisory levels. It is important to note that the status of the knowledge of perchlorate exposure and risks continues to evolve and it will be essential to periodically re-evaluate what is known about this chemical. However, the reference dose recommended by National Research Council and adopted by the U.S. Environmental Protection Agency will protect the health of even the most sensitive groups of people over a lifetime of exposure.

ACKNOWLEDGMENTS

The author wishes to acknowledge all of the researchers who contributed to the perchlorate database and the Department of Defense, Perchlorate Study Group and NASA for funding perchlorate research. Special thanks go to Teresa R. Sterner for assistance with the preparation of this chapter.

REFERENCES

1. Motzer, W.E. Perchlorate: Problems, detection, and solutions. Environ. Forensics. 2001; 2:301-311.

2. Urbansky, E.T. Perchlorate chemistry: implications for analysis and remediation. Bioremed. J. 1998; 2:81-95.

3. Gibbs, J.P., Ahmad, R., Crump, K.S., Houck, D.P., Leveille, T.S., Findley, J.E., Francis, M. Evaluation of a population with occupational exposure to airborne ammonium perchlorate for possible acute or chronic effects on thyroid function. J Occup. Environ. Med. 1998; 40:1072-1082.

4. Braverman, L.E., He, X., Pino, S., Cross, M., Magnani, B., Lamm, S.H., Kruse, M.B., Engel, A., Crump, K.S., and Gibbs, J.P. The effect of perchlorate, thiocyanate, and nitrate on thyroid function in workers exposed to perchlorate long-term. The Journal of Clinical Endocrinology & Metabolism. 2005; 90(2):700-706.

5. Greenspan, F.S. The role of fine-needle aspiration biopsy in the management of palpable thyroid nodules. Am. J. Clin. Pathol. 1997; 108:S26-30.

6. Scanlon, M.F, Issa, B.G., Dieguez, C. Regulation of growth hormone secretion. Horm. Res. 1996; 46(4-5):149-54.

7. Capen, C.C. Mechanistic data and risk assessment of selected toxic end points of the thyroid gland. Toxicol. Pathol. 1997; 25:39-48.

8. Wolff, J. Perchlorate and the thyroid gland. Pharmacolog. Rev. 1998; 50:89-105.

9. Yu, K.O., Narayanan, L., Mattie, D.R., Godfrey, R.J., Todd, P.N., Sterner, T.R., Mahle, D.A., Lumpkin, M.H., Fisher, J.W. 2002. The pharmacokinetics of perchlorate and its effect on the hypothalamus- pituitary-thyroid axis in the male rat. Toxicol. Appl. Pharmacol. 2002; 182:148-159.

10. Greer, M.A., Goodman, G., Pleus, R.C., Greer, S.E. Health effects assessment for environmental perchlorate contamination: The dose response for inhibition of thyroidal radioiodine uptake in humans. Environ. Health Perspect. 2002; 110:927-937.

11. Merrill, E.A., Clewell, R.A., Gearhart, J.M., Robinson, P.J., Sterner, T.R., Yu, K.O., Mattie, D.R., Fisher, J.W. PBPK predictions of perchlorate distribution and its effect on thyroid uptake of radioiodide in the male rat. Toxicol. Sci. 2003; 73:256-269.

12. Merrill, E.A., Clewell, R.A., Robinson, P.J., Jarabek, A.M., Gearhart, J.M., Sterner, T.R., Fisher, J.W. PBPK model for radioactive iodide and perchlorate kinetics and perchlorate-induced inhibition of iodide uptake in humans. Toxicol. Sci. 2005; 83:25-43.

13. Clewell, R.A., Merrill, E.A., Yu, K.O., Mahle, D.A., Sterner, T.R., Mattie, D.R., Robinson, P.J., Fisher, J.W., Gearhart, J.M. Predicting fetal perchlorate dose and inhibition of iodide kinetics during gestation: A physiologically-based pharmacokinetic analysis of perchlorate and iodide kinetics in the rat. Toxicol. Sci. 2003a; 73:235-255.

14. Clewell, R.A., Merrill, E.A., Yu, K.O., Mahle, D.A., Sterner, T.R., Fisher, J.W., Gearhart, J.M. Predicting neonatal perchlorate dose and inhibition of iodide uptake in the rat during lactation using physiologically-based pharmacokinetic modeling. Toxicol. Sci. 2003b; 74:416-436.

15. Caldwell, D.J., King, J.H., Kinkead, E.R., Wolfe, R.E., Narayanan, L., Mattie, D.R. Results of a fourteen day oral-dosing toxicity study of ammonium perchlorate. In: 1995 JANNAF Safety and Environmental Protection Subcommittee Meeting: Volume 1, Tampa, FL. Joint Army, Navy, NASA, Air Force (JANNAF) Interagency Propulsion Committee Publication 634. Columbia, MD: Chemical Propulsion Information Agency. 1995.

16. Siglin, J.C., Mattie, D.R., Dodd, D.E., Hildebrandt, P.K., Baker, W.H. A 90-Day drinking water toxicity study in rats of the environmental contaminant ammonium perchlorate. Toxicol. Sci. 2000; 57:61-74.

17. Keil, D., Warren, D.A., Jenny, M., EuDaly, J., Dillard, R. Effects of ammonium perchlorate on immunotoxicological, hematological, and thyroid parameters in B6C3F1 female mice. Medical University of South Carolina, Charleston, SC. Final report of contract DSWA01-97-0008. 1999

18. Narayanan, L. Consultative letter, AFRL-HE-WP-CL-2000-0034. Thyroid hormone and TSH co-laboratory study report [memorandum with attachments to Annie Jarabek]. Wright-Patterson Air Force Base, Dayton, OH: Air Force Research Laboratory. June 15, 2000.

19. York, R.G., Lewis, E., Brown, W.R., Girard, M.F., Mattie, D.R., Funk, K.A. and Strawson, J.S. 2005a. Refining the Effects Observed in a Developmental Neurobehavioral Study of Ammonium Perchlorate Administered Orally in Drinking Water to Rats. I - Thyroid and Reproductive Effects (In Press).

20. York, R.G., Barnett, Jr, J., Girard, M.F., Mattie, D.R., Bekkedal, M., Garman, R.H. and Strawson, J.S. 2005b. Refining the Effects Observed in a Developmental Neurobehavioral Study of Ammonium Perchlorate Administered Orally in Drinking Water to Rats. II - Behavioral and Neurodevelopment Effects (In Press).

21. U.S. Environmental Protection Agency. Disposition of Comments and Recommendations for Revisions to "Perchlorate Environmental Contamination: Toxicological Review and Risk Characterization External Review Draft (January 16, 2002)." National Center for Environmental Assessment, Washington, DC. October 27, 2003a.

22. York, R.G., Brown, W.R., Girard, M.F., Dollarhide, J.S. Oral (drinking water) developmental toxicity study of ammonium perchlorate in New Zealand white rabbits. Int. J. Toxicol. 2001a; 20:199-205.

23. York, R.G., Funk, K.A., Girard, M.F., Mattie, D., Strawson, J.E. Oral (drinking water) developmental toxicity study of ammonium perchlorate in Sprague-Dawley rats. Int. J. Toxicol. 2003; 22:453-464.

24. York, R.G., Barnett, J., Brown, W.R., Garman, R.H., Mattie, D.R., Dodd, D. A rat neurodevelopmental evaluation of offspring, including evaluation of adult and neonatal thyroid, from mothers treated with ammonium perchlorate in drinking water. Int. J. Toxicol. 2004; 23:191-214.

25. Bekkedal, M.Y.V., Carpenter, T., Smith, J., Ademujohn, C., Maken, D., Mattie, D.R. A neurodevelopmental study of the effects of oral ammonium perchlorate exposure on the motor activity of pre-weanling rat pups. Naval Health Research Center Detachment (Toxicology), Wright-Patterson AFB, OH. TOXDET-00-03. 2000.

26. Bekkedal, M.Y.V., Arffsten, D., Mattie, D. An evaluation of neurobehavioral tests used to assess the neurodevelopmental effects of early ammonium perchlorate exposure. J. Toxicol. Environ. Health A. 2004; 67:835-844.

27. York, R.G., Brown, W.R., Girard, M.F., Dollarhide, J.S. Two-generation reproduction study of ammonium perchlorate in drinking water in rats evaluates thyroid toxicity. Int. J. Toxicol. 2001b; 20:183-197.

28. Soldin, O.P., Braverman, L.E., Lamm, S.H. Perchlorate clinical pharmacology and human health: a review. Ther. Drug Monit. 2001; 23:316-331.

29. Lamm, S.H., and M. Doemland. Has perchlorate in drinking water increased the rate of congenital hypothyroidism? J. Occup. Environ. Med. 1999; 41(5):409-411.

30. Li, Z., F.X. Li, D. Byrd, G.M. Deyhle, D.E. Sesser, M.R. Skeels, and S.H. Lamm. Neonatal
    thyroxine level and perchlorate in drinking water. J. Occup. Environ. Med. 2000; 42(2):200-205.

31. Li, F.X., D.M. Byrd , G.M. Deyhle, D.E. Sesser, M.R.Skeels, S.R.Katkowsky, and S.H. Lamm. Neonatal Thyroid-Stimulating Hormone Level and Perchlorate in Drinking Water. Teratology. 2000; 62(6):429-431.

32. Li, F.X., L. Squartsoff, and S.H. Lamm. Prevalence of thyroid diseases in Nevada counties with respect to perchlorate in drinking water. J. Occup. Environ. Med. 2001; 43(7):630-634.

33. Chang, S., C. Crothers, S. Lai, and S. Lamm. Pediatric neurobehavioral diseases in Nevada counties with respect to perchlorate in drinking water: An ecological inquiry. Birth Defects Res. Part A Clin. Mol. Teratol. 2003; 67(10):886-892.

34. Brechner, R.J., Parkhurst, G.D., Humble, W.O., Brown, M.B., Herman, W.H. Ammonium perchlorate contamination of Colorado River drinking water is associated with abnormal thyroid function in newborns in Arizona. J. Occup. Environ. Med. 2000; 42:777-782.

35. Schwartz, J. Gestational Exposure to Perchlorate is Associated with Measures of Decreased Thyroid Function in a Population of California Neonates. M.S. Thesis, University of California, Berkeley. 2001.

36. Kelsh, M.A., P.A. Buffler, J.J. Daaboul, G.W. Rutherford, E.C. Lau, J.C. Cahill, A.K. Exuzides, A.K. Madl, L.G. Palmer, and F.W. Lorey. Primary congenital hypothyroidism, newborn thyroid function, and environmental perchlorate exposure among residents of a southern California community. J. Occup. Environ. Med. 2003; 45(10):1116-1127.

37. Lamm, S. H. Perchlorate exposure does not explain differences in neonatal thyroid function
    between Yuma and Flagstaff. [Letter]. J. Occup. Environ. Med. 2003; 45(11):1131-1132.

38. Buffler, P.A., M.A. Kelsh, E.C. Lau, C.H. Edinboro, and J.C. Barnard. 2004. Epidemiologic Studies of Primary Congenital Hypothyroidism and Newborn Thyroid Function Among California Residents, Final Report. April 2004, Berkeley, CA.

39. Morgan, J.W., and R.E. Cassady. 2002. Community cancer assessment in response to long-time exposure to perchlorate and trichloroethylene in drinking water. J. Occup. Environ. Med. 44(7):616-621.

40. Brabant, G., Bergmann, P., Kirsch, C.M., Kohrle, J., Hesch, R.D., von zur Muhlen, A. Early adaptation of thyrotropin and thyroglobulin secretion to experimentally decreased iodine supply in man. Metabolism. 1992; 41:1093-1096.

41. Braverman, L.E., X. He, S. Pino, B. Magnani, and A. Firek. The effect of low dose perchlorate on thyroid function in normal volunteers [abstract]. Thyroid. 2004; 14(9):691.

42. Lawrence, J.E., Lamm, S.H., Pino, S., Richman, K., Braverman, L.E. The effect of short-term low-dose perchlorate on various aspects of thyroid function. Thyroid. 2000; 10:659-663.

43. Lawrence, J., Lamm, S., Braverman, L.E. Low dose perchlorate (3 mg daily) and thyroid function. Thyroid 2001; 11:295.

44. Lamm, S.H., Braverman, L.E., Li, F.X., Richman, K., Pino, S., Howearth, G. Thyroid health status of ammonium perchlorate workers: A cross- sectional occupational health study. J. Occup. Environ. Med. 1999; 41:248-260.

45. Crump, C., Michaud, P., Tellez, R., Reyes, C., Gonzalez, G., Montgomery, E.L., Crump, K.S., Lobo, G., Becerra, C., Gibbs, J.P. Does perchlorate in drinking water affect thyroid function in newborns or school-age children? J. Occup. Environ. Med. 2000; 42:603-612.

46. Tellez, R.T., Chacon, P.M., Abarca, C.R., Blount, B.C., Van Landingham, C.B., Crump, K.S., and Gibbs, J.P. Chronic Environmental Exposure to Perchlorate Through Drinking Water and Thyroid Function During Pregnancy and the Neonatal Period. Thyroid. 2005; (In Press).

47. Hill, R.N., Erdreich, L.S., Paynter, O.E., Roberts, P.A., Rosenthal, S.L., Wilkinson, C.F. Thyroid follicular cell carcinogenesis. Fundam. Appl. Toxicol. 1989; 12:629-697.

48. National Research Council. 2005a. *Health Implications of Perchlorate Ingestion.* Washington, D.C.: National Academies Press.

49. Barnes, D.G., Dourson, M.L. Reference dose (RfD): Description and use in health risk assessments. Reg. Toxicol. Pharmacol. 1988; 8:471-486.

50. Dourson, M.L. Methods for establishing oral reference doses. In: Risk Assessment of Essential Elements. W. Mertz, C. O. Abernathy, and S. S. Olin, eds. Washington, D.C.: ILSI Press. Ch. 51-61. 1994.

51. U.S. Environmental Protection Agency. A Review of the Reference Dose and Reference Concentration Processes. Risk Assessment Forum. Washington, DC. May 2002. EPA/630/P-02/002A.

52. U.S. Environmental Protection Agency. Integrated Risk Information System (IRIS). Glossary of Terms. Office of Research and Development. Washington, DC. 2003b. Available online: http://www.US EPA.gov/iris.

53. Haber, L.T., Dollarhide, J.S., Maier, A., Dourson, M.L. Noncancer Risk Assessment: Principles and Practice in Environmental and Occupational Settings. In Patty's Toxicology. E. Bingham, B. Cohrssen and C.H. Powell Ed. 5th ed, Wiley and Sons, Inc. New York, NY. p. 169-232. 2001.

54. University of Nebraska. Perchlorate State of the Science Symposium. University of Nebraska Medical Center, Omaha, Nebraska. September 29th to October 1, 2003. Available online: http://www.unmc.edu/coned.

55. Strawson, J., Zhao,Q., Dourson, M. Reference dose for perchlorate based on thyroid hormone change in pregnant women as the critical effect. Regul. Toxicol. Pharmacol. 2004; 39(1):44-65.

56. Capen, C.C.   Toxic responses of the endocrine system.   In: Casarett and Doull's Toxicology The Basic Science of Poisons. 6th edition. C. D. Klaassen, ed. New York: McGraw-Hill. Ch. 21. pp. 724. 2001.

57. Toxicology Excellence for Risk Assessment (*TERA*).   Use of human data in risk assessment. Comments submitted to U.S. EPA. February 19, 2002.

58. Meek, M., Renwick, A., Ohanian, E., Dourson, M., Lake, B., Naumann, B., Vu, V. Guidelines for application of compound specific adjustment factors (CSAF) in dose/concentration response assessment. Comm. Toxicol. 2001; 7:575-590.

59. Mattie, D.R., Sterner, T.R., Merrill, E.A., Clewell, R.A., Zhao, Q., Strawson, J.E., Dourson, M.L.   Use of Human and Animal PBPK Models in Risk Assessment for Perchlorate. The Toxicologist. 2004; 78(S-1): 361.

60. BRT. Ammonium perchlorate: Effect on immune function. Quality assurance audit: study no. BRT 19990524 – plaque forming cell (PFC) assay; study no. BRT 19990525 – local lymph node assay (LLNA) in mice. Burleson Research Technologies, Inc., Raleigh, NC. Study # BRT 19990524 & 19990525. 2000.

61. Kessler, F.J., Krüskemper, H.J.   (1966).   Experimentelle Schilddrüsentumoren durch mehrjährige Zufuhr von Kaliumperchlorat [Experimental thyroid tumors caused by long-term potassium perchlorate administration]. Klin. Wochenschr. 1966; 44:1154-1156.

Chapter 9

Using Biomonitoring to Assess Human Exposure to Perchlorate

Benjamin C. Blount and Liza Valentín-Blasini[*]

Division of Laboratory Sciences, National Center for Environmental Health, Centers for Disease Control and Prevention, Atlanta, GA 30341

INTRODUCTION

Perchlorate (ClO_4^-) is an inorganic anion used primarily as an oxidant in solid rocket fuel. Perchlorate salts are also used in a variety of products such as road flares, air bags, explosives, and pyrotechnics.[1] It can also form naturally in the atmosphere,[2] producing ClO_4^- in precipitation. This natural process is thought to geologically concentrate ClO_4^- in certain regions such as west Texas[2] and northern Chile.[3] A combination of human and nonhuman activities has led to the widespread presence of ClO_4^- in the environment. Concerns about this led the U.S. Environmental Protection Agency (EPA) to include perchlorate on the Drinking Water Candidate Contaminant List[4] and to monitor perchlorate levels in public water systems. As of August 2005, ClO_4^- was detected in 5.4% of public water utility samples in 34 different states, with levels ranging from the method detection limit (4 µg/L) to a maximum of 420 µg/L.[5] The majority of this drinking water contamination is likely due to polluted source water, although in certain rare instances electrolytic ClO_4^- formation can occur *in situ* in water distribution systems.[6] Dietary perchlorate exposure is also likely, because of the contamination of vegetable[7] and forage[8–10] crops irrigated with perchlorate-containing water or fertilized with Chilean nitrate.[3] Although ClO_4^- contamination is a national problem, it is most severe in California, Nevada, and Arizona, where it routinely occurs in the lower Colorado River at parts-per-billion levels.[11]

The prevalence of trace levels of ClO_4^- in the environment almost certainly leads to human exposure. Environmental perchlorate exposure is of potential health concern because large clinical and experimental doses of ClO_4^- competitively inhibit iodide uptake,[12,13] and sustained inhibition of iodide

---

[*] The findings and conclusions in this report are those of the authors and do not necessarily represent the views of the Centers for Disease Control and Prevention.

uptake could potentially lead to hypothyroidism. The thyroid plays a crucial role in homeostasis and neurological development. Hypothyroidism can cause metabolic problems in adults and abnormal development in children.[14] Severe hypothyroidism due to iodine deficiency during pregnancy is a cause of cretinism, a permanent cognitive impairment of the developing fetus.[14] Even mild hypothyroidism during pregnancy is associated with measurable cognitive deficits in children.[15,16] The key question is whether or not exposure to environmental amounts of perchlorate impairs thyroid function, especially in sensitive subpopulations such as pregnant women.

ASSESSING HUMAN EXPOSURE TO PERCHLORATE

To assess the health effects of exposure we must first accurately quantify human exposure to ClO_4^-. Human exposure assessment has been traditionally based solely on limited information about the levels of a toxicant in the environment (e.g., in water, soil, air, or food). Based on these environmental data, human exposure has been predicted using models to account for variables that can affect the transfer of contaminants from the environment to individuals. These variables include proximity to exposure source(s), toxicant movement in the environment, absorption variables, genetic factors, personal habits, lifestyle, nutritional status, and other factors. All of these variables contribute to significant uncertainty about human exposure data based solely on environmental measurements. For this reason, model-based exposure assessment is typically considered one of the weakest links in the process of examining potential linkage between environmental exposure and health effects.

The accuracy and precision of exposure assessments can be improved by using environmental monitoring in conjunction with human biological monitoring, or biomonitoring.[17] Biomonitoring evaluates exposure by directly measuring an environmental toxicant or its metabolite(s) in human matrices, such as urine, blood, saliva, or other tissue.[18] The direct measurement of a biomarker of exposure provides confirmation that an exposure has occurred, as well as information about the magnitude of the exposure. This direct exposure information is much more precise than the assumption that all individuals in a certain geographic area have the same degree of perchlorate exposure. As shown in Figure 1, individual biomarker data are directly comparable with individual health effects. Biomonitoring thus decreases the uncertainty about the relationships between these two variables. For these reasons the NAS recommends direct assessment of perchlorate exposure in studies of human populations.[19]

Exposure and health effects pathway

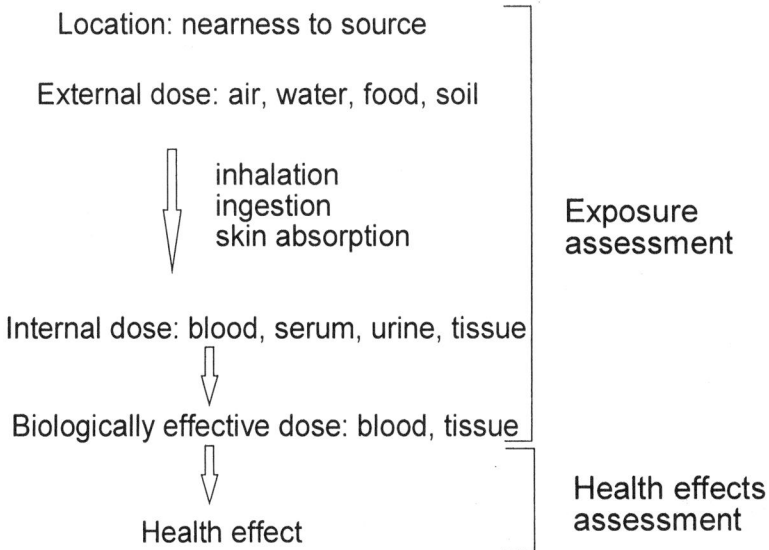

Location: nearness to source

External dose: air, water, food, soil

inhalation
ingestion
skin absorption

Exposure
assessment

Internal dose: blood, serum, urine, tissue

Biologically effective dose: blood, tissue

Health effect

Health effects
assessment

Figure 1. Exposure and health effects pathway. Environmental monitoring and biologic monitoring provide complementary exposure assessment information.

Biomarkers of exposure are directly related to the internal dose of an environmental toxicant. Most importantly, by measuring the biomarker levels in one human tissue, the amount of toxicant delivered to the site of the potential health effect (target tissue) can be estimated on the basis of pharmacokinetic calculations.[20] These calculations are relatively uncomplicated for ClO_4^- because the majority of a perchlorate dose is excreted unmetabolized in the urine.[21,22] For example, quantification of ClO_4^- in the urine of a study participant can be used to estimate the levels of the contaminant in that person's blood. Blood perchlorate levels provide a reasonable estimate of the ClO_4^- levels flowing through the thyroid capillary bed to the sodium-iodide symporter in the basolateral surface of the thyroid.[19] We can therefore more confidently derive the biologically effective dose of ClO_4^- delivered to the target tissue by directly measuring the contaminant in an individual than by using environmental determinations alone.

On the basis of available data describing exposure and health effects, the National Academy of Sciences defined a reference dose (RfD) for ClO_4^- of 0.0007 mg/kg-day.[19] An RfD is defined as "an estimate, with uncertainty spanning perhaps an order of magnitude, of a daily oral exposure to the human population (including sensitive subgroups) that is likely to be without an appreciable risk of deleterious effects during a lifetime."[23] The NAS-recommended RfD was subsequently adopted by the EPA. This benchmark can be compared with urinary perchlorate biomonitoring data by extrapolating measured ClO_4^- excretion rates to total daily excretion based on urine volume[24] or creatinine.[25] Thus, biomonitoring data can be used to assess the degree of risk/safety of a ClO_4^- exposure. The most important consideration is that direct measurement of human exposure markedly decreases uncertainty in assessment of human health risks.

ATTRIBUTES OF EFFECTIVE BIOMONITORING METHODS

Effective biomonitoring methods must address issues ranging from appropriate biomarker selection to adequate analytical specificity. For ClO_4^- the choice of an appropriate biomarker of exposure is straightforward. Perchlorate is not significantly metabolized in humans[21] and therefore direct measurement of the parent compound is the logical choice. Perchlorate does appear to be metabolized in the rumen of cattle,[8] but not in the human digestive system. One must also select the appropriate biological matrix in which perchlorate will be measured. Matrix selection is based on availability, pharmacokinetics, analytical issues, and the target population. For example, blood serum is the matrix closest to the target tissue. Perchlorate in serum is therefore the closest measure of the biologically effective dose of perchlorate delivered to the thyroid stroma and the thyroid sodium-iodide symporter.[19] However, ClO_4^- levels in blood tend to be much lower than urinary levels[12] because of efficient renal clearance of this anion. Additionally, blood collection is an invasive procedure compared with collection of urine or milk. Human milk is an excellent matrix for assessing ClO_4^- exposure of both a lactating woman and her breast-fed infant. However this matrix is only available from lactating women; thus another matrix must be chosen for exposure assessment in the general population. Therefore we chose urine as the matrix to use for ClO_4^- biomonitoring in the general population. In non-lactating women, most of a perchlorate dose (<95%) is excreted in the urine.[21] Thus, total ClO_4^- dose can be estimated from urinary perchlorate excretion. Additionally, urine is easily collected, non-pathogenic, and residual samples can be obtained from large population-based studies.

Analytical sensitivity is crucial for measuring picogram to nanogram quantities of ClO_4^- in biological samples, especially given the limited volumes of sample typically available (<1 mL). In 1997, improved analytical methods were used to quantify perchlorate contamination of the lower Colorado River.[26] Recent improvements utilizing mass spectrometry now allow for detection of ClO_4^- in a wide range of foods[27] and biological samples.[28,29] Methods of lesser sensitivity have not found measurable background levels of ClO_4^- in non-occupationally exposed Americans.[12,22,30,31] By applying more sensitive biomonitoring methods several groups have reported widespread perchlorate exposure in the general population.[10,29,32,33]

Selective methods are needed to quantify trace levels of perchlorate anion in highly variable and complex biological matrices. The specificity of the assay for ClO_4^- is improved by using a combination of selective cleanup,[34] chromatography,[35] and/or detection.[28] We recently developed a selective analytical method capable of quantifying ClO_4^- in the presence of much higher levels of chloride, sulfate and phosphate anions in human urine.[29] This method permits a high degree of selectivity by using a combination of selective ion chromatography and tandem mass spectrometry. Perchlorate is typically detected in water using conductivity, but more complex matrices require a more selective method, such as mass spectrometry. Comparison of a human urine sample analyzed by both tandem mass spectrometry and conductivity detection (Figure 2) demonstrates the magnitude of improved selectivity achieved for ClO_4^- quantification in a difficult matrix. Note that the signal-to-noise ratio for the perchlorate quantification ion trace (~970) greatly exceeds that for the conductivity signal (~19) that is composed of both the ClO_4^- analyte and the internal standard. Clearly tandem mass spectrometry improves measurement of perchlorate in difficult matrices, especially when coupled with ion chromatography.

Analytical accuracy is necessary for comparing biomonitoring results between different studies, methods and laboratories, and for estimating accurate exposure doses. We must be confident that our reported value is close to the true value. In the urine biomonitoring method of Valentín-Blasini et al.,[29] absolute assay accuracy is verified by the blind analysis of certified perchlorate reference solutions (AccuStandard, New Haven, CT) prepared to final concentration in synthetic urine. Analysis of four sets of these proficiency testing solutions across a one year time period yielded an average percent difference of –5.2% (CI, –7.2% to –3.2%). Matrix spike experiments indicated that the method was similarly accurate in quantifying ClO_4^- spiked into urine samples.

The analytical precision of biomonitoring methods is important for establishing a degree of certainty in the measured amounts. The CDC perchlorate biomonitoring method evaluates assay precision primarily using quality control pools of matrix containing known amounts of perchlorate. Two quality control pools are analyzed in each analytical batch with unknown samples; subsequent results must meet the accuracy and precision specifications outlined in the Westgard rules.[36] Over a 1-year period we analyzed these two quality control pools multiple times ($n = 117$) with an interday precision of 2.8% relative standard deviation at 71± 2.0 μg/L and 3.0% relative standard deviation at 4.7±0.14 μg/L. We also reevaluated the reproducibility of the assay by reanalyzing 5% of the samples, yielding an average relative percent difference of 1.5% (CI, 1.1–2.0%). Precise data such as these are necessary for verifying method validity.

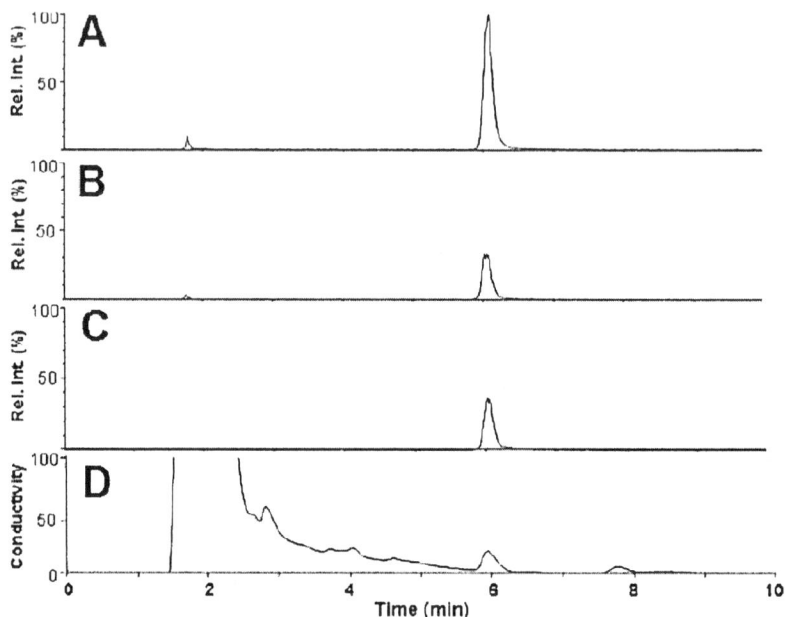

Figure 2. Comparison of a human urine sample analyzed by both tandem mass spectrometry and conductivity. Panel A shows MS/MS signal, and panel D shows conductivity trace for the same run. Panels A and C show the ClO_4^- quantification ion transition of 99–83 m/z, and the confirmation ion transition of 101-85 m/z, respectively. Panel B shows the ^{18}O-labeled ClO_4^- internal standard ion transition 107–89 m/z. Panel D shows the conductivity signal measured in microsiemens. This urine sample contains 10 μg/L perchlorate and 4 μg/L internal standard.

Biomonitoring methods should provide data on multiple analytes relevant to the toxicological interpretation of exposure. Perchlorate exposure can compromise iodine uptake by competitive inhibition of the sodium-iodide symporter (NIS). Thiocyanate and nitrate anions are also competitive inhibitors of NIS.[37] Thus the interpretation of the physiological impact of a given ClO_4^- exposure depends on the relative levels of iodide, thiocyanate, and nitrate. The CDC perchlorate biomonitoring program is now validating

IC-MS/MS methods that will quantify these 4 physiologically relevant anions that interact with NIS. By quantifying all of these anions in a single sample analysis, our method will provide a snapshot of the iodine import milieu experienced by NIS. Using this approach the CDC perchlorate biomonitoring program will provide complementary data for an improved toxicological interpretation of a given ClO_4^- exposure.

Biomonitoring methods should also be sufficiently rugged and have a high sample throughput. Through the use of an online divert valve and rugged chromatography we developed a biomonitoring method with improved throughput (75 unknowns/day).[29] A rugged method is one that can tolerate day-to-day variability without failing. By developing a rugged perchlorate biomonitoring method with sufficient throughput, we are able to study ClO_4^- exposure and health effects in large populations. By examining exposure in thousands of individuals we can improve the statistical power of studies to detect associations between perchlorate exposure and health effects.

BIOMONITORING APPLICATIONS

We have applied perchlorate biomonitoring techniques in two recently completed studies. The first was a study of 61 urine samples collected from anonymous healthy adult donors in an area with no known ClO_4^- contamination (Atlanta, Georgia).[29] Urine samples were collected in February 2004, immediately chilled on ice, frozen within 24 hours, and stored at $-20°C$ until analyzed. As shown in Table 1, ClO_4^- was detectable in all of the urine samples from this convenience population, with a log-normal distribution and a median concentration of 3.2 µg/L (7.8 µg ClO_4^-/g creatinine). Urinary ClO_4^- levels were adjusted for urinary creatinine to partially compensate for variable hydration and lean body mass. Because of the anonymous sampling design of this study, the source(s) of trace levels of perchlorate exposure in this population cannot be identified.

Table 1. Geometric means and medians for urinary ClO_4^- in selected populations

population	tap water ClO_4^- (µg/L)	n	Urinary ClO_4^- (µg/L)		Urinary ClO_4^- (µg/g creatinine)	
			geometric mean	median	geometric mean	median
Atlanta[29]	0.2	61	3.70	3.2	7.01	7.8
Antofagasta[39]	0.4	100*	13.8	15	19.8	21
Chañaral[39]	5.8	90*	39.3	35	44.0	37
Taltal[39]	114	98*	88.0	98	111	120

*n value represents up to 3 urine samples from 60 study participants at each of these sites.

We also applied our method to residual urine samples from a study of Chilean women drinking tapwater that contained differing levels of naturally occurring perchlorate.[39] The Chilean study collected 3 spot urine samples from approximately 60 pregnant women in each of 3 cites: Antofagasta (tapwater ClO_4^- approximately 0.4 µg/L), Chañaral (tapwater ClO_4^- approximately 5.8 µg/L), and Taltal (tapwater ClO_4^- approximately 114 µg/L). The primary findings in the Chilean study are reported elsewhere.[39] Our analysis of the urine samples indicated widespread and varied ClO_4^- exposure in all women from the three locations (Table 1). As with the urine collected in Atlanta, all Chilean samples contained measurable levels of ClO_4^- (median = 35 µg/L) and were adjusted for urinary creatinine content (median = 43 µg/g creatinine). The median levels of urinary perchlorate increased with increasing levels of ClO_4^- in the tapwater (Antofagasta, 21 µg/g creatinine; Chañaral, 37 µg/g creatinine; Taltal, 120 µg/g creatinine). Urinary ClO_4^- levels varied dramatically (>100-fold) within each city population. This variability likely resulted from varied water use and dietary habits.

Our initial measurements indicate that perchlorate exposure is widespread, albeit at trace levels. The toxicological impact of chronic trace level ClO_4^- exposure is uncertain. The National Research Council of the National Academy of Sciences recently defined a reference dose for perchlorate at 0.7 µg/kg-day.[17] Only one sample from the Atlanta convenience population contained ClO_4^- at levels in excess of the amount expected to be excreted by an individual exposed at the reference dose.[29] The potential health effects of trace level exposure needs to be addressed in future studies that combine perchlorate biomonitoring with assessment of thyroid function.

Urinary perchlorate is an effective biomarker for ClO_4^- exposure. Not only is urine collection less invasive than blood sampling, but urinary ClO_4^- levels tend to be much higher than serum levels due to efficient renal clearance. These advantages will allow us to assess perchlorate exposure in large population-based studies (e.g., National Health and Nutrition Examination Survey [NHANES])[38] and report the data as part of the *National Report on Human Exposure to Environmental Chemicals*.[40] In fact, the CDC perchlorate biomonitoring program has quantified ClO_4^- in 2820 urine samples collected from NHANES during 2001–2002.[32] These data will provide information for risk modeling and provide a reference range for comparisons with results from other potentially-exposed population groups. Additional studies of the NHANES data will evaluate the sources and health implications of perchlorate exposure in the general population.

CONCLUSIONS

Scientifically valid exposure assessment is crucial to risk assessment, risk management, and prevention of environmental disease. Biomonitoring is an excellent tool for evaluating human exposure to perchlorate from all sources, and often provides complementary information to environmental data. Analytical methods for perchlorate biomonitoring must be adequately sensitive and selective to detect sub-parts-per-billion levels in complex matrixes. Through improved exposure assessment we can better evaluate the relevance of environmental perchlorate to human exposure and health.

REFERENCES

1. Mendiratta, S. K.; Dotson, R. L.; and Brooker, R.T. Perchloric acid and perchlorates. 1996, *18*, 157-170. New York, John Wiley & Sons. Kirk-Othmer encyclopedia of chemical technology. Kroschwitz JI and Howe-Grant M.

2. Dasgupta, P. K.; Martinelango, P. K.; Jackson, W. A.; Anderson, T. A.; Tian, K.; Tock, R. W.; Rajagopalan, S. The origin of naturally occurring perchlorate: the role of atmospheric processes. *Environ.Sci.Technol.* 2005, *39*,1569-75.

3. Urbansky, E. T.; Brown, S. K.; Magnuson, M. L.; Kelty, C. A. Perchlorate levels in samples of sodium nitrate fertilizer derived from Chilean caliche. *Environ.Pollut.* 2001, *112*, 299-302.

4. Environmental Protection Agency *Federal* Registry 1998, *63*, 10274.

5. Unregulated Contaminant Monitoring Regulation (UCMR) data from public water systems. US EPA. 2004. http://www.epa.gov/safewater/ucmr/data.html [Accessed October 11, 2005].

6. Jackson, A.; Arunagiri, S.; Tock, R.; Anderson, T. A.; Rainwater, K. Electrochemical generation of perchlorate in municipal drinking water systems. *Journal of the American Water Works Association* 2004, *96*, 103-08.

7. Yu, Lu; Canas, Jaclyn E; Cobb, George P.; Jackson William A.; Anderson, T. A. Uptake of perchlorate in terrestrial plants. *Ecotoxicology and Environmental Safety* 2004, *58*, 44-49.

8. Capuco, A. V.; Rice, C. P.; Baldwin, R. L. Fate of dietary perchlorate in lactating dairy cows. *Proceedings of the National Academy of Sciences* 2005, *102(45):*16152-16157.

9. Kirk, A. B.; Smith, E. E.; Tian, K.; Anderson, T. A.; Dasgupta, P. K. Perchlorate in milk. *Environ.Sci.Technol.* 2003, *37*, 4979-81.

10. Kirk, A. B.; Martinelango, P. K.; Tian, K.; Dutta, A.; Smith, E. E.; Dasgupta, P. K. Perchlorate and iodide in dairy and breast milk. *Environ.Sci.Technol.* 2005, *39*, 2011-17.

11. Hogue, C. Environmental Pollution: Rocket-Fueled River. *Chemical & Engineering News* 2003, *81*, 37-46.

12. Greer, M. A.; Goodman, G.; Pleus, R. C.; Greer, S. E. Health effects assessment for environmental perchlorate contamination: the dose response for inhibition of thyroidal radioiodine uptake in humans. *Environ.Health Perspect.* 2002, *110*, 927-37.

13. Wyngaarden, J. B; Stanbury, J. B.; Rapp, B. The effects of iodide, perchlorate, thiocyanate and nitrate administration upon the iodide concentrating mechanism of the rat thyroid. *Endocrinology* 1953, *52*, 568-74.

14. Utiger, R. D. and Braverman, L. E. *Werner and Ingbar's The Thyroid: A fundamental and clinical text.* 8th ed.; Lippincott Williams & Wilkins: Philadelphia, PA, 2000.

15. Haddow, J. E.; Palomaki, G. E.; Allan, W. C.; Williams, J. R.; Knight, G. J.; Gagnon, J.; O'Heir, C. E.; Mitchell, M. L.; Hermos, R. J.; Waisbren, S. E.; Faix, J. D.; Klein, R. Z. Maternal thyroid deficiency during pregnancy and subsequent neuropsychological development of the child. *N Engl.J Med.* 1999, *341*, 549-55.

16. Klein, R. Z.; Sargent, J. D.; Larsen, P. R.; Waisbren, S. E.; Haddow, J. E.; Mitchell, M. L. Relation of severity of maternal hypothyroidism to cognitive development of offspring. *J Med.Screen.* 2001, *8*, 18-20.

17. NAS. Monitoring Human Tissues for Toxic Substances. 1991. Washington, National Academy Press. http://www.nap.edu/catalog/1787.html [Accessed October 11, 2005].

18. Pirkle, J. L.; Needham, L. L.; Sexton, K. Improving exposure assessment by monitoring human tissues for toxic chemicals. *J.Expo.Anal.Environ.Epidemiol.* 1995, *5*, 405-24.

19. NAS. Health Implications of Perchlorate Ingestion. 2005. Washington. D.C, National Research Council, National Academy Press. http://www.nap.edu/catalog/11202.html [Accessed October 11, 2005].

20. Merrill, E. A.; Clewell, R. A.; Robinson, P. J.; Jarabek, A. M.; Gearhart, J. M.; Sterner, T. R.; Fisher, J. W. PBPK model for radioactive iodide and perchlorate kinetics and perchlorate-induced inhibition of iodide uptake in humans. *Toxicol.Sci.* 2005, *83*, 25-43.

21. Anbar, M.; Guttmann, S.; Lweitus, Z. *Int J Appl Radiat Isot* 1959, *7*, 87-96.

22. Lawrence, J. E.; Lamm, S. H.; Pino, S.; Richman, K.; Braverman, L. E. The effect of short-term low-dose perchlorate on various aspects of thyroid function. *Thyroid* 2000, *10*, 659-63.

23. US EPA. 2005 Glossary of Terms. America's Children and the Environment. http://www.epa.gov/envirohealth/children/background/glossary.htm#r [Accessed October 11, 2005].

24. Fenske, R. A.; Kissel, J. C.; Lu, C.; Kalman, D. A.; Simcox, N. J.; Allen, E. H.; Keifer, M. C. Biologically based pesticide dose estimates for children in an agricultural community. *Environ.Health Perspect.* 2000, *108*, 515-20.

25. Mage, D. T.; Allen, R. H.; Gondy, G.; Smith, W.; Barr, D. B.; Needham, L. L. Estimating pesticide dose from urinary pesticide concentration data by creatinine correction in the Third National Health and Nutrition Examination Survey (NHANES-III). *J Expo.Anal.Environ.Epidemiol.* 2004, *14*, 457-65.

26. Okamoto, H. S.; Rishi, D. K.; Steeber, W. R.; Baumann, F. J.; Perera, S. K. Using ion chromatography to detect perchlorate. *J.Am.Water Works Assoc.* 1999, *91*, 73-84.

27. Krynitsky, A. J.; Niemann, R. A.; Nortrup, D. A. Determination of perchlorate anion in foods by ion chromatography-tandem mass spectrometry. *Anal.Chem.* 2004, *76*, 5518-22.

28. Martinelango, P. K.; Anderson, J. L.; Dasgupta, P. K.; Armstrong, D. W.; Al Horr, R. S.; Slingsby, R. W. Gas-phase ion association provides increased selectivity and sensitivity for measuring perchlorate by mass spectrometry. *Anal.Chem.* 2005, *77*, 4829-35.

29. Valentin-Blasini, L.; Mauldin, J. P.; Maple, D.; Blount, B. C. Analysis of perchlorate in human urine using ion chromatography and electrospray tandem mass spectrometry. *Anal.Chem.* 2005, *77*, 2475-81.

30. Braverman, L. E.; He, X.; Pino, S.; Cross, M.; Magnani, B.; Lamm, S. H.; Kruse, M. B.; Engel, A.; Crump, K. S.; Gibbs, J. P. The effect of perchlorate, thiocyanate, and nitrate on thyroid function in workers exposed to perchlorate long-term. *J Clin.Endocrinol.Metab* 2005, *90*, 700-06.

31. Gibbs, J. P.; Narayanan, L.; Mattie, D. R. Crump et al. study among school children in chile: subsequent urine and serum perchlorate levels are consistent with perchlorate in water in taltal. *J.Occup.Environ.Med.* 2004, *46*, 516-17.

32. Blount, B. C.; Valentin-Blasini, L.; Mauldin, J. P.; Pirkle, J. L.; Osterloh, J. (personal communication).

33. Pearce, E. N.; Braverman, L. E.; Blount, B. C.; Valentin-Blasini, L. (personal communication).

34. Ellington, J. J.; Evans, J. J. Determination of perchlorate at parts-per-billion levels in plants by ion chromatography. *J.Chromatogr.A* 2000, *898*, 193-99.

35. Anderson, T. A.; Wu, T. H. Extraction, cleanup, and analysis of the perchlorate anion in tissue samples. *Bull.Environ.Contam Toxicol.* 2002, *68*, 684-91.

36. Westgard, J. O.; Barry, P. L.; Hunt, M. R.; Groth, T. A multi-rule Shewhart chart for quality control in clinical chemistry. *Clin.Chem.* 1981, *27*, 493-501.

37. Tonacchera, M.; Pinchera, A.; Dimida, A.; Ferrarini, E.; Agretti, P.; Vitti, P.; Santini, F.; Crump, K.; Gibbs, J. Relative potencies and additivity of perchlorate, thiocyanate, nitrate, and iodide on the inhibition of radioactive iodide uptake by the human sodium iodide symporter. *Thyroid* 2004, *14*, 1012-19.

38. CDC. 2004 National Health and Nutrition Examination Survey. National Center for Health Statistics, Hyattsville, MD. http://www.cdc.gov/nchs/nhanes.htm [Accessed October 11, 2005].

39. Tellez, R. T.; Chacon, P. M.; Abarca, C. R.; Blount, B. C.; Landingham, C. B.; Crump, K. S.; Gibbs, J. P. Long-term environmental exposure to perchlorate through drinking water and thyroid function during pregnancy and the neonatal period. *Thyroid* 2005, *15*, 963-75.

40. CDC 2005 National Report on Human Exposure to Environmental Chemicals. http://www.cdc.gov/exposurereport/ [Accessed October 11, 2005].

.

Chapter 10

Recent Advances in Ion Exchange for Perchlorate Treatment, Recovery and Destruction

Baohua Gu and Gilbert M. Brown

Environmental Sciences and Chemical Sciences Divisions
Oak Ridge National Laboratory, Oak Ridge, TN 37831

INTRODUCTION

Ion-exchange technology has been used for drinking water and wastewater treatment for over a half century.[1-4] Ion exchange is a process by which ions of interest in the solution phase, such as perchlorate (ClO_4^-), are selectively removed by exchange with ions associated with a synthetic solid phase, often referred to as "resin." Usually, synthetic resins are made from styrenic and acrylic polymeric organic materials (or small beads) that contain surface functional groups, such as positively charged quaternary ammonium, which attract negatively charged anions such as ClO_4^- in solution. Depending on their effectiveness and affinity to sorb ClO_4^-, resins may be categorized into selective and non-selective. The selective resin sorbs ClO_4^- strongly and effectively in the presence of competing anions such as sulfate (SO_4^{2-}), chloride (Cl^-), nitrate (NO_3^-), and bicarbonate (HCO_3^-), which usually exist in relatively high concentrations in groundwater and surface water. Non-selective resins sorb anions somewhat indiscriminately or by the mass reaction based on their charge and concentration. Therefore, divalent SO_4^{-2} anions usually are more strongly sorbed than those monovalent anions on non-selective resins.[5]

Resin regeneration, secondary waste production, perchlorate destruction, and their associated costs are major issues related to the use of either selective or non-selective ion exchange for water treatment. Until recently, selective resins usually are not regenerated but are incinerated after a single use, resulting in relatively high capital and operational costs. Although the non-selective resin can be regenerated, it requires frequent regenerations because of its low selectivity and sorption capacity, thus producing a large quantity of secondary waste brine solution that must be disposed of properly.[6, 7]

Recent technological development has led to numerous new products and discoveries such as highly selective ion-exchange resins (e.g., the Purolite A-530E and Amberlite PWA-2 resins), novel regeneration techniques for selective ion exchangers, and perchlorate recovery or destruction technologies, leading to practically zero secondary waste production. These combined treatment technologies hold great promises for treating perchlorate-contaminated water with greatly reduced capital and operational costs. This chapter is therefore intended to provide a detailed description of ion-exchange processes and mechanisms, new technological advances in ion-exchange and resin regeneration, and resin performance issues at varying environmental conditions such as the feed ClO_4^- concentration, the presence or absence of competing ions and other co-contaminants (e.g., naturally occurring uranium). Perchlorate recovery, destruction, and waste minimization are also discussed, and comparisons are made with respect to the pros and cons of using various ion-exchange and regeneration techniques such as conventional non-selective ion exchange and brine regeneration, the single-use fixed bed versus the highly selective, regenerable ion-exchange systems.

NATURE OF ION EXCHANGE

Selectivity

Most ion exchange resins are macroporous, as shown schematically in Figure 1. An anion-exchange resin, as received from the manufacturer, contains positively charged surface functional groups sorbed with counter ions, usually Cl⁻ anions. When it is exposed to a solution containing counter ions of a different kind (e.g., ClO_4^-), some counter ions in solution will enter the

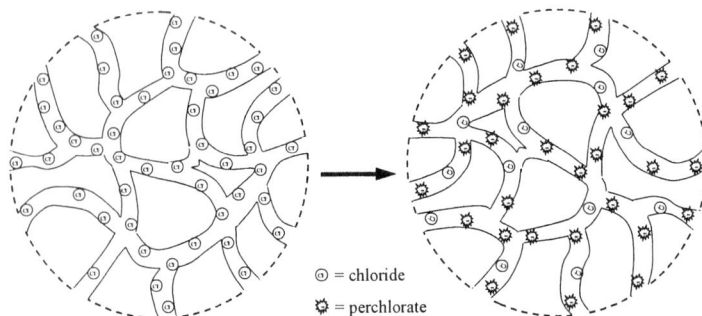

⊙ = chloride
✿ = perchlorate

Figure 1. Schematic illustration of the binary ion exchange between perchlorate and chloride ions on a synthetic resin bead. The resin is initially saturated with chloride as counter ions.

ion exchanger, in exchange for Cl^- from the resin bead. At equilibrium, both the exchanger and the liquid will contain both kinds of counter ions, although they are usually not in the same ratio. Such a binary exchange reaction (e.g., between ClO_4^- and Cl^-) may be written as:

$$R\text{-}Cl^- + ClO_4^- \rightleftharpoons R\text{-}ClO_4^- + Cl^-$$

where $R\text{-}Cl^-$ and $R\text{-}ClO_4^-$ represent Cl^- and ClO_4^- ions associated with the resin exchanger. Since both Cl^- and ClO_4^- are monovalent anions, the thermodynamic equilibrium constant (K) can be expressed as

$$K = \frac{\{R - ClO_4^-\}\{Cl^-\}}{\{R - Cl^-\}\{ClO_4^-\}}$$

where { } refers to the activity of these ionic species. Ion exchangers do prefer certain counter ions to other ions in solution. The K value is thus called *selectivity coefficient* and is a measure of the preference of counter ions to other ions in solution. A more practical description of this equilibrium behavior is so called *distribution or partitioning coefficient* (K_d), which is defined as

$$K_d = \frac{[R - ClO_4^-]}{[ClO_4^-]} = mL/g \text{ of exchanger}$$

where $[R\text{-}ClO_4^-]$ and $[ClO_4^-]$ represent concentrations of sorbed ClO_4^- on resin beads and those free ClO_4^- ions in the solution phase, respectively. Therefore, a high K_d value indicates that the resin has a relatively high selectivity or efficiency for sorbing ClO_4^- from the solution phase. An advantage of using K_d is its simplicity, because it can be calculated regardless the presence of any other counter ions in the system. Natural groundwater usually contains multiple counter ions with different valence states, such as SO_4^{2-}, Cl^-, NO_3^-, CO_3^{-2}, and HCO_3^-. Obviously, the disadvantage in using K_d is that it is not a constant; K_d values will change or depend upon, among other factors, the initial concentration of the target counter ion, solution ionic composition, nature of the counter ions (e.g., charge-to-size ratio and hydration energy),[8] surface functional groups, total anion-exchange capacity (AEC), and structural properties of resin matrixes.[9, 10] The above concept can best be illustrated by examining the sorption isotherms and kinetics of ClO_4^- on resins with varying surface functional groups and background electrolyte concentrations, as described in detail below.

Sorption Equilibrium

A series of quaternary ammonium strong-base anion-exchange resins were first studied for sorption of ClO_4^- from aqueous solution. These resins have the same polystyrene backbone structure and similar bead sizes but different trialkylammonium functional groups. The trimethyl resin (A-500) was obtained from Purolite, whereas the triethyl, tripropyl, tributyl, and trihexyl resins were obtained from Wandong Chemical Plant, China. All resins were used as received with Cl⁻ as the counter ions. The sorption experiments were performed in two simulated groundwater solutions, one of which has higher background electrolyte concentrations than the other.[6, 11] These studies were designed to evaluate how trialkylammonium functional groups, ClO_4^- concentrations, and competing ions influence the sorption equilibrium of ClO_4^- (Figure 2 and Table 1). Results indicate that the tributyl resin appeared to be the best to sorb ClO_4^- at low equilibrium concentrations as indicated by its initial steep slope of the isotherm, which is equivalent to K_d by assuming a linear relationship between the amount of ClO_4^- sorbed and that left in solution (at <0.2 mg/L ClO_4^-). This initial slope of the sorption isotherm decreased systematically with a decrease in the size of the trialkyl chain length from tributyl, tripropyl, triethyl, to trimethyl. In other words, the selectivity for ClO_4^- sorption increased with an increase in the size and consequently the hydrophobicity of the trialkylammonium functional group. The resin with the longest trihexyl groups behaved differently from other resins: although its initial slope of the isotherm is comparable with that of the tributyl resin, it flattened out quickly as solution perchlorate concentration

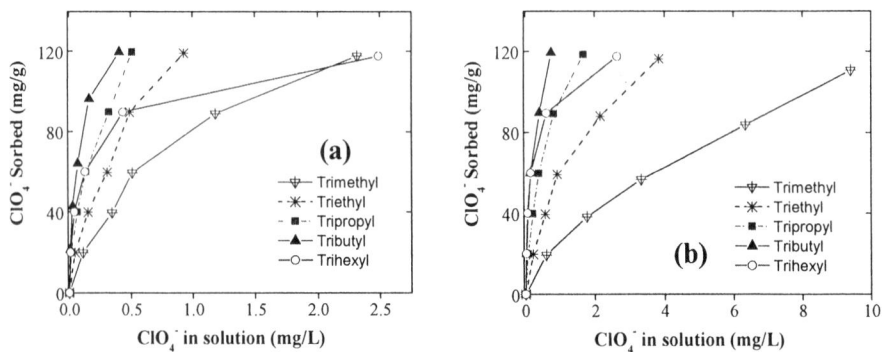

Figure 2. Perchlorate sorption isotherms on resins with varying surface trialkylammonium functional groups in simulated electrolyte background solutions of (a) 0.2-mM $NaNO_3$, 0.2-mM $Mg(NO_3)_2$, 0.3-mM K_2SO_4, and 1-mM $NaHCO_3$ and (b) 1-mM $NaClO_4$, 10-mM NaCl, 1-mM $CaCl_2$, 2-mM K_2SO_4, and 2-mM $NaHCO_3$. The initial ClO_4^- concentration varied from 0 to 120 mg/L. The resin-to-solution ratio was 1:1000 (w/v), and the equilibration time was 24 h.

increased (Figure 2). This observation can be explained by the fact that (i) an increase in the alkyl chain length surrounding the anion exchange site results in steric congestion and therefore a lower total AEC and (ii) a concomitant decrease occurs in the wettability of the resin, and thus the rate of exchange decreases.

Table 1. Perchlorate distribution coefficients $(K_d)^*$ on synthetic resins with varying surface trialkylammonium functional groups and experimental conditions (i.e., different background electrolyte concentrations and reaction times). The initial added ClO_4^- concentration was 100 mg/L.

Resin	20-min K_d (mL/g)	1-h K_d (mL/g)	24-h K_d (mL/g)	168-h K_d (mL/g)
Low background electrolyte solution [#]				
Trimethyl (Purolite A-500)	no data	21,500	51,100	52,800
Triethyl (WBR-100)	7,200	66,300	107,000	115,000
Tripropyl (WBR-102)	5,300	80,800	243,000	250,000
Tributyl (WBR-004)	3,400	39,800	668,000	713,000
Trihexyl (WBR-006)	56	179	487,000	2,100,000
High background electrolyte solution [#]				
Trimethyl (Purolite A-500)	1,350	7,090	12,100	12,600
Triethyl (WBR-100)	3,360	19,500	37,500	39,000
Tripropyl (WBR-102)	3,720	32,100	79,300	84,700
Trihexyl (WBR-006)	78	400	128,000	471,000

[*] K_d values were rounded to the nearest thousandth when the solution concentration is near the detection limit, or when the degree of sorption is >99%, which results in a great uncertainty in calculated K_d values (because K_d is calculated as the ratio between the amount of ClO_4^- sorbed per unit mass of resin to that left in the equilibrium solution). When the degree of sorption is relatively low, the measured ClO_4^- concentration and the calculated K_d are usually within an error of <3%.

[#] The low background electrolyte solution consisted of 1-mM $NaClO_4$, 0.2-mM $NaNO_3$, 0.2-mM $Mg(NO_3)_2$, 0.3-mM K_2SO_4, and 1-mM $NaHCO_3$, whereas the high background electrolyte solution consisted of 1-mM $NaClO_4$, 10-mM $NaCl$, 1-mM $CaCl_2$, 2-mM K_2SO_4, and 2-mM $NaHCO_3$. Experiments were performed with 0.1-g resin (dry weight equivalent) in 100 mL of the test solution.

The shape of the sorption isotherms appeared to be similar in both low and high electrolyte background solutions (Figure 2). However, as can be expected, lower amounts of ClO_4^- were sorbed at high electrolyte concentrations than at low electrolyte concentrations, thereby leaving more ClO_4^- ions in the solution phase (Figure 2b) and resulting in lower K_d values (Table 1). Figure 2 also indicates that the sorption isotherms of perchlorate are not linear, particularly at relatively high equilibrium ClO_4^- concentrations. In other words, the slope of the sorption isotherm, which determines the K_d, depends on the equilibrium ClO_4^- concentration because K_d is not a true equilibrium constant, as stated earlier. As shown in Table 2, at a lower added ClO_4^- concentration (10 mg/L), the K_d values increase substantially for resins functionalized with triethyl up to trihexyl groups. In comparison with data shown in Table 1 (at the low background electrolyte concentration), K_d values increased from one to four-fold for all resins studied. However, sorption on the trimethyl resin (A-500) is an exception, primarily because of its relatively low selectivity for ClO_4^- ions, which do not effectively compete with other anions in the background solution.

Table 2. Perchlorate distribution coefficients $(K_d)^*$ on synthetic resins with varying surface trialkylammonium functional groups at the low background electrolyte solution. (See Table 1.) The initial added ClO_4^- concentration was 10 mg/L.

Resin	1-h K_d (mL/g)	24-h K_d (mL/g)	168-h K_d (mL/g)
Trimethyl (Purolite A-500)	17,600	41,600	46,900
Triethyl (WBR-100)	141,000	251,000	377,000
Tripropyl (WBR-102)	233,000	657,000	905,000
Tributyl (WBR-004)	45,400	737,000	1,400,000
Trihexyl (WBR-006)	5,800	2,900,000	>3,300,000

Surface functional groups on the resin affect not only the sorption of ClO_4^- but also other anions such as nitrate and sulfate.[5, 12] As illustrated in Figure 3, the Purolite A-500 resin, functionalized with small trimethylammonium groups, is particularly good to sorb multi-valent anions such as sulfate. The amount of SO_4^{2-} sorbed on the A-500 resin was the highest and remained relatively constant within the concentration range of perchlorate studied. The sorption of nitrate was lower and decreased as the perchlorate concentration in solution increased. On the other hand, the sorption of SO_4^{2-} was the lowest on the WBR-006 resin functionalized with large trihexylammonium groups. Only small amounts of SO_4^{2-} were sorbed initially when added ClO_4^- concentration was low, and the sorption decreased rapidly and became negligible at higher ClO_4^- concentrations. Nitrate was more selectively sorbed than SO_4^{2-} in this case, but its sorption also decreased rapidly and became negligible on the trihexyl resin when the added ClO_4^- concentration increased. These results suggest that the trihexyl resin actually rejects SO_4^{2-} partly because of its surface hydrophobicity (with a six-carbon chain) and a relatively high hydration energy of sulfate (ΔG^0, $-1,103$ kJ/mol).[8, 10, 13] The hydration energy of these anions is in the order of sulfate < nitrate ($\Delta G^0 = -314$ kJ/mol) < perchlorate ($\Delta G^0 = -205$ kJ/mol). Therefore, the nature of these counter ions (i.e., charge-to-size ratio and hydration energy) greatly influences their selectivity and competition with ClO_4^- ions for sorption onto the resin.

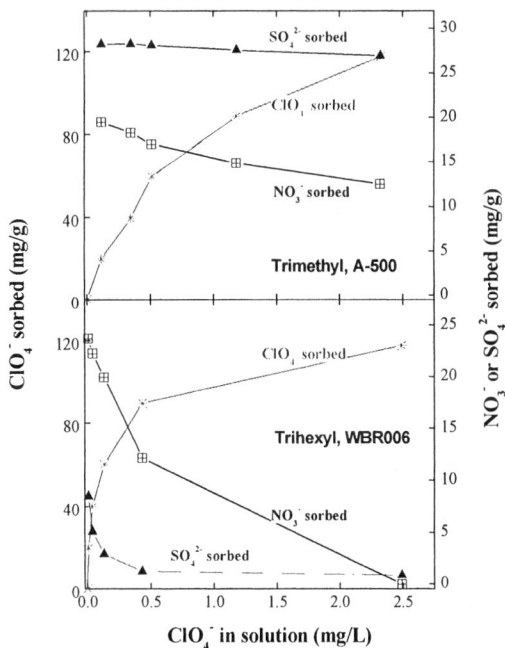

Figure 3. Sorption of perchlorate, nitrate, and sulfate on the Purolite A-500 (trimethyl) and WBR-006 (trihexyl) resins under the same experimental conditions described in Figure 2a. The nitrate and sulfate were added as background electrolytes and thus kept constant throughout the experiment.

Sorption Kinetics

The fact that the sorption isotherms of the trihexyl resin (WBR-006) behaved differently from other resins (Figure 2) is due mainly to its size and hydrophobicity of the trihexyl group. Additionally, the 24-h equilibration time (used for the sorption experiment) occurred to be insufficient for the anion-exchange reaction to reach equilibrium. Tables 1 and 2 and Figure 4 clearly show the dependence of K_d on reaction time. Data indicate that, for the trihexyl resin, the K_d value increased dramatically over time; it was the lowest initially (within 1 h) and became the greatest among all resins studied after one week of equilibration. All these resins showed a time-dependent sorption of perchlorate, particularly within the first 24 h, because of the macroporous structure of resin beads and the diffusion and exchange processes of counter ions. With the exception of the trihexyl resin, K_d values become relatively constant after 24 h of equilibration (Table 1 and Figure 4). For the trihexyl resin, the reaction did not even reach equilibrium at the end of the kinetic experiment (after 1 week of equilibration) (Figure 4).

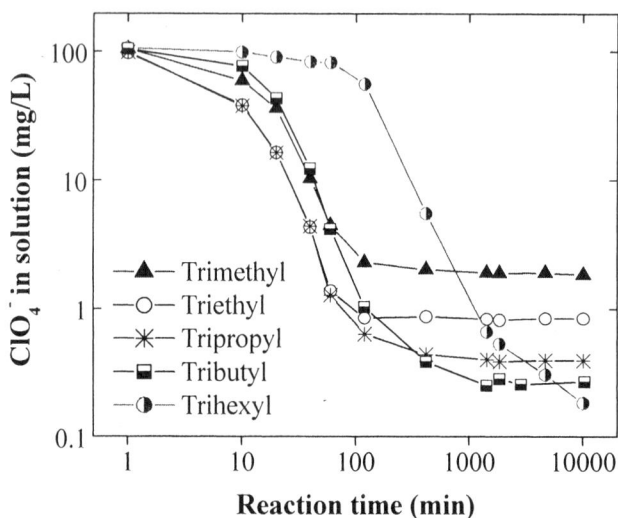

Figure 4. Perchlorate sorption kinetics on resins with varying surface trialkylammonium functional groups under the same experimental conditions described in Figure 2, except that the initial added ClO_4^- concentration was kept constant at 100 mg/L.

At a given time of equilibration, for example at 20 min, K_d values decreased consistently with an increase in the size of the trialkyl chain length, but this order was reversed after one week of equilibration (Table 1). For the 1-h K_d, it increased consistently up to propyl, but for the 24-h K_d, it increased up to butyl. These results indicate that a high selectivity of the resin can thus be achieved with the larger trialkyl ammonium group but at the expense of slow exchange kinetics.

These observations have led to the design of the bifunctional anion-exchange resins,[9, 10] in which both short and long chains of trialkylammonium groups are functionalized on resin surfaces to provide excellent selectivity for ClO_4^-

sorption and, at the meantime, good exchange kinetics, as will be discussed in detail below. Similar observations have been made for the sorption of radioactive pertechnetate (TcO_4^-) anions on various synthetic resins.[9] It was reported that K_d values for the sorption of TcO_4^- steadily increased up to the tributyl resin in a 24-h equilibration period; the trihexyl resin gave a substantially lower value because sorption on the trihexyl resin is not at equilibrium at 24 h; the K_d continues to climb with equilibration time, surpassing that of the butyl resin, giving the highest K_d of all resins tested.

SELECTIVE AND NON-SELECTIVE RESINS

On the basis of their selectivity, resins generally may be categorized into selective and non-selective towards perchlorate sorption. Generally speaking, the Type-I polystyrenic strong-base anion-exchange resins (with quaternary ammonium functional groups attached to the aromatic ring of styrene or divinylbenzene cross-linking agents) have natural bias for sorbing poorly hydrated anions such as ClO_4^-, as illustrated in Figure 2 and Tables 1–2. This is because the polystyrene divinylbenzene matrix is considered to be nonpolar (or hydrophobic) compared with matrices containing oxygen molecules, such as those polyacrylic matrixes, in which the styrene monomer is replaced by the acrylic monomer in the polymer chain. Similarly, the Type-II anion-exchange resins are formed by the replacement of one of the trialkyl groups with an ethanol group, resulting in a slightly hydrophilic character with lower basic strength. Therefore, resin with polyacrylic backbones and Type-II resins usually are much less selective than Type-I polystyrenic resins (Table 3). Increasing the length of the trialkyl group from methyl to hexyl increases the hydrophobicity as well as charge-separation distances of the resin. Therefore, as the trialkyl chain length increases, the resin selectivity for ClO_4^- increases (Tables 1–2), but it decreases the selectivity for hydrated divalent anions such as SO_4^{2-} (Figure 3).

However, a drawback of increasing the alkyl chain length is a greatly reduced sorption kinetics, as discussed earlier (Tables 1–2). Therefore, a balance must be maintained between the selectivity and reaction kinetics, which has led to the development of a new class of highly efficient bifunctional anion-exchange resins at the Oak Ridge National Laboratory (ORNL).[9, 10] The bifunctional resins have two quaternary ammonium groups, one having long chains for higher selectivity and one having shorter chains for improved reaction kinetics. The resin initially was developed to remove extremely low levels of radioactive TcO_4^- (<1 ppb Tc) from contaminated groundwater at the U.S. Department of Energy's Paducah Gaseous Diffusion Plant site in Kentucky.[9,10,14] A pilot-scale field test successfully demonstrated that the bifunctional resin was able to treat >700,000 bed volumes (BV) of the contaminated groundwater (running at a flow rate of ~6 BV min^{-1}) before a breakthrough occurred, despite the fact that the concentration of TcO_4^- was ~6 orders of magnitude lower than that of the competing anions such as Cl^-, SO_4^{2-}, NO_3^-, and HCO_3^-.[10] Due to the chemical similarities between TcO_4^- and ClO_4^- oxyanions, the bifunctional resin was also found to be highly selective for perchlorate sorption. Table 3 lists the relative performances of these resins in comparison with non-selective polyacrylic resins. Note that both Purolite D-3696 and WBR-106 are experimental bifunctional resins (with triethyl and trihexyl groups) made in small batches in laboratory. The Purolite A-530E is a commercially available, scale-up version of the D-3696 resin. The WBR-112L resin is a custom-made batch of the triethyl and tributyl bifunctional resin used for field groundwater treatment. Evidently, all these bifunctional resins exhibit orders-of-magnitude higher K_d's than those polyacrylic resins (Purolite A-830 and A-850). In comparison with those mono-functional tributyl and trihexyl resins (Tables 1 and 2), the bifunctional resins also show much faster reaction kinetics with a greatly increased 1-h K_d. This increased initial reaction kinetics is crucial under continuous flow-through conditions because of a short contact time used in such practices (usually less than a few minutes during water treatment), as will be discussed in detail below.

Table 3. Comparison of time-dependent distribution coefficient (K_d) of perchlorate on various synthetic resins under similar experimental conditions as described in Table 1, except that the added initial ClO_4^- concentration was 10 mg/L. The Purolite A-830 and A-850 are polyacrylic resins, the Amberlite PWA2 is the polystyrenic resin made by Rohm and Haas, and Purolite A530E, D-3696, WBR-112L and 106 are bifunctional resins.

Resin	1 h K_d (mL/g)	24 h K_d (mL/g)	168 h K_d (mL/g)
Purolite A-830	360	140	180
Purolite A-850	1,300	1,700	1,800
Purolite 530E	237,000	568,000	627,000
Purolite D-3696 *	165,000	1,877,000	1,842,000
WBR-112L	195,000	482,000	493,000
WBR-106	468,000	>3,000,000	>3,000,000
Amberlite PWA2	103,000	779,000	920,000

* Data from Gu et al.[6]

It is worth mentioning that, in addition to the bifunctional resin, the potential market for perchlorate removal and water treatment has also led to the development of other highly selective anion-exchange resins such as the

Amberlite PWA2 resin, recently marketed by Rohm and Haas. The PWA2 is a polystyrenic resin, and its detailed composition and surface functional groups remain proprietary.[15] Its initial reaction kinetics and the 1-h K_d appeared to be much slower than those of the bifunctional resins (Table 4), although its 24-h and 168-h surpassed the Purolite A-530E resin. A high initial K_d is particularly important to the removal of trace quantities of ClO_4^- under flow-through conditions, as will be discussed in the following section. Additionally, we found that, unlike the Purolite A-530E resin, the spent resin (PWA2) cannot be regenerated once loaded with perchlorate, even by washing with a combination of ferric chloride and hydrochloric acid solution.[16]

COLUMN FLOW-THROUGH OPERATIONS

Breakthroughs and Kinetic Effect

The performance of resins to remove ClO_4^- from contaminated water is also evaluated in continuous column flow-through systems. Unlike the batch ion-exchange process, fresh feed of perchlorate and competing ions is continuously added to the resin bed or column. As ion exchange progresses, the perchlorate concentration at the inlet of the resin bed loads to equilibrium with the concentration in the feed solution. Consequently, the solution and resin farther down the column equilibrate at lower concentrations of perchlorate so that the concentration at the outlet of the column can be very low or below the detection limit. Therefore, the column approach to ion exchange can remove perchlorate to a much lower concentration with less resin exchanger than the batch or fluidized bed-contacting techniques. In other words, for the concentration in the effluent solution to approach zero, its concentration must also approach zero in the resin exchanger. In a batch-type system, this is only possible if the amount of resin is infinite or through repeated contacts. In a column flow-through system, however, the effluent concentration depends only on the concentration of perchlorate in resins near the outlet, which may be at only trace levels during much of the loading process. As loading of perchlorate continues, a concentration gradient forms along the column flow direction; more of the resin bed will be in equilibrium with the feed solution, and the perchlorate concentration gradient will move further down the column, eventually resulting in the breakthrough of perchlorate (Figure 5).

Figure 5 illustrates the breakthrough of perchlorate in small laboratory packed resin columns. The data are plotted as the ratio of effluent concentration (C) to the influent concentration (C_0) of ClO_4^- against the number of bed volumes (BVs) of water that has passed through the column. The resin (Purolite A-530E) is usually wet-packed in columns as slurries and, in this case, a high-precision HPLC pump was used for feeding the ClO_4^- solution at a constant flow rate. A relatively high influent ClO_4^- concentration (1–10 mg/L) and a flow rate of ~17 BV/min were used in these experiments. At an influent ClO_4^- concentration of 1 mg/L (Figure 5a), no breakthrough of ClO_4^- was observed until about 2,000 BVs of solution was fed through the column, and <0.6% breakthrough occurred with an input of ~5,000 BVs of the test solution. Obviously, due to limited sorption capacity of the resin, the breakthrough will depend on the influent ClO_4^- concentration. At a high influent ClO_4^- concentration (10 mg/L), the breakthrough occurred more quickly, and C/C_0 increased rapidly (Figure 5b) because of a very high flow rate (~17 BV/min) used in such an accelerated laboratory test.

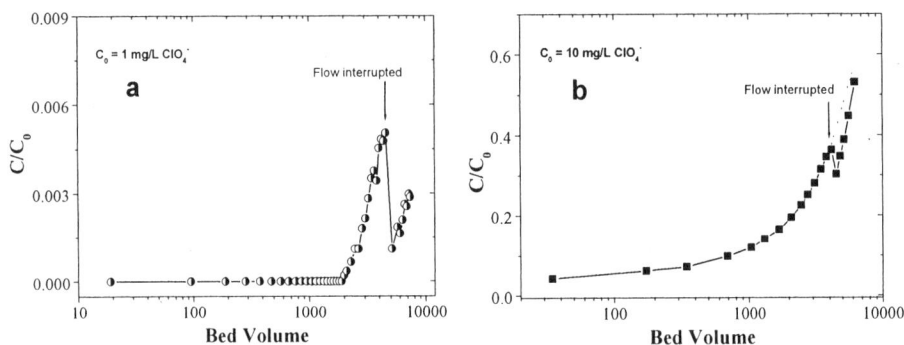

Figure 5. Effluent perchlorate concentration profiles in continuous column flow-through operations at two different influent concentrations (a = 1 and b = 10 mg/L ClO_4^-). The column (10×22 mm) was packed with Purolite A-530E resin, and the flow was halted overnight and restarted during column loading.

Breakthrough curves from these laboratory column experiments also provide information about the kinetic equilibrium behavior of the resin. The rate-limiting step can be inferred from the results of the "flow-interruption" test, in which the flow was halted overnight and restarted during column loading (Figure 5). The behavior of C/C_0 after the column was restarted is correlated to the mass-transfer mechanism. In other words, if the exchange rate is controlled by diffusion in the resin particles, diffusion of ClO_4^- ions within resin beads (Figure 1) would continue even after flow is stopped. Highly concentrated ClO_4^- ions at the outer layers of the resin bead will diffuse toward the particle centers, thereby leveling the concentration gradient across the resin bead. The net result was a sharp decrease in C/C_0 when the column was restarted (Figure 5a, b): the C/C_0 decreased from about 0.5 to 0.1% at the influent concentration of 1 mg/L, and from ~37 to 30% at the influent concentration of 10 mg/L when the flow was interrupted overnight. This translates to an increased sorption of ClO_4^- on the resin bed. As the run continues, the concentration gradients reestablish, and the breakthrough curve should slowly approach the shape it would have had without flow interruption.

Therefore, for resins with slow reaction kinetics such as the tributyl and trihexyl resins (Tables 1 and 2), a slow flow rate could substantially increase the resin performance to remove ClO_4^- from contaminated water. For example (Figure 6), at a flow rate of ~1.7 BV/min, no perchlorate breakthrough was observed until ~1,000 BVs of the test solution was passed through the tributyl resin column (WBR-004). In contrast, about 15% breakthrough of ClO_4^- occurred immediately (<40 BVs) when the column was operated at ~17 BV/min in the same laboratory accelerated breakthrough experiments. At such a high flow rate, one also would expect the trihexyl resin to perform poorly in removing ClO_4^- (Figure 7) despite the fact that its 24-h and 168-h K_d's are the highest among various monofunctional anion-exchange resins (Tables 1 and 2). On the other hand, the tripropyl and triethyl resins performed the best in these accelerated laboratory column

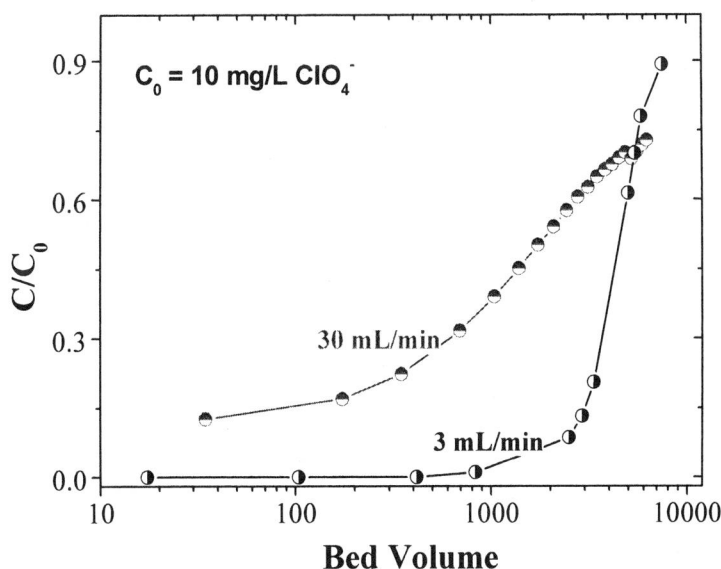

Figure 6. Effect of flow rate on the breakthrough of perchlorate in a packed column (10×22 mm) of the tributyl resin (WBR-004). The influent concentration was 10 mg/L ClO_4^- in a background solution of 3 mM $NaHCO_3$, 1 mM NaCl, 0.5 mM $CaCl_2$, 0.5 mM Na_2SO_4, 0.5 mM KNO_3, and 0.25 mM $MgCl_2$.

flow-through tests (Figure 7). Although the trimethyl resin (Purolite A-500) gives relatively fast reaction kinetics, it did not perform as well as the triethyl and tripropyl resins because of its relatively low selectivity (Table 2). It only outperformed the tributyl and trihexyl resins under the given experimental conditions.

It is noted, however, that the flow hydrodynamics could complicate the breakthrough by affecting the elution profile. Hydrodynamic effects include dispersion and channeling. Dispersion may be caused by molecular diffusion and eddy currents. The liquid tends to travel at different rates through various pore spaces within the resin bed so that the concentration in solution also changes. Unlike molecular diffusion, eddy dispersion is proportional to the flow rate and occurs mostly at high flow rates. Channeling is much the same as eddy dispersion but occurs on macroscopic scales. Liquid traverses the path of least resistance (or the lowest pressure drop). Therefore, if packing irregularities or resin shrinkage have caused uneven distribution of resins, water will bypass the resin, resulting in an early breakthrough.

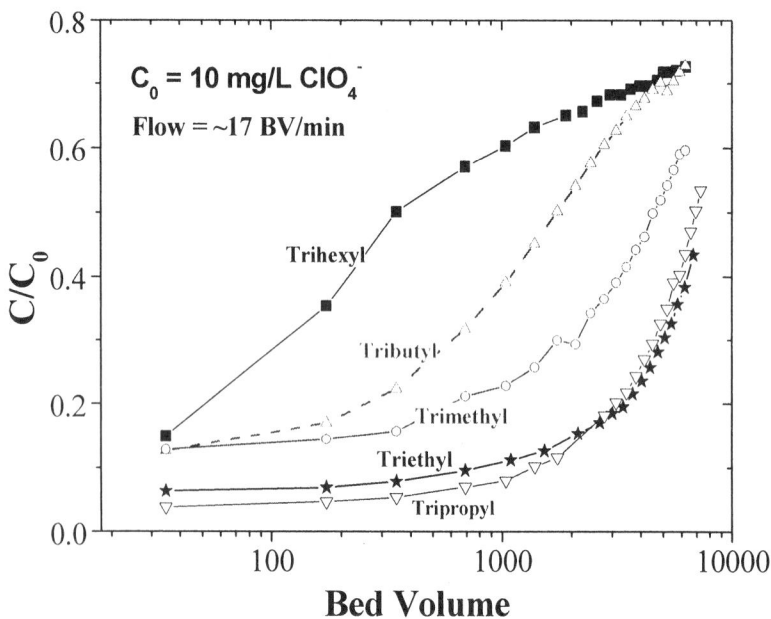

Figure 7. Comparisons of the breakthrough of perchlorate in packed columns (10×22 mm) of various monofunctional synthetic resins (Purolite A-500, WBR-100, WBR-102, WBR-004, and WBR-006) under the same experimental conditions as described in Figure 6.

Perchlorate Breakthroughs on Selected Ion-Exchange Resins

The column flow-through experiments also have been used to evaluate the performance of various commercially available selective anion-exchange resins such as the Amberlite PWA2 and Sybron SR-7 in comparison with the bifunctional resin, Purolite A-530E (Figure 8). Note that a nitrate-selective resin, Purolite A-520E, and an initial batch of the bifunctional resin, Purolite D3696,[6] also were included in Figure 8 for comparison. Results indicate that the bifunctional resin performed the best initially, and less than 10% breakthrough of perchlorate was observed after about 1,000 BV of test solution passed through the column at a high flow rate (~17 BV/min). Similar to the nitrate-selective resin (A-520E), a relatively high initial breakthrough (~10%) occurred at ~40 BV for the PWA2 resin, although its C/C_0 became lower than that of the bifunctional resin at the later stage because of its relatively high anion-exchange capacity.

These observations are consistent with the fact that the bifunctional resin shows relatively higher 1-h K_d but lower 24-h and 168-h K_d values than the PWA2 resin (Table 4). As has been reported previously,[12] a high initial K_d is thought to be particularly important to the removal of trace quantities of ClO_4^- from solution at a high flow rate, for example, in the treatment of drinking water containing only a few parts per billion of ClO_4^-. A high K_d value becomes less important at relatively high ClO_4^- concentrations because of decreased competition between ClO_4^- ions and other anions in solution for sorption on the resin. The K_d value also decreases as the ClO_4^- concentration in solution increases, as described earlier. At relatively high ClO_4^- concentrations, the bifunctional resins show limited or no advantages over other resins in sorbing ClO_4^-. The sorption kinetics and ion-exchange capacity are more important parameters to consider when competing anions are relatively low in solution. As can be expected, the monofunctional tripropyl resin, Sybron SR-7, performed the worst among these polystyrenic resins in removing ClO_4^- from the test solution. Nearly 18% breakthrough of ClO_4^- occurred when the first data point was taken at ~30 BV (Figure 8). However, it is emphasized again that these data were collected under accelerated laboratory experiments at a high flow rate (~17 BV/min) and a high influent ClO_4^- concentration (10 mg/L). The performances of these resins are still orders of magnitude better than those of polyacrylic and Type-II resins (e.g., Purolite A-830 and 850) in removing ClO_4^- from contaminated water at lower concentrations and lower flow rates (usually <2 BV/min).

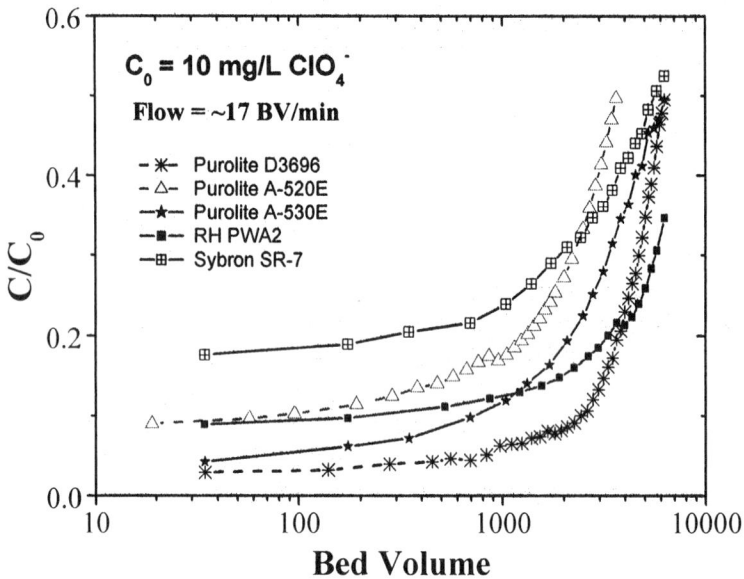

Figure 8. Perchlorate breakthrough curves on the bifunctional resin (Purolite A-530E), the Amberlite PWA2, and Sybron SR-7 resins in laboratory accelerated flow-through experiments, as described in Figure 6.   The breakthrough of a nitrate-selective resin, Purolite A-520E, and an experimental bifunctional resin, Purolite D3696, also were plotted for comparison.[6]

RESIN REGENERATION AND REUSE

Conventional Brine Regeneration

The use of highly selective ion-exchange resins increases the efficiency in removing ClO_4^- and may significantly reduce the operational cost during water treatment because a fixed resin bed can be left for unattended operation for months or even years, depending on the influent ClO_4^- concentration and the flow rate. An even more cost-effective approach would require the regeneration or recycling of the resin bed to further reduce the capital cost. The conventional wisdom is that strong brine solutions (e.g., 12% NaCl) are the only cost-effective means of regenerating an anion-exchange resin. Unfortunately, the exceptionally high affinity of ClO_4^- for the Type-I strong-base anion-exchange resins has made the regeneration little success even by washing with large excess volumes of concentrated brine solutions.[7, 13, 16, 17] The increasing order of affinity of singly charged ions for these resins is known to be perchlorate > nitrate > chloride > bicarbonate. For example, Tripp and Clifford[17] reported that, even with a relatively non-selective resin and with counterflow of the brine, regeneration required a large excess of NaCl. Similarly, Batista et al.[7] found that many bed volumes of concentrated brine were able to remove only a small portion of the loaded ClO_4^- from the Sybron SR-7 resin, and heating the ClO_4^--laden resins during regeneration had only limited success. Therefore, this brine regeneration technique is applicable only for Type-II resins or the Type-I resin with polyacrylic backbones.[7, 13, 17, 18]

The difficulties of brine regeneration of Type-I strong-base anion-exchange resins can be understood if we examine the distribution coefficients as a function of the brine concentration using pertechnetate as an analog for perchlorate sorption on the Purolite D3696, A-520E, and A-850 resins (Figure 9).[13] These measurements were made by equilibrating a sample of resin with a solution containing 6 μM TcO_4^- and the indicated NaCl concentration for 24h. Results clearly show that the least-selective polyacrylic resin, Purolite A-850, is the only resin that has a small enough K_d in 3 M NaCl to be effectively regenerated with a strong NaCl solution. The observed K_d in 3 M NaCl is ~17 ml/g, and the implications are that a given bed volume of resin will have to be contacted with approximately 10 BVs of 3 M NaCl or about 30 BVs of 1 M NaCl to remove sorbed TcO_4^- ions. The Purolite D3696 resin clearly is the most selective resin, and its distribution coefficient remains exceptionally high in 3 M NaCl so that it becomes impractical to be regenerated with a brine solution. Even with the less-selective resin (Purolite A-520E), the K_d is so high in 3 M NaCl that

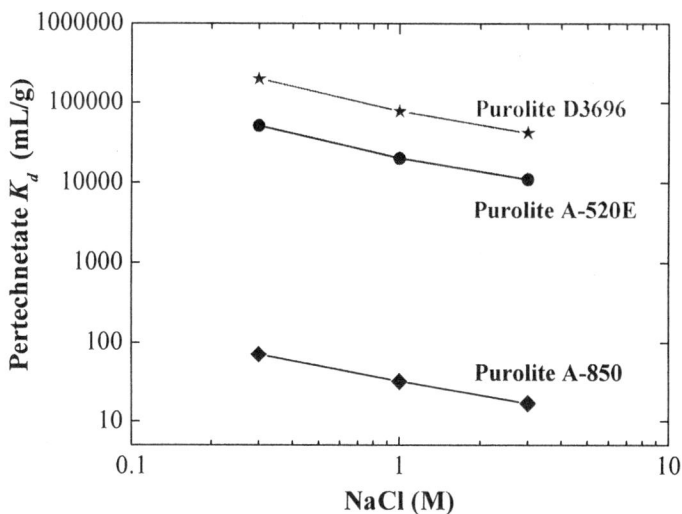

Figure 9. Distribution coefficients for pertechnetate (TcO_4^-) sorption on various resins in a 24-h equilibration period as a function of NaCl concentration (with permission from Kluwer Publishers).[13]

hundreds or thousands of bed volumes of brine will be required to completely remove TcO_4^- from the resin. This translates into high operating cost and waste-disposal problems associated with the brine containing TcO_4^- or ClO_4^-. Therefore, the high costs for resin regeneration and regenerant disposal would render the ion-exchange technology unattractive unless the two issues of resin regeneration and final waste disposal could be resolved.[7]

New Paradigm of Resin Regeneration

The exceptionally high selectivity of perchlorate on Type-I styrenic anion-exchange resins necessitates the development of a new paradigm of resin regeneration that can be used cost-effectively for practical applications.[16] As a general strategy, the logical approaches are (i) the use of reduction techniques, such as those described in Chapter 16, in which perchlorate ions are chemically or electrochemically reduced to chloride, and (ii) the use of displacement reactions in which the displacing anion could later be decomposed or be readily washed off the resin bed. These displacing anions are usually large, poorly hydrated anions like ClO_4^- and also must be strongly sorbed by the resin. Any of the regeneration schemes should not incur excessive cost, nor should their decomposition produce environmental challenges of their own.

Among various candidate anions studied, the most attractive candidate meeting these requirements has been identified as the tetrachloroferrate anion ($FeCl_4^-$), formed in a ferric chloride ($FeCl_3$) solution in the presence of an excess amount of hydrochloric acid (HCl) or chloride.[16, 19] The chemical equilibrium can be written as:

$$FeCl_3(aq) + Cl^- \rightleftharpoons FeCl_4^- \qquad (1)$$

Similar to the ClO_4^- ion, the $FeCl_4^-$ is a large, poorly hydrated anion and known as one of the most strongly extracted anions from HCl solutions by either liquid-liquid solvent extraction[20] or anion exchange.[21, 22] It is therefore anticipated to effectively displace ClO_4^- from the resin, particularly in the presence of an organic solvent. On the other hand, the $FeCl_4^-$ ion has a much desired chemical property - it decomposes in water or dilute acidic solutions according to the following rapidly established chemical equilibria, Eq. 2–5 (water of hydration not shown in the equation):

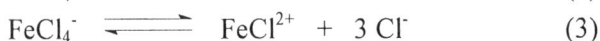

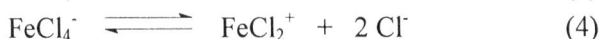

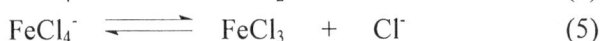

$$FeCl_4^- \rightleftharpoons Fe^{3+} + 4\,Cl^- \qquad (2)$$
$$FeCl_4^- \rightleftharpoons FeCl^{2+} + 3\,Cl^- \qquad (3)$$
$$FeCl_4^- \rightleftharpoons FeCl_2^+ + 2\,Cl^- \qquad (4)$$
$$FeCl_4^- \rightleftharpoons FeCl_3 + Cl^- \qquad (5)$$

Accordingly, by decreasing Cl^- concentration, the $FeCl_4^-$ anion converts to positively charged or neutral Fe(III) species such as Fe^{3+}, $FeCl^{2+}$, $FeCl_2^+$, and $FeCl_3$, which are readily desorbed from the resin by charge repulsion. Therefore, the resin is regenerated to its original state with Cl^- as the counter ion by charge balance. Because the rinse water does not contain ClO_4^-, it can be readily disposed of after proper neutralization by either passing it through a carbonate gravel filter bag or simply adding soda to the rinse water. Both $FeCl_3$ and HCl are relatively inexpensive, and their use in treatment of drinking water is well established. This methodology thus potentially offers a cost-effective means to regenerate strong-base anion-exchange resins loaded with ClO_4^-. This regeneration technology has been demonstrated successfully in field operations, as will be described in detail in Chapter 11, and it is being marketed by the Calgon Carbon Corporation.

Both laboratory and field regeneration studies have been carried out to evaluate the efficacy of using the $FeCl_3$-HCl technique to regenerate various resins that are loaded with ClO_4^-. The time-dependent desorption of ClO_4^- by the regenerant solution was first examined by Gu et al.[16] using the bifunctional D-3696 resin and the monofunctional A-520E resin that were sorbed with ClO_4^- at ~50% of their exchange capacity. It was reported that perchlorate ions are rapidly desorbed upon contacting with a solution consisting of 0.35 M $FeCl_3$, 2 M HCl, and 35% ethanol; about 60% of the sorbed ClO_4^- can be displaced within 30 min, and more than 96% of the ClO_4^- is displaced in 8 h. Desorption of ClO_4^- from loaded resin beds was then evaluated in laboratory using the same packed resin columns as those used in the column breakthrough experiments (Figures 5–8).[16] The amount of ClO_4^- desorbed during regeneration was monitored so that a mass balance could be determined. It is found that the sorbed ClO_4^- could be rapidly and effectively desorbed from the resin bed. About 98% of the ClO_4^- was eluted off the resin bed with about 10 BVs of the regenerant solution. Examination of the cumulative ClO_4^- desorption indicated a near 100% recovery or desorption of ClO_4^- after elution with ~15 BVs of the same regenerant solution (0.35 M $FeCl_3$, 2 M HCl, and 35% ethanol). As indicated earlier, because the $FeCl_4^-$ is a poorly hydrated anion and known as one of the most strongly extracted anions from HCl solutions by solvent extraction,[20] ethanol was used to enhance the extraction efficiency of ClO_4^- from the resin bed. However, it was eliminated later in the process because of potential health implications and environmental concerns related to the use of ethanol during large-scale field applications. Instead, a more concentrated solution of $FeCl_3$ (1 M) and HCl (4 M) was used in follow-on studies. This mixed solution of $FeCl_3$ and HCl was found to effectively regenerate the resin bed; nearly 100% of sorbed ClO_4^- could be eluted with as little as ~2 BVs of the regenerant solution in a small-scale field experiment.[11]

Figure 10 illustrates the effectiveness of using the $FeCl_3$ (1 M) and HCl (4 M) solution to elute ClO_4^- from the bifunctional resin bed (WBR-112L). The sorbed ClO_4^- was rapidly desorbed, and it resulted in a peak effluent concentration of ~35,000 mg/L (or ~350 mM) at ~2.5 BV. Within ±5% error, nearly 100% of ClO_4^- was recovered after eluting with about 5 BVs of the regenerant solution. Although our initial studies were focused on the regeneration of highly selective bifunctional resins, this same technique is equally applicable for the regeneration of other Type-I polystyrenic resins, as illustrated in Figure 11. Similarly, the sorbed ClO_4^- could be rapidly and quantitatively eluted off the resin bed by washing with as little as 3–5 BVs of the regenerant solution. Note that the difference between peak effluent concentrations of WBR-102 and WBR-004 (Figure 11) was attributed to different initial loadings of ClO_4^- on the resin bed. Regardless of the initial loading, the sorbed ClO_4^- could be recovered by eluting with 3–5 BVs of the $FeCl_3$ and HCl solution. In fact, even less of the regenerant is needed if a perfect plug-flow condition is maintained and a sufficient equilibrium time is given during the regeneration. Potential channeling and nonequilibrium exist in these laboratory small column experiments (Figures 10 and 11). As indicated earlier, only 2 BVs were found to be necessary for regenerating a relatively large column (5×30 cm) in a small-scale field experiment.[11] More recently in a large-scale field test (with a 100-gal resin bed), we achieved nearly 98% elution of the sorbed ClO_4^- by washing with ~1 BV of the

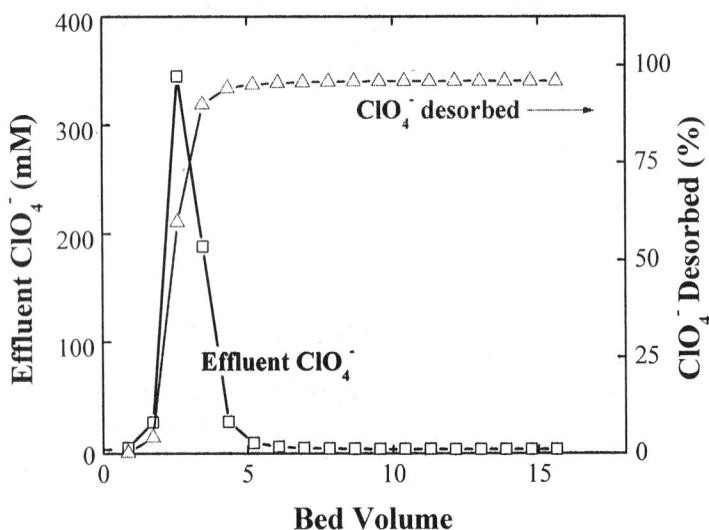

Figure 10. Regeneration of the bifunctional resin (WBR-112L) that had been treated with ClO_4^- at an influent concentration of 10 mg/L in a laboratory column flow-through experiment. The regenerant solution consisted of 1 M $FeCl_3$ and 4 M HCl.

regenerant solution, as will be described in detail in Chapter 11. Although additional regenerant solution is used to continue eluting residual ClO_4^- in the resin, these additional BVs of regenerant could be reused for next regeneration cycles without any significant changes in chemical property or efficiency.[6] Therefore, the high efficiency of the new $FeCl_3$-HCl regeneration scheme opens the door to the use of new, highly selective ion-exchange materials for perchlorate with enhanced treatment efficiency and longevity but reduced operational and capital costs in comparison with techniques of single use, throwaway resins.

Studies also have been performed to evaluate the resin performance after repeated treatment and regeneration cycles. No significant deterioration of resin performance was noted after seven regeneration cycles in laboratory flow-through experiments.[6] This conclusion has also been confirmed in our recent field studies (Chapter 11).

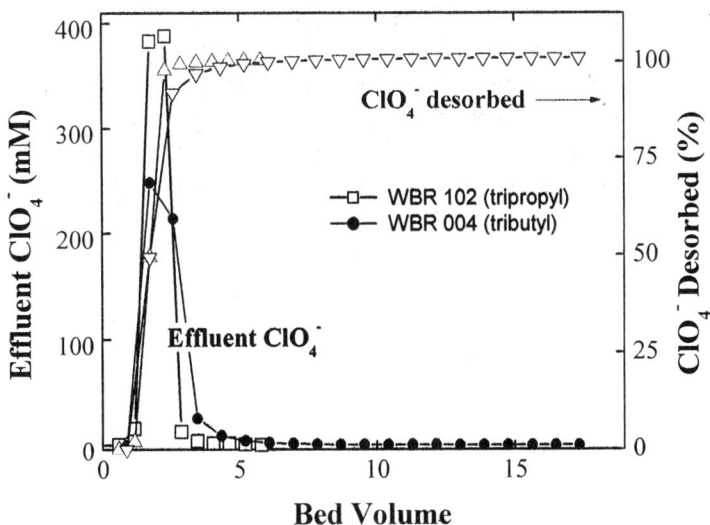

Figure 11. Regeneration of the monofunctional resins (WBR-102 and 004) that had been previously loaded with ClO_4^- in small packed columns (shown in Figure 7). The regenerant solution consisted of 1 M $FeCl_3$ and 4 M HCl.

However, there is one exception – although this new technique is found to be highly efficient in regenerating both the bifunctional and monofunctional Type-I strong-base anion-exchange resins (Figures 10 and 11), it failed to regenerate the new, highly selective Amberlite PWA2 resin (Figure 12). Under the same experimental conditions, <10% of the sorbed ClO_4^- was desorbed from the resin bed after eluting with more than 20 BVs of the

FeCl$_3$-HCl regenerant solution. The effluent ClO$_4^-$ concentration was very low (<2.5 mM), and no elution peaks were observed. Perchlorate appeared to be slowly and continuously released into the effluent with an increased input of the regenerant solution. This is in contrast to rapid desorption and sharp elution peaks observed within 1–5 BVs for the regeneration of the bifunctional resins (WBR-112L and Purolite D-3696)[11,16] and the monofunctional WBR-102 and WBR-004 resins. No satisfactory explanations could be given for these observations, primarily because of a lack of the chemical and structural information about the PWA2 resin (proprietary).[15]

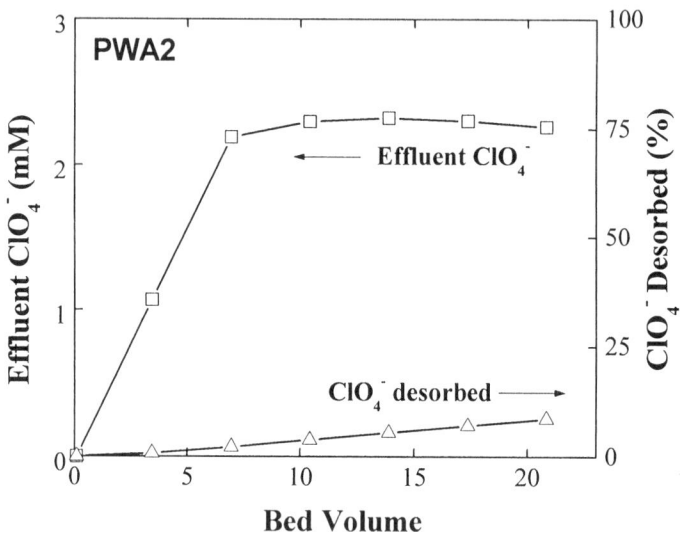

Figure 12. Regeneration of the Amberlite PWA2 resin that had been previously loaded with ClO$_4^-$ in a small packed column (shown in Figure 8). The regenerant solution consisted of 1 M FeCl$_3$ and 4 M HCl.

Regenerant Minimization and Perchlorate Recovery

Regeneration using the FeCl$_3$-HCl technique leads to highly concentrated perchlorate in the first 1–2 BVs of the regenerant solution, as described earlier. Our recent field studies show that the peak eluent ClO$_4^-$ concentration could reach as high as ~113,000 mg/L (or ~1.1 M ClO$_4^-$), and more than 98% of sorbed ClO$_4^-$ was eluted with only 1 BV of the regenerant solution (Chapter 11). This amount of waste production, i.e., 1 BV, is trivial in comparison with hundreds or thousands of BVs of brine wastes produced during the regeneration of Type-II or Type-I polyacrylic resins such as those used in the ISEP+^{TM} systems.[7, 18] This assumes the treatment of >100,000 BVs of contaminated water at concentrations of 50 μg/L or less.[6] In other words, only about 0.001% of wastes are produced compared with the amount of water treated. Venkatesh et al.[18] reported the production of about 0.75% and 1.75% of brine wastes by using the ISEP+ system for treating waters containing about 50–80 and 1200 μg/L ClO_4^-, respectively.

The waste regenerant solution may be further reduced to half BV because very little or no ClO_4^- is eluted in the first half BV of the regenerant solution if we examine the elution curve carefully. Therefore, perchlorate is concentrated in the second half BV of the regenerant solution. More importantly, because perchlorate is so concentrated but Fe^{3+} depleted (due to its sorption on resin) in the first BV of the regenerant solution, the eluted ClO_4^- could be readily recovered and recrystallized as pure solid phases such as $KClO_4$ due to its relatively low solubility in aqueous solution (K_{sp}=1.05×10^{-2} at 20°C). At the concentration of ~100,000 mg/L ClO_4^-, >90% of ClO_4^- can be recovered as pure solids. This translates into significant cost savings and waste reduction because the recovered perchlorate salt could be reused or disposed of. Moreover, the first BV of the regenerant solution can be reused after proper makeup with additional $FeCl_3$ and HCl because of a low ClO_4^- concentration in solution following its recovery.

The first BV of the regenerant solution with concentrated ClO_4^- also can be treated to destroy ClO_4^- either by thermal decomposition,[23] biodegradation of the neutralized solution, or electrochemical decomposition as described in Chapter 16. Thermal decomposition of ClO_4^- in the $FeCl_3$ and HCl solution is accomplished by the reduction of ClO_4^- ions with ferrous ion as a reducing agent at a elevated temperature (usually <200°C).[23] This process degrades ClO_4^- into chloride and water and oxidizes Fe^{2+} to Fe^{3+} ion, which is one of the essential gradients of the regenerant solution, so it can be recycled. Detailed discussion of perchlorate destruction is given in this and Chapter 16. Alternatively, the waste regenerant solution may be neutralized with NaOH so that ClO_4^- will stay in the neutralized brine solution for further treatment by techniques such as biodegradation. Recent studies have shown that certain species of microorganisms are capable of degrading ClO_4^- anaerobically in highly saline water.[24-26] For example, a mixed culture inoculated from marine sediment was found to reduce ClO_4^- and nitrate simultaneously in a synthetic medium in the presence of 30 g/L NaCl with the addition of some trace metals, Na_2S, and phosphate.[25]

OTHER NOVEL REGENERATION TECHNIQUES

It is worth mentioning that, in search for novel techniques to regenerate highly selective strong-base anion-exchange resins loaded with ClO_4^-, several candidate ions [including salicylate, tetrafluoroborate (BF_4^-), and thiocyanate (SCN^-)] also were found to effectively displace sorbed ClO_4^-, thereby regenerating the resin bed. They also are large anions with relatively low hydration energy and thus strongly sorbed by the Type-I polystyrenic resins. As such, a major issue by using these ions for resin regeneration is that a second or third displacement step is needed to regenerate the resin to its original state with Cl^- as the counter ion. With the exception of salicylate, regeneration using either tetrafluoroborate or thiocyanate could be costly and impractical. Therefore, only a salicylate regeneration technique is described below.

Salicylic acid is a naturally forming weak organic acid and known to be extracted from plants such as willow trees. It forms an anion in an alkaline solution and was first reported to displace pertechnetate anions loaded on activated carbon (similar to an anion exchanger).[27] Once it is sorbed on the resin, it could be readily desorbed by rinsing with a dilute HCl acid, in which it converts to a neutral species (salicylic acid) as illustrated in the following chemical equilibriums:

$$R–ClO_4^- \;+\; Sal^- \;\rightarrow\; R–Sal^- \;+\; ClO_4^- \qquad (6)$$

$$R–Sal^- \;+\; HCl \;\rightarrow\; R–Cl^- \;+\; Salicylic\ acid \qquad (7)$$

where $R–ClO_4^-$, $R–Sal^-$ and $R–Cl^-$ are sorbed ClO_4^-, salicylate, and chloride ions on the resin, respectively. Laboratory studies were performed to determine the regeneration efficiency of the bifunctional resin, Purolite D-3696, using the salicylate displacement technique. An effort was made to obtain a mass balance and to determine the minimum amount of regenerant that is needed to complete regeneration of the resin. Similarly to the regeneration by the $FeCl_3$-HCl (Figures 10-11), these studies were performed in small laboratory columns; the resin was first loaded with ClO_4^- by monitoring its breakthrough curves, followed by the regeneration using the salicylate regenerant solution consisting of 0.8 M Na-salicylate and 0.4 M NaOH in 20% methanol. The release of ClO_4^- is plotted as a function of the number of BV of the salicylate solution passed through the column (at a flow rate of 0.1 mL/min). Results indicate that perchlorate could be rapidly desorbed by washing with the salicylate solution; >80% of ClO_4^- was desorbed within the first 10 BVs (Figure 13). However, unlike the elution by the $FeCl_3$-HCl solution, a significant tailing or slow desorption occurred, and a nearly complete desorption of ClO_4^- (~97%) was achieved only after washing with ~60 BVs of the salicylate solution.

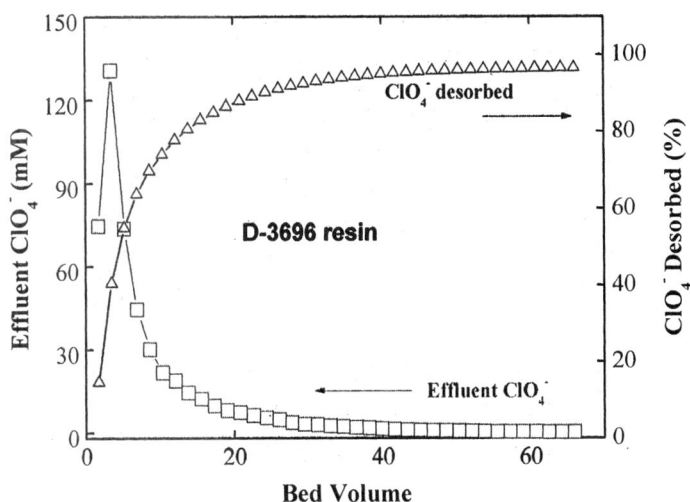

Figure 13. Regeneration of the Purolite D-3696 resin by salicylate displacement technique. The regenerant solution consisted of 0.8 M Na-salicylate and 0.4 M NaOH in 20% methanol.

These observations suggest that salicylate potentially could be used to regenerate strong-base anion-exchange resins loaded with ClO_4^-, although additional studies may be required to optimize the conditions such as the concentrations of salicylate, NaOH, and methanol. Note that an organic solvent is used to enhance the solubility of salicylate and the swelling of the resin copolymers, thereby increasing the desorption efficiency. Once perchlorate is completely desorbed, the sorbed salicylate could be readily rinsed off with a dilute acid (e.g., 0.3 M HCl in 40% methanol). As shown in Figure 14, salicylate was rapidly desorbed from the resin bed. More than 80% desorption occurred within the first 10 BVs, and nearly 100% desorption was observed after washing with ~30 BVs of dilute HCl (Figure 14). Therefore, the resin was regenerated to its original Cl^- form (as chloride displaces salicylate).

To validate the regeneration efficiency, repeated column breakthrough experiments also were performed. Results (Figure 15a) show that both the initial breakthrough curve and that after regeneration virtually overlap, suggesting successful regeneration of the bifunctional resin bed by the salicylate displacement technique. We also evaluated the applicability of this technique to regenerate monofunctional strong-base anion-exchange resins,

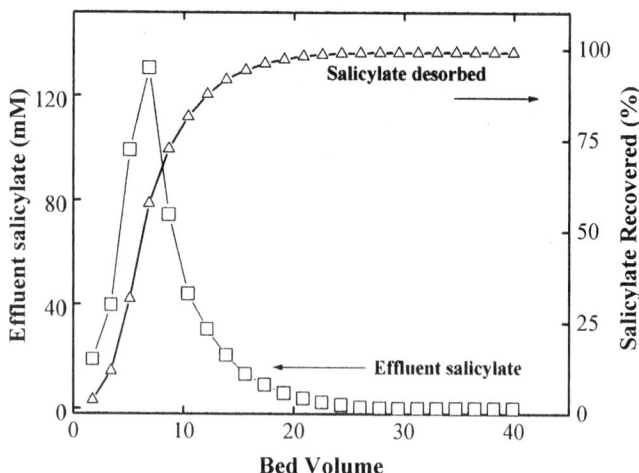

Figure 14. Desorption of salicylate from the bifunctional D-3696 resin by
0.3 M HCl in 40% methanol.

such as Purolite A-520E and Amberlite PWA2, by repeated regeneration and
breakthrough experiments. Similarly, results indicate that perchlorate could
be rapidly displaced by salicylate from the Purolite A-520E resin, but it
failed to regenerate the PWA2 resin. A nearly complete regeneration of the
A-520E resin but less than 30% regeneration of the PWA2 resin were
observed after eluting with ~25 BVs of the salicylate regenerant solution
(data not shown). The successful regeneration of the A-520E resin is
evidenced by the fact that no significant deterioration of the resin occurred
after repeated regeneration and breakthroughs (three cycles) (Figure 15b).

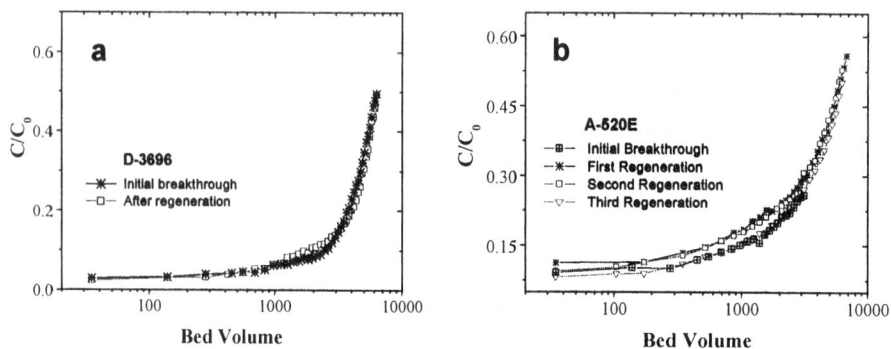

Figure 15. Breakthroughs of perchlorate in columns of (a) Purolite D-3696 and (b) A-520E
resins after repeated regeneration using salicylate displacement technique. Experimental
conditions are the same as used in Figures 7 and 8.

An advantage of using salicylate as a regenerant is that salicylate can be recovered for reuse or waste minimization after each regeneration cycle. This is accomplished by acidification of the alkaline salicylate regenerant solution, which results in the precipitation of salicylic acid due to its low solubility in aqueous solution. However, a disadvantage is that a complete regeneration of a spent resin bed may take many more bed volumes of the salicylate solution than the $FeCl_3$-HCl regenerant solution. Additionally, salicylate is not stable in alkaline solution and may undergo oxidation with notable color changes over an extended period (days or weeks).

PERCHLORATE DESTRUCTION AND REGENERANT RECYCLING

As described earlier, regeneration using the $FeCl_3$-HCl technique produces a very small quantity of secondary wastes (<0.001% of treated water) on the basis of the treatment of >100,000 BVs of contaminated water at concentrations of 50 μg/L or less using bifunctional resins,[6] and the concentrated ClO_4^- in the first BV of the spent regenerant solution may be recovered as pure solid phase $KClO_4$ and reused. An alternative approach is to destroy ClO_4^- in the first BV of the spent regenerant solution so that the regenerant could be recycled to further reduce the waste disposal costs.[23, 28] This can be accomplished by the reduction of ClO_4^- using ferrous iron (usually $FeCl_2$) as a reducing agent (Eq. 8).

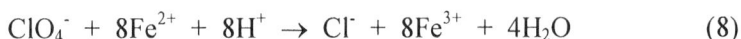

$$ClO_4^- + 8Fe^{2+} + 8H^+ \rightarrow Cl^- + 8Fe^{3+} + 4H_2O \qquad (8)$$

The reaction products are chloride, ferric ion (Fe^{3+}), and water. While ClO_4^- is reduced or destroyed, ferrous Fe^{2+} is oxidized to Fe^{3+}, which is one of the gradients of the $FeCl_3$-HCl regenerant solution and thus perfectly replenishes or "regenerates" this solution. This is because Fe^{3+} ions in the first BV of spent regenerant solution are depleted due to their sorption on the resin bed. The amount of ferrous ion required for ClO_4^- destruction depends on the concentrations of ClO_4^- in the spent regenerant solution. Usually, a 20–30% excess amount of Fe^{2+} ion is used because of the presence of other oxidants such as nitrate, which is also reduced or degraded into nitrogen in the process. After appropriate makeups to desired concentrations (with respect to Fe^{3+} and HCl), the regenerant solution can thus be reused. Obviously, the most important advantage of this perchlorate destruction process is the elimination of the need for disposal of hazardous regenerant wastes containing ClO_4^-. When combined with the selective ion exchange (for water treatment) and the $FeCl_3$-HCl regeneration techniques, essentially no secondary wastes are produced by the entire treatment process.

Detailed reaction kinetics between ferrous Fe^{2+} and ClO_4^- has been reported by Gu et al.[23] The reaction was found to be slow at temperatures below 100°C despite its favorable thermodynamics. This is because a high activation energy is required for the degradation of ClO_4^- to occur in aqueous solutions. However, the reaction accelerates as temperature is increased (Figure 16). For example, at 195°C, a complete reduction of ClO_4^- (with an initial concentration of ~9,000 mg/L) occurred in <1 h, and the reaction half-life was only ~8 min. The reaction followed the pseudo-first order rate law, and the reaction rate increased nearly three orders of magnitude when the temperature was increased from 110°C to 195°C. The estimated activation energy of the reaction is about 120 kJ/mol.[23]

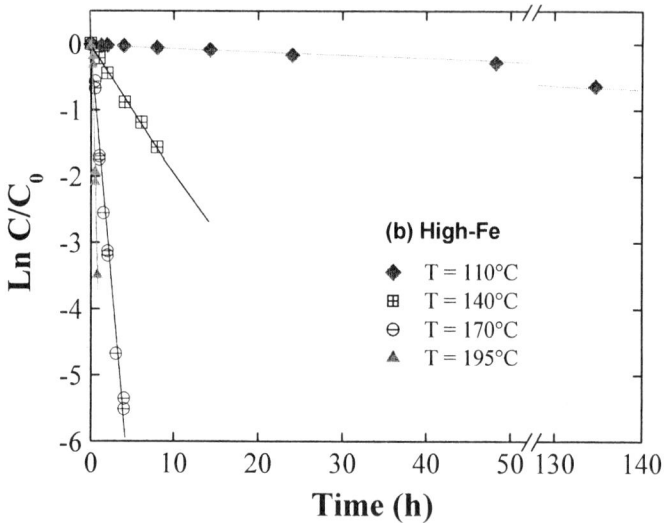

Figure 16. Perchlorate reduction kinetics by Fe^{2+} in $FeCl_3$-HCl solution at varying temperatures. The initial ClO_4^- concentration was about 9,000 mg/L[23] (with permission from the American Chemical Society).

On the basis of these kinetics studies, either a batch-type or flow-through reactor can be made to destroy ClO_4^- in spent regenerant solutions at an elevated temperature (<200°C). A relatively small reactor unit is usually needed because only a small volume (~0.5–1 BV) of spent regenerant solution needs to be treated per water treatment–regeneration cycle. When selective ion-exchange resins are used, the treatment phase can usually last ~6 months or longer if the treatment system is running at 1 BV/min at an influent ClO_4^- concentration of ~50 µg/L or less.[6] Additionally, a complete destruction of ClO_4^- is unnecessary because the presence of small quantities of ClO_4^- in the $FeCl_3$-HCl solution does not significantly affect its regeneration efficiency, as described previously.[16] Both a laboratory prototype flow reactor and a large-scale perchlorate destruction unit have been constructed and tested for their effectiveness in degrading ClO_4^-. The large-scale unit was made of tantalum (Ta) metal coil (Figure 17), which is inert to HCl and dilute perchloric acid at temperatures below ~230°C. The unit includes a corrosion-resistant pump capable of withstanding a backpressure of ~300 psi or higher, a Ta coil immersed in a large heating pot at one end and a small cooling pot at the other, and a needle valve to regulate the flow rate and pressure. As the mixture of ferrous Fe^{2+} and spent regeneration solution is pumped through the unit, perchlorate is destroyed in the heated Ta coil. This unit is capable of processing about 50 gal/day spent $FeCl_3$-HCl solution at a temperature of ~200°C. Detailed field test results will be presented in the case studies. In general, an average perchlorate destruction rate of 95% is achieved at an influent concentration of ~6,000 to 10,000 mg/L at an operating temperature of about 195°C. Laboratory studies also indicate that, at an input concentration of 9,000 mg/L ClO_4^-, >95% of the ClO_4^- was degraded after passing through the flow reactor at varying flow rates and a temperature above 170°C.[23] More than 98% of the ClO_4^- degradation can be achieved with a residence time ranging from 2 to 7 h.

Figure 17. A flow-through perchlorate-destruction unit is built with tantalum metal coil immersed in a heating oil bath (large pot) and then cooled in a small water bath at exit. Perchlorate is destroyed as it passes through the heated coil, and the spent $FeCl_3$-HCl regenerant solution is renewed and recycled.

Studies also have shown that the reaction rates and the perchlorate destruction efficiency depend on system pressure.[23,28] The higher the temperature and pressure and the longer the residence time, the more ClO_4^- is degraded. For example, at 170°C, the estimated reaction half-life is about 3.6

h. Theoretical calculations suggest that about 94% degradation would occur at a residence time of ~2 h. When the system pressure was increased to 300 psi, the reaction accelerated and resulted in the degradation of ~98.2% of ClO_4^- in a residence time of ~2 h. Nevertheless, both laboratory and field studies demonstrated the feasibility of the perchlorate destruction technology; it could be a valuable addition to the resin regeneration technology using the $FeCl_3$-HCl solution.

Figure 18 shows an integrated water treatment system using selective ion exchange, resin regeneration, and perchlorate destruction. The system could potentially result in tremendous savings in capital and operational costs and, more importantly perhaps, in minimized secondary waste production. In this process, three resin beds are used, with two in the lead-and-lag configuration and the third in reserve. When the lead resin bed is exhausted, it is regenerated (as a reserve), and the lag resin bed becomes the lead, and so on. The resin regeneration and perchlorate destruction can be performed on-site or off-site, depending on site-specific conditions and regulatory requirements. In the case of large-scale applications with multi-treatment systems in operation, a centralized or an off-site regeneration and destruction station is preferred to further reduce overall capital and operational costs. This is because the regeneration and perchlorate destruction process usually

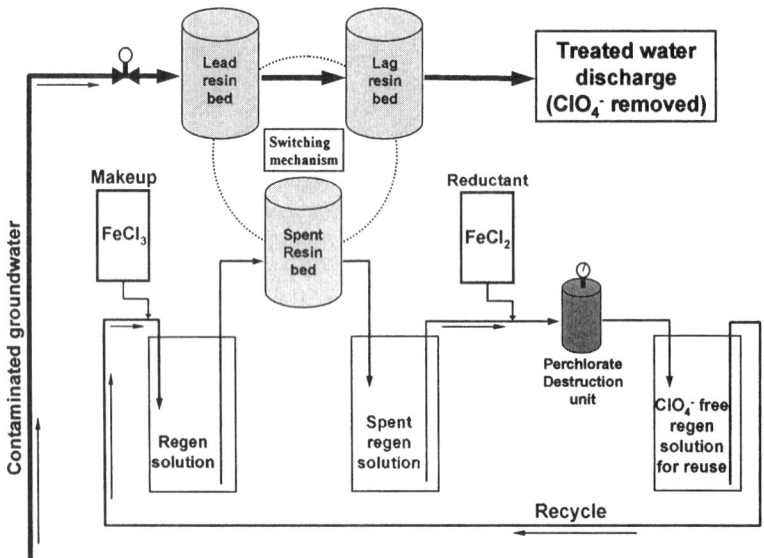

Figure 18. Schematic illustration of an integrated water treatment using selective ion exchange, resin regeneration, and perchlorate destruction system. Spent resin is regenerated using the $FeCl_3$-HCl solution, and perchlorate is destroyed in spent regenerant so that both resin and regenerant solution can be recycled, resulting in minimized waste production.

takes about 1–2 weeks whereas the water treatment phase can usually last months or years, depending on the influent ClO_4^- concentration and the flow rate. Additionally, a centralized regeneration and destruction station may be built to avoid the handling of hazardous hydrochloric acid at each treatment site.

Another perchlorate destruction system has been developed at Calgon Carbon Corporation to degrade ClO_4^- in the NaCl brine solution used in the ISEP+ system.[18] Similarly, nitrate is also degraded in the process, and the system is called perchlorate and nitrate destruction module (PNDM). The module uses a catalytic chemical reduction process for treating ClO_4^- and nitrate in the waste brine. In this process, ammonium is added as a reductant. The system is operated at 250°C and is relatively energy intensive because of a higher operating temperature required and a much higher waste brine volume (~0.75% of treated water) produced by the ISEP+ system[18] than by the $FeCl_3$-HCl regeneration, as described earlier. The PNDM also includes a nanofiltration unit to remove sulfate present in the brine. Pilot studies indicate that the PNDM can effectively reduce both ClO_4^- and nitrate present in the regeneration brine waste, and that the treated regenerant stream, after appropriate makeups, can be reused or recycled to regenerate polyacrylic resins used in the ISEP+ system.

ION-EXCHANGE SELECTION ISSUES AND COST ANALYSIS

When evaluating the potential usefulness of an ion-exchange system for the remediation of perchlorate contaminated water, several important factors must be considered. They include:

- Site-specific water chemistry and water quality
- The feed perchlorate concentration and discharge requirements
- Generation of secondary waste streams
- Presence of co-contaminants such as uranium and organic solvents
- Total capital and operational costs.

These factors may determine the choice of ion-exchange media (e.g., selective or non-selective resins), whether or not to regenerate the resin (i.e., single use or multiple use of the resin bed), regeneration frequency and methodology, or the use of other treatment technologies such as biodegradation.

Effect of Water Chemistry and Water Quality

The water chemistry and quality can have a significant impact on the treatment effectiveness of ion-exchange systems. In particular, the presence of competing anions such as nitrate, sulfate, chloride, and carbonate greatly limit the treatment efficiency and longevity of ion-exchange systems, particularly the conventional non-selective ion exchange. These competing anions are usually present in orders-of-magnitude higher concentrations than the ClO_4^- concentration in water. Divalent anions such as SO_4^{2-} are particularly strong competitors for sorption onto the Type-II or Type-I acrylic resins because of their relatively low selectivity for ClO_4^- sorption (Figure 3 and Table 3). As such, most of the exchange sites (>99%) in the resin bed usually are sorbed with competing anions rather than ClO_4^- ions by mass reaction because of a relatively low concentration of perchlorate in the contaminated water. Therefore, for treatment of water with high competing anion concentrations but low ClO_4^- concentrations (e.g., at sub-ppb ranges found in drinking water), selective ion exchange is preferred. The order of selectivity of these anions on selective ion exchangers is $ClO_4^- \gg NO_3^- > Cl^-$ > SO_4^{2-}, and sulfate is the least strongly sorbed ion on selective ion exchangers such as the bifunctional resin.[5, 6, 12]

High levels of dissolved organic matter and suspended solids in water may cause fouling and clogging of the resin bed, resulting in an increased pressure drop and/or a restricted flow. Dissolved organic matter is common in groundwater or drinking water, as it is derived from soil, plants, animals and/or microorganisms.[29-31] Its concentration typically ranges from 0.05 to >10 mg/L. It is a mixture of various natural organic compounds (e.g., organic acids, carbohydrates) and usually carries multiple negative charges as a result of the deprotonation of carboxyl and hydroxyl functional groups of the organics. It is thus strongly sorbed by anion-exchange resins, particularly the gel-type, polyacrylic resin. Therefore, more frequent regeneration and/or replacement of the exchange resin may result. However, previous studies indicate that dissolved organic matter does not appear to significantly impact the removal of ClO_4^- by using highly selective ion-exchange resins, even after repeated treatment and regeneration cycles.[6, 11] The use of strong acidic HCl solution (with $FeCl_3$) appears to be especially helpful in cleaning the resin bed during the regeneration process.

Suspended solids such as fine sands and colloidal clay particles usually can be dealt with, for example, by adding an in-line particle filter (e.g., sand filters or membrane filters). However, high levels of dissolved calcium (Ca^{2+}) and ferrous ions in water sometimes could be troublesome because they may not be filtered out by pre-filtration. Calcium carbonate may precipitate out within the resin bed because of changes in temperature and partial pressure as groundwater is pumped above ground. The solubility of calcium carbonate in water generally decreases in the temperature range of 25–95°C.[32] Calcium also may form precipitates (e.g., $CaSO_4$) with sorbed sulfate within the resin bed. Depending on the concentrations, the precipitation of both $CaCO_3$ and $CaSO_4$ could cause bridging of resin particles, restricting flow and increasing the pressure drop. The ferrous ion concentration in water usually is very low under oxic and suboxic conditions. Significant amounts of dissolved ferrous ion may be found in groundwater only under anaerobic or reducing conditions. It could form precipitates or coatings of ferric oxyhydroxides in the resin bed upon contact with air. Under these circumstances, the water may be aerated and filtered to remove iron before it is routed through the resin bed.

Effect of Perchlorate Concentration in Water

Selection of ion-exchange technology also is influenced by the initial ClO_4^- concentration in water and the treatment target concentration. Selective ion-exchange resins could perform orders of magnitude better than those non-selective polyacrylic and monofunctional polystyrenic resins at low ClO_4^- concentrations (e.g., <1 mg/L). On the other hand, at relatively high ClO_4^- concentrations but low competing anions, the selectivity becomes unimportant because of decreased competition between ClO_4^- and other anions in water. Under these conditions, selective ion exchange only shows limited advantages over conventional non-selective resins for sorbing ClO_4^-. As described earlier, the selectivity or distribution coefficient, K_d, is a function of the ClO_4^- concentration, and it decreases as the ClO_4^- concentration in solution increases because the sorption isotherms are not linear at high concentrations.[12] The above concept may be best illustrated in Figure 19. For example, at a relatively high ClO_4^- concentration (~5 mg/L), both the monofunctional A-500 and the bifunctional A-530E resins are able to sorb ClO_4^- at >100 mg/g. At the concentration of >50 mg/L, even the polyacrylic resin, Purolite A-850, is able to sorb ~100 mg/g ClO_4^- (data not shown).[12] However, at the concentration of 0.2 mg/L, the A-530E resin sorbed ~75 mg/g ClO_4^- in comparison with only ~12 and <1 mg/g ClO_4^- sorbed by the A-500 and A-850 resin, respectively.

These observations demonstrate that a high K_d for ClO_4^- sorption is particularly important to the removal of trace quantities of ClO_4^- from solution, for example, in drinking water containing only a few parts per billion of ClO_4^-. As noted earlier, the bifunctional resin initially was developed for removing trace quantities of radioactive pertechnetate anion at parts per trillion concentrations from contaminated water.[9, 10, 13] Therefore, highly selective resins are best suited for treating groundwater or drinking water that is contaminated with low levels of ClO_4^-. At high ClO_4^-

concentrations (e.g., >50 mg/L), ion exchange may not be a cost-effective approach for water treatment, simply because of its limited sorption capacity, causing the resin bed to be replaced or regenerated frequently. Under these circumstances, bioremediation might be a better choice, particularly if the water quality is unfavorable for the use of ion exchange.

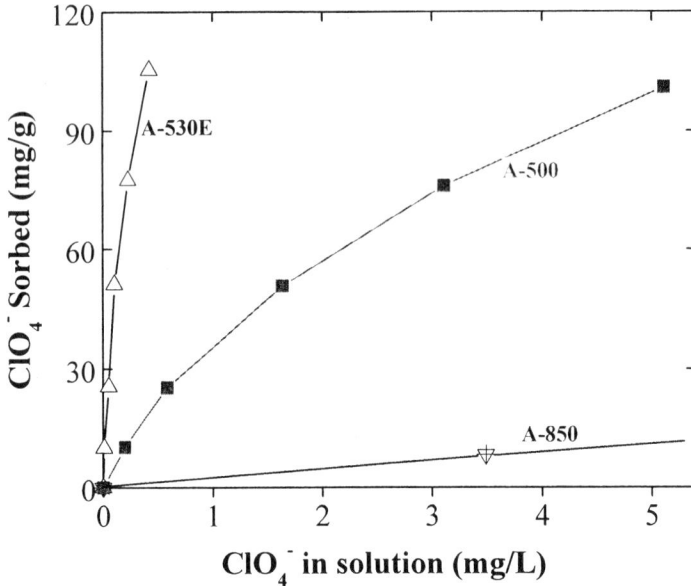

Figure 19. Sorption isotherms of ClO_4^- by Purolite A-530E, A-500, and A-850 resins in a background solution of 3 mM $NaHCO_3$, 1 mM NaCl, 1 mM Na_2SO_4, and 1 mM KNO_3.[12]

Effect of Uranium and Co-contaminant Sorption

The naturally occurring uranium [as U(VI)] in water could be a concern in certain regions of the United States, thus affecting the selection of ion-exchange techniques for treatment of perchlorate-contaminated water. Uranium-238 is among the longest-lived radionuclides (with a half-life of ~4.5×10^9 years) and poses health risks to humans, particularly at relatively high concentrations.[33] The most stable oxidation state of uranium in oxic and suboxic environments is U(VI). Most groundwater contains carbonates and bicarbonates that complex with U(VI) to form stable anionic species[33-35] such as $UO_2(CO_3)_2^{2-}$ and $UO_2(CO_3)_3^{4-}$. These highly charged U(VI) anionic species are thus retained and concentrated by anion-exchange resins as a result of electrostatic interactions, even if the U(VI) concentration in water may be low (e.g., below the drinking water MCL of 20 µg/L).[33] In fact, the sorption of U(VI) on anion-exchange resins is known, and previous studies have indicated that many strong-base anion-exchange resins show varying degrees of sorption of U(VI) in water.[12, 33, 36-38] The use of anion-exchange resins to remove U(VI) has been a common water purification practice in many regions of the world where the concentration of naturally occurring U(VI) is high.[33, 36, 37, 39]

From the human health perspective, it is certainly desirable to remove both uranium and ClO_4^- from contaminated water. However, the sorption and concentration of uranium by the resin is a potential concern, because mixed wastes containing both radioactive uranium and ClO_4^- are generated either in the spent resin itself or in the solution used for resin regeneration. Fortunately, various synthetic resins show different sorption affinities for U(VI), and the sorbed U(VI) usually can be eluted by a dilute acid wash.[12] This is because U(VI)-carbonates are unstable in acid and thus dissociate into UO_2^{2+} and CO_2 according to the following equilibrium (Eqs. 9 and 10):

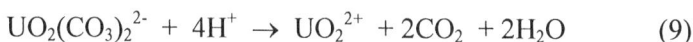

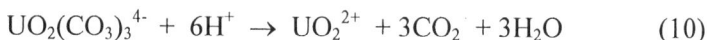

$$UO_2(CO_3)_2^{2-} + 4H^+ \rightarrow UO_2^{2+} + 2CO_2 + 2H_2O \qquad (9)$$

$$UO_2(CO_3)_3^{4-} + 6H^+ \rightarrow UO_2^{2+} + 3CO_2 + 3H_2O \qquad (10)$$

Therefore, U(VI)-carbonates convert to positively charged UO_2^{2+} ions, which are readily desorbed from the anion-exchange resin due to charge repulsion. These properties of U(VI)-carbonates can thus be used to separate uranium from sorbed ClO_4^- for secondary waste segregation and minimization.[12] There are two options to get rid of uranium during water treatment: (1) the use of a small pre-treatment resin canister, which is specific to sorb uranium but not ClO_4^-, and (2) a pre-wash of the spent resin bed with a dilute acid before regeneration using the $FeCl_3$-HCl technique. Because perchlorate is so strongly sorbed (on Type-I styrenic resins), it will remain on the resin while uranium is desorbed.

The sorption and desorption behavior of both uranium and ClO_4^- has been studied in great detail recently by Gu et al.[12] using a number of synthetic resins including Purolite A530E, A520E, A-500, and A-850, WBR-109, and Dowex 1-X8. The gel-type, polyacrylic resin (A-850) was found to have the highest sorption affinity for uranium (K_d up to ~600,000) but the lowest affinity for ClO_4^- sorption, as described earlier. This high affinity of the A-850 resin for uranium is attributed to its relatively high anion-exchange capacity and its aliphatic, gel-type acrylic backbone structure, which makes the anion- exchange sites more mobile or flexible than those in resins with a rigid polystyrene backbone structure. As a result, multicharged anionic species such as $UO_2(CO_3)_2^{2-}$ and $UO_2(CO_3)_3^{4-}$ are sorbed more strongly (due to electrostatic forces) by gel-type acrylic resins than by polystyrenic resins.[8] Similarly, the A-850 resin showed the highest sorption for SO_4^{2-}. On the other hand, the bifunctional A-530E resin exhibits a low affinity and kinetics for uranium and SO_4^{2-} sorption because of the use of both small and large trialkylamine functional groups, which result in a larger charge separation and thus weaker sorption of multicharged anionic species. Indeed, the bifunctional resins are designed to preferentially sorb poorly hydrated and singly charged anions such as ClO_4^-, as discussed earlier.

As can thus be expected, the sorbed uranium would be more readily desorbed from the bifunctional resin than from the A-850 resin. As shown in Figure 20, a single wash (once with 0.01 M HCl) resulted in the desorption of ~75% of the sorbed U(VI) from the bifunctional resin A530E but only ~20% from the A-850 resin and <10% from the Dowex 1-X8 and Purolite A-500 resins. Similarly in a column flow-through experiment, uranium was found to be readily eluted off the A530E resin, and ~100% of the U(VI) can be desorbed or recovered by elution with 0.01 M HCl.[12] However, <80% of U(VI) was eluted from the A-850 resin column even after elution with ~250 BVs of the acid. These observations demonstrate the effectiveness of dilute acid wash for uranium desorption and segregation from sorbed ClO_4^- in the resin bed. For treatment of water containing both uranium and ClO_4^-, the bifunctional resin is thus preferred; a dilute acid prewash before regeneration can prevent the production of mixed hazardous wastes because the dilute acid is ineffective in desorbing ClO_4^- from the spent resin bed. The acid washing solution containing uranium may be further neutralized, and uranium can be precipitated (as hydroxides) for waste minimization during field remediation processes.

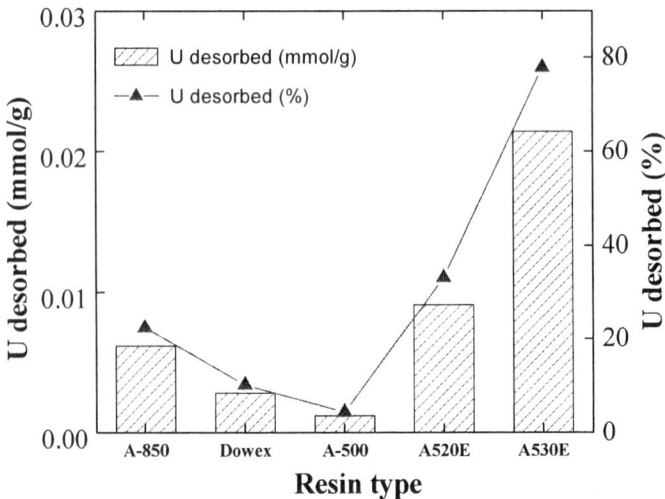

Figure 20. Uranium desorption from various synthetic resins by a single wash with 0.01 M HCl.[12]

Note that uranium also may be bio-reduced, precipitated, and concentrated in bioreactors during bioremediation of contaminated water containing both uranium and ClO_4^-. No data are currently available, although a variety of anaerobic microorganisms are known to rapidly reduce uranium, and the reduced U(IV) species are only sparingly soluble.[40-43] In fact, biological reduction of uranium has been proposed as one of the promising remediation technologies for immobilizing uranium in soil and thus impeding its migration in groundwater.

Sorption of other co-contaminants such as arsenic and organic solvents on synthetic resins are not well studied to date. However, it is generally known that arsenite [As(III)] is poorly sorbed on anion-exchange resins. Although arsenate [As(V)] can be sorbed by anion-exchange resins, the sorption is relatively weak.[44, 45] Other anions, especially SO_4^{2-}, strongly compete with arsenate for sorption sites. Therefore, unless the arsenic concentration is high in water, it is not expected to be strongly retained in the resin bed during water treatment. Some organic solvents may be retained by the ion-exchange resin, given the hydrophobic nature of the resin materials. Further studies are obviously needed in order to determine the sorption affinity and capacity of resins for various organic solvents. However, it is safe to say that multiple contaminants and/or high concentrations of contaminants may warrant the use of multiple technologies to completely treat a contaminant plume.

Field Applications and Cost Considerations

Currently, Calgon's ISEP+ continuous ion-exchange system using acrylic resins, the selective and regenerable ion exchange using $FeCl_3$-HCl regeneration, and the older fixed-bed, single-use system using Type-I selective resins are among the most commonly used ion-exchange techniques for treating water contaminated with ClO_4^-. Major pros and cons of using these different treatment techniques are given in Table 4. As stated earlier, fixed-bed processes usually consist of three or more beds in a sequential or parallel operation where at least one of the beds is polishing the water that has been treated by the lead or upstream beds (Figure 18). Normally two or more beds are in the adsorption mode, with one bed held in reserve. When the bed that has been in adsorption mode the longest becomes saturated, it is removed from service and replaced by a new or regenerated one. The continuous ISEP+ system divides the resin into multiple smaller beds.[18] All the ion-exchange process steps are carried out continuously, so that regenerant flows and concentrations are virtually constant ("merry-go-round" operation). Continuous ion exchange allows techniques that reduce waste volumes to be used and can support the addition of processes for the destruction of ClO_4^- by the PNDM system. However, because low selective acrylic resins are used in the ISEP+ systems, the whole treatment process

Table 4. Comparisons of advantages and disadvantages of using different ion-exchange and regeneration techniques of (1) conventional non-selective ion-exchange and brine regeneration, (2) single-use, fixed-bed system, and (3) highly selective and regeneration by FeCl₃-HCl techniques.

Non-selective, brine regeneration	Single-use, fixed-bed system	Selective and regenerated by FeCl₃-HCl
○ Effective and able to remove ClO_4^- below the detection limit	○ Effective and able to remove ClO_4^- below detection limit	○ Highly efficient and last >5 times longer
○ Fast reaction, high flow rate, and simple operations	○ Fast reaction, high flow rate, and simple operations	○ Particularly suited to remove ClO_4^- at low concentrations but high competing ions
○ Practical and economical	○ Simplicity and low capital cost	○ No changes to water quality and no remineralization
● Competition by other anions (e.g., NO_3^-, Cl^-, $SO_4^=$, HCO_3^-)	● High operational cost for resin replacement and disposal	○ Fast reaction, high flow rate, and simple operations
● Frequent regeneration and production of large quantities of secondary brine wastes	● Competition by other anions (e.g., NO_3^-, Cl^-, $SO_4^=$, HCO_3^-)	○ Low regenerant volume, 1 BV per regeneration cycle
● High capital cost for continuous ion-exchange & regeneration system	● Suspended solids causing decreased efficiency and clogging	○ Perchlorate recovered or destroyed, and regenerant recycled
● Changes to water quality and remineralization may be needed		○ Low capital and minimized operational and waste disposal costs
● Suspended solids causing decreased efficiency and clogging		● Handling of acidic regenerant solution
		● Resin cost about twice more than non-selective resins
		● Suspended solids causing decreased efficiency and clogging

generates a significant amount of brine wastes, about 0.75% to 1.75% of treated water, depending on the ClO_4^- and other competing ion concentrations in the feed water.[18,46,47] Additionally, the capital costs associated with the ISEP+ continuous ion-exchange system have been reported to be more than $4 million dollars for a 2,500-gpm treatment system at California's La Puente Valley County Water District.[47]

The use of highly selective, regenerable ion-exchange systems developed at ORNL is a great enhancement of the ion-exchange process and could result in significant cost savings. As described earlier, the combined technology of highly selective ion exchange, resin regeneration using $FeCl_3$-HCl, and perchlorate destruction is relatively new and is being tested in field at flow rates up to 150 gpm. Although no cost information is available for a full-scale treatment system, results to date are promising because (i) the selective bifunctional resin could last 5 to 10 times longer than most of the Type-I styrenic resins during water treatment phase (assuming the feed ClO_4^- concentrations of 50 µg/L or less), (ii) the spent resin can be effectively regenerated with as little as 1 BV of the $FeCl_3$-HCl solution, and (iii) perchlorate in the first BV of regenerant solution can be recovered or destroyed and the regenerant recycled. Additionally, the resin price has dropped significantly during the last few years down to ~$200/cu. ft and thus is comparable to some of the selective ion-exchange resins, due to increased production by Purolite. Therefore, even without the destruction process, the economic advantage of this combined treatment process could be evaluated by considering the cost of ~1 BVs of HCl and $FeCl_3$ solution versus

(a) costs of ~5–10 BVs of fresh resin plus its disposal (loaded with ClO_4^-) if a relatively low selective strong base anion-exchange resin is used but not regenerated, or

(b) costs of the treatment and/or disposal of hundreds of BVs of concentrated NaCl brine solution if non-selective Type-II or Type-I acrylic resins are used and regenerated.[47]

There are a few vendors who supply potable water grade $FeCl_3$ as a 40% solution and food grade hydrochloric acid (36-38%) at commodity chemical prices. Midland Resources, Inc. (Lawrence, Kansas), can supply potable-water-grade $FeCl_3$ solution in a 3,700-gal tank truck for about $4,000. This material is NSF certified, and it is ordinarily used as a flocculating agent or for the precipitation of phosphates and sulfides in drinking water applications. BASF Company (Geismar, Louisiana) produces HCl as a by-product and sells a 36% food-grade HCl in a 20-ton tank truck at ~$50/ton, although shipping costs could be as high as the material cost itself. In other words, the actual cost of HCl is likely on the order of $100/ton delivered. Since HCl is about 5 kg/gal, the tank truckload will contain about 4,000 gal HCl. Therefore, to make 2,500 gal regenerant solution containing 1 M $FeCl_3$ (~15%) and 4 M HCl (~12%), it requires the use of 1,000-gal $FeCl_3$ at 40% and ~840-gal HCl at 36%, and the remainder is water. This translates into a total cost of <$1,500 per 2,500-gal regenerant solution. Obviously, this is only a fractional cost (<5%) in comparison with purchasing a 2,500-gal single-use resin bed, which could cost >$30,000 (assuming the resin cost is ~$100 per cu. ft for many of the Type-I styrenic resins). Significant labor costs also must be accounted for when using low-selective resins that require more frequent replacement of resin beds. Again, depending on water quality (i.e., competing anions) and the perchlorate feed concentration, the bifunctional A-530E resin could last 5–10 times longer than many of the commercially available resins currently in use, with the exception of the PWA2 resin. However, because the PWA2 resin cannot be regenerated, its cost ($60,000–$75,000 for a 2,500-gal bed at ~$200 per cu. ft) is still more than an order of magnitude higher than the cost to make up 2,500 gal of the $FeCl_3$-HCl regenerant solution.

Additional cost savings also should be considered by precipitating or recovering perchlorate solids or destroying ClO_4^- ions in the concentrated regenerant solution and thus recycling the regenerant solution, as described earlier. The only drawback to using the $FeCl_3$-HCl regeneration is the use of hydrochloric acid. As recommended previously, a centralized or off-site regeneration station will be a valuable addition when several treatment facilities are in simultaneous operation. It will not only reduce the overall operational costs but also avoid the use of acid in the field. Handling of corrosive materials such as HCl is a routine process in industry, but a centralized service minimizes the environmental impact, and Calgon Carbon Corporation is in the planning stage of offering such services.

ACKNOWLEDGMENTS

We thank Y.K. Ku and H. Yan for their technical assistance in laboratory. This project was partially supported by the Environmental Security Technology Certification Program (ESTCP) of the U.S. Department of Defense, the Corporate Environmental Safety and Health of Lockheed Martin Corporation, the Environmental Management Restoration Branch at Edwards Air Force Base, and the U.S. Department of Energy (DOE). Oak Ridge National Laboratory is managed by UT-Battelle, LLC, under contract DE-AC05-00OR22725 with DOE.

REFERENCES

1. Nachod, F. C., Schubert, J. *Ion exchange technology*; Academic Press: New York, 1956.

2. Arden, T. V. *Water purification by ion exchange*; Plenum Press: New York, 1968.

3. Kunin, R. *Ion exchange resins, 2nd Ed.*; Krieger Pub.: Malabar, Fla., 1972.

4. Helfferich, F. *Ion Exchange*; Dover Publications, New York, 1995.

5. Gu, B., Ku, Y., Jardine, P. M. Sorption and binary exchange of nitrate, sulfate, and uranium on an anion-exchange resin. *Environ Sci. Technol.*, 2004; *38*:3184-3188.

6. Gu, B., Brown, G. M., Alexandratos, S. D., Ober, R., Dale, J. A., Plant, S. In *Perchlorate in the Environment*; Urbansky, E. T., Ed.; Kluwer/Plenum: New York, 2000, pp 165-176.

7. Batista, J. R., McGarvey, F. X., Vieira, A. R. In *Perchlorate in the Environment*; Urbansky, E. T., Ed.; Kluwer/Plenum: New York, 2000, pp 135-145.

8. Moyer, B. A., Bonnesen, P. V. In *Supramolecular chemistry of anions*; Bianchi, A., Bowman-James, K., Garcia-Espana, E., Eds.; VCH: New York, 1997.

9. Bonnesen, P. V., Brown, G. M., Bavoux, L. B., Presley, D. J., Moyer, B. A., Alexandratos, S. D., Patel, V., Ober, R. Development of bifunctional anion exchange resins with improved selectivity and sorptive kinetics for pertechnetate. 1. Batch-equilibrium experiments. *Environ. Sci. Technol.*, 2000; *34*:3761-3766.

10. Gu, B., Brown, G. M., Bonnesen, P. V., Liang, L., Moyer, B. A., Ober, R., Alexandratos, S. D. Development of novel bifunctional anion-exchange resins with improved selectivity for pertechnetate sorption from contaminated groundwater. *Environ. Sci. Technol.*, 2000; *34*:1075-1080.

11. Gu, B., Ku, Y., Brown, G. M. Treatment of perchlorate-contaminated water using highly-selective, regenerable ion-exchange technology: a pilot-scale demonstration. *Remediation*, 2002; *12*:51-68.

12. Gu, B., Ku, Y., Brown, G. Sorption and desorption of perchlorate and U(VI) by strong-base anion-exchange resins. *Environ. Sci. Technol.*, 2005; *39*:901-907.

13. Brown, G. M., Bonnesen, P. V., Moyer, B. A., Gu, B., Alexandratos, S. D., Patel, V., Ober, R. In *Perchlorate in the Environment*; Urbansky, E. T., Ed.; Kluwer/Plenum: New York, 2000, pp 155-164.

14. Alexandratos, S. D., Hussain, L. A. Bifunctionality as a means of enhancing complexation kinetics in selective ion exchange resins. *Ind. Eng. Chem. Res.*, 1995; *34*:251-258.

15. Barrett, J. H., Lundquist, E. G., Miers, J. A., Pafford, M. M., Carlin Jr., W. H. High selectivity perchlorate removal resins and methods and systems using same. *US Patent Application, 10/831,988*, 2004.

16. Gu, B., Brown, G. M., Maya, L., Lance, M. J., Moyer, B. A. Regeneration of perchlorate (ClO_4^-)-loaded anion exchange resins by novel tetrachloroferrate ($FeCl_4^-$) displacement technique. *Environ. Sci. Technol.*, 2001; *35*:3363-3368.

17. Tripp, A. R., Clifford, D. A. In *Perchlorate in the Environment*; Urbansky, E. T., Ed.; Kluwer/Plenum: New York, 2000, pp 123-134.

18. Venkatesh, K. R., Klara, S. M., Jennings, D. L., Wagner, N. J. In *Perchlorate in the Environment*; Urbansky, E. T., Ed.; Kluwer/Plenum: New York, 2000, pp 147-153.

19. Brown, G. M., Gu, B., Moyer, B. A., Bonnesen, P. V. Regeneration of anion exchange resins by sequential chemical displacement. *US Patent, 6,448,299*, 2002.

20. Marcus, Y., Kertes, A. S. *Ion exchange and solvent extraction of metal complexes*; Wiley Interscience: New York, 1969.

21. Diamond, R. M., Whitney, D. C. In *Ion Exchange, Vol. 1*; Marinsky, J. A., Ed.; Marcel Dekker: New York, 1966, pp 277-351.

22. Kraus, K. K., Nelson, F. *Proc. Int. Conf. Peaceful Uses Atomic Energy, Geneva*, 1956; *7*:113-135.

23. Gu, B., Dong, W., Brown, G. M., Cole, D. R. Complete degradation of perchlorate in ferric chloride and hydrochloric acid under controlled temperature and pressure. *Environ. Sci. Technol.*, 2003; *37*:2291-2295.

24. Logan, B. E., Wu, J., Unz, R. F. Biological perchlorate reduction in high salinity solutions. *Water Res.*, 2001; *35*:3034-3038.

25. Cang, Y., Roberts, D. J., Clifford, D. A. Development of cultures capable of reducing perchlorate and nitrate in high salt solutions. *Wat. Res.*, 2004; *38*:3322-3330.

26. Gingras, T. M., Batista, J. R. Biological reduction of perchlorate in ion exchange regenerant solutions containing high salinity and ammonium levels. *J. Environ. Mon.*, 2002; *4*:96-101.

27. Gu, B., Dowlen, K. E., Liang, L., Clausen, J. L. Efficient separation and recovery of technetium-99 from contaminated groundwater. *Sep. Technol.*, 1996; *6*:123-132.

28. Gu, B., Cole, D. R., Brown, G. M. Destruction of perchlorate in ferric chloride and hydrochloric acid solution with control of temperature, pressure, and chemical reagents. *US Patent 6,800,203*, 2004.

29. Chen, J., Gu, B., LeBoeuf, E. J., Pan, H., Dai, S. Spectroscopic characterization of the structural and functional properties of natural organic matter fractions. *Chemosphere*, 2002; *48*:59-68.

30. Chin, Y. P., Traina, S. J., Swank, C. R., Backhus, D. Abundance and properties of dissolved organic-matter in pore waters of a fresh-water wetland. *Limnol. Oceanography*, 1998; *43*:1287-1296.

31. Gu, B., Schmitt, J., Chen, Z., Liang, L., McCarthy, J. F. Adsorption and desorption of natural organic matter on iron oxide: mechanisms and models. *Environ. Sci. Technol.*, 1994; *28*:38-46.

32. Hatch, G. B. Evaluation of scaling tendencies. *Mater. Prot. Perform.*, 1973; *12*:49-50.

33. Clifford, D. A., Zhang, Z. Modifying ion exchange for combined removal of uranium and radium. *J. AWWA*, 1994; *86*:214-227.

34. Gu, B., Liang, L., Dickey, M. J., Yin, X., Dai, S. Reductive precipitation of uranium(VI) by zero-valence iron. *Environ. Sci. Technol.*, 1998; *32*:3366-3373.

35. Gu, B., Brooks, S. C., Roh, Y., Jardine, P. M. Geochemical reactions and dynamics during titration of a contaminated groundwater with high uranium, aluminum, and calcium. *Geochim. Cosmochim. Acta*, 2003; *67*:2749-2761.

36. Huikuri, P., Salonen, L. Removal of uranium from Finnish groundwaters in domestic use with a strong base anion resin. *J. Radioanal. Nucl. Chem.*, 2000; *245*:385-393.

37. Vaaramaa, K., Pulli, S., Lehto, J. Effects of pH and uraniium concentration of the removal of uranium from drinking water by ion exchange. *Radiochim. Acta*, 2000; *88*:845-849.

38. Zhang, Z., Clifford, D. A. Exhausting and regenerating resin for uranium removal. *J. AWWA*, 1994; *86*:228-241.

39. Vaaramaa, K., Lehto, J., Jaakkola, T. Removal of U-234,U-238, Ra-226, Po-210 and Pb-210 from drinking water by ion exchange. *Radiochim. Acta*, 2000; *88*:361-367.

40. Lovley, D. R., Phillips, E. J. P., Gorby, Y. A., Landa, E. R. Microbial reduction of uranium. *Nature*, 1991; *350*:413-416.

41. Liu, C. X., Gorby, Y. A., Zachara, J. M., Fredrickson, J. K., Brown, C. F. Reduction kineitcs of Fe(III), Co(III), U(VI), Cr(VI), and Tc(VII) in cultures of dissimilatory metal-reducing bacteria. *Biotechnol. Bioeng.*, 2002; *80*:637-649.

42. Gu, B., Chen, J. Enhanced microbial reduction of Cr(VI) and U(VI) by different natural organic matter fractions. *Geochim. Cosmochim. Acta*, 2003; *67*:3575-3582.

43. Fredrickson, J. K., Kostandarithes, H. M., Li, S. W., Plymale, A. E., Daly, M. J. Reduction of Fe(III), Cr(VI), U(VI), and Tc(VII) by *Deinococcus radiodurans* R1. *Appl. Environ. Microbiol.*, 2000; *66*:2006-2011.

44. Bissen, M., Gremm, T., Koklu, U., Frimmel, F. H. Use of the anion-exchange resin amberlite IRA-93 for the separation of arsenite and arsenate in aqueous samples. *Acta Hydrochim. Hydrobiol.*, 2000; *28*:41-46.

45. Kim, J., Benjamin, M. M., Kwan, P., Chang, Y. J. A novel ion exchange process for As removal. *J. AWWA*, 2003; *95*:77-85.

46. Venkatesh, K. R. Removal of perchlorate and other contaminants from groundwater at JPL, a pilot study. *Final report to Jet Propulsion Laboratory, Pasadena, CA*, 1999:NAS7.000218, SSIC No. 009661.

47. Calgon Carbon Corporation. Full-scale ISEP(R) groundwater treatment plant. 2000: http://www.perchloratenews.com/case-study4.html.

Chapter 11

Field Demonstration using Highly Selective, Regenerable Ion Exchange and Perchlorate Destruction Technologies for Water Treatment

Baohua Gu and Gilbert M. Brown

Environmental Sciences and Chemical Sciences Divisions
Oak Ridge National Laboratory, Oak Ridge, TN 37831

INTRODUCTION

Currently, the two most commonly used ion-exchange technologies for perchlorate (ClO_4^-) removal in water are (i) selective but non-regenerable Type-I styrenic anion-exchange resins and (ii) nonselective Type-II or Type-I acrylic anion-exchange resins with NaCl brine regeneration (described in Chapter 10). In the first case, the spent resin cannot be regenerated, hence is discarded after it reaches its sorption capacity. It is not cost-effective because the resin bed must be replaced every month or so, depending on the feed ClO_4^- concentration. Additionally, the spent resin containing hazardous ClO_4^- has to be disposed of properly. In the second case, the Type-II or Type-I acrylic resins can be regenerated by flushing with NaCl brine solution. However, because of its extremely low selectivity, these resins indiscriminately remove practically all anions other than chloride. This low selectivity results in a low capacity to sorb ClO_4^-, and thus the resin has to be regenerated frequently (e.g., hourly or daily), resulting in the production of large volumes of secondary brine waste containing ClO_4^-. These factors contribute to the relatively high capital and operating costs of conventional ion-exchange technologies, which are discussed in numerous reports and publications elsewhere.[1-7] Although bioremediation is technically feasible and has been demonstrated successfully[1,2,8,9] (see Chapter 14), the method could be ineffective or costly for treatment of large plumes with low ClO_4^- concentrations (e.g., in drinking water with <0.1 mg/L ClO_4^-) because a highly reducing environment is required for the biodegradation of ClO_4^-.[8,10] The challenge is to sustain enough biomass to create continuous, reducing conditions at a high flow rate without adding substantial amounts of electron donors and/or acceptors. Many groundwater constituents such as nitrate, dissolved O_2, and Fe^{3+}/Mn^{4+} are also known to be preferred electron acceptors for bioreduction, and they must be reduced before the reduction of

ClO_4^- can occur. Additionally, post-treatment is usually required to remove added nutrients, other ingredients, and/or potential pathogens.[8,10]

This chapter focuses on field demonstrations of a recently developed selective ion-exchange, resin regeneration and perchlorate destruction technology for water treatment at two sites in California [namely the Edwards Air Force Base (AFB) site and an undisclosed Site-1 in Northern California]. The new technology involves the use of highly selective bifunctional resins for ClO_4^- sorption, the novel $FeCl_3$-HCl regeneration method, and perchlorate destruction techniques.[2,9,11,12] These treatment technologies make it possible to recycle both the resin and the regenerant solution, leading to practically no production of secondary wastes containing ClO_4^- and significantly improved cost savings.

The overall objective of the demonstrations was to evaluate treatment efficiency and longevity of the bifunctional resin after repeated regeneration and water treatment, and to determine the efficacy and feasibility of the perchlorate destruction and spent regenerant recycling. To date, the system has been subjected to two treatment and two regeneration cycles at the Site-1, and four treatment and three regeneration cycles at the Edwards AFB site over a period of more than two years. Regeneration has proven successful in restoring the ion-exchange sites and in cleaning up the resin bed, and limited data suggest no significant deterioration of the resin's sorption capacity for ClO_4^- after repeated regeneration. The eluted ClO_4^- also was found to be highly concentrated in the first bed volume (BV) of the regenerant solution and could be effectively destroyed, enabling solution recycling. These treatment technologies thus hold considerable promise for treating perchlorate-contaminated water with greatly enhanced treatment efficiency and longevity, and reduced operating and capital costs in comparison with conventional ion-exchange techniques.

CASE STUDY 1

Site–1 Background and Treatment System

Case Study 1 was performed at an undisclosed site in northern California. The site has been involved in manufacturing and testing rocket engines using liquid and solid propellants for military and commercial use. Some wastes from these manufacturing activities were disposed of in surface impoundments, landfills, and open burn areas. As a result of these former disposal practices, groundwater has been contaminated with perchlorate, and remedial actions are currently in progress at the site. In late 2003, a decision was made to test and demonstrate the highly-selective, regenerable ion-exchange system developed at the Oak Ridge National Laboratory (ORNL) to remove and destroy ClO_4^- from contaminated groundwater.[2,11,12] More specifically, the project was designed to demonstrate:

i) the efficiency and longevity of the bifunctional resin to remove ClO_4^- after repeated water treatment and regeneration cycles; and

ii) the destruction of ClO_4^- in the spent regenerant solution so that regenerant could be recycled and waste minimized.

The treatment system consisted of a primary resin canister for water treatment and a backup resin canister for polishing (Figure 1). Each canister has a capacity of ~270 gal but was filled with only about 120 gal of the bifunctional resin, Purolite A-530E. Lateral distributors were installed both at the top and bottom of the canister for even distribution of the water flow. Groundwater was pumped directly from the extraction well to the resin canister (from top to bottom) after passing through a series of bag filters (5-μm size) to remove suspended particles and sediments. The system has been running typically at ~150 gpm or about 1.25 BV/min.

Groundwater quality data are provided in Table 1. The groundwater can generally be characterized as having low to moderate concentrations of competing anions such as nitrate, chloride and sulfate. The groundwater (influent) ClO_4^- concentration was relatively constant during the water treatment phase. The first breakthrough study was performed with a contaminated groundwater source containing ~250 μg/L ClO_4^-, but subsequent studies used groundwater containing ~870 μg/L ClO_4^- in order to accelerate its breakthrough and to maximize the number of water treatment and regeneration cycles on the primary resin canister within limited demonstration time. The effluent ClO_4^- concentrations were monitored daily during treatment. Analyses were performed according to EPA Method 314.0

Table 1. Groundwater geochemical properties at Site-1 in California.

Major anions and pH		Major cations	
Cl^- (mg/L)	10	Fe (mg/L)	0.3
NO_3^- (mg/L)	10	Ca (mg/L)	27
SO_4^{2-} (mg/L)	15	Mg (mg/L)	16
ClO_4^- (μg/L)	250–900	Na (mg/L)	13
pH	7.3	K (mg/L)	2.1

using an ion chromatograph equipped with Dionex AS16 analytical and AG16 guard columns. The detection limit was about 1 μg/L.

Figure 1. Perchlorate groundwater treatment system at Site-1 in California. Each resin canister was filled with ~120 gal of the bifunctional resin, Purolite A-530E, and typically ran at ~150 gpm.

Because of the relatively high resin sorption capacity and the longevity of each treatment cycle, studies were focused on only the primary resin canister to maximize the number of water treatment and regeneration cycles within the limited demonstration time. When ClO_4^- breakthrough was observed, the system was taken offline, and the primary resin bed was regenerated using the $FeCl_3$-HCl displacement technique.[11] Spent regenerant solution was collected in three separate tanks (~2 BVs each). The effluent in the first tank usually contained the highest amount of eluted ClO_4^- and thus was run through the perchlorate-destruction system described in Chapter 10,[12] whereas the rest of the spent regenerant solution was recycled for subsequent regeneration cycles.

Field Breakthrough

The first field breakthrough study was initiated in early January 2004 and lasted approximately 12 weeks. The influent groundwater ClO_4^- concentration was about 250 µg/L, and the system was operated at 100–150 gpm or about 1 BV/min. No perchlorate breakthrough was observed until approximately 55,000 BVs of contaminated groundwater had been treated in the resin bed (Figure 2). During the 12-week treatment period, about 7.2 kg of ClO_4^- (equivalent to about 8.5 kg ammonium perchlorate) were removed by the resin bed. Previous studies using the bifunctional resin have reported the treatment of ~100,000 BVs of groundwater before ClO_4^- breakthrough was observed.[2] The lower volume of groundwater treated in this field study (Figure 2) is primarily caused by a ClO_4^- feed concentration five times greater (250 µg/L) than that in previous studies (50 µg/L).[2] As has been described in Chapter 10, the amount of groundwater treated (or the amount of ClO_4^- removed) depends not only on the anion-exchange capacity (AEC) of the resin, but also on the selectivity (or distribution coefficient, K_d) of the resin for ClO_4^- and the influent feed concentration. This relationship may be

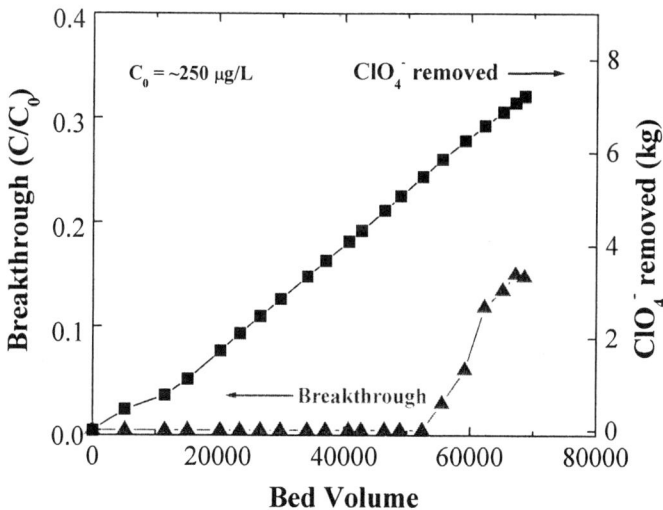

Figure 2. Perchlorate breakthrough and mass removal by the ion-exchange resin bed (~120 gal, Purolite A-530E) in a field groundwater treatment study in California (Site 1). The influent ClO_4^- concentration ranged from 240 to 260 µg/L, and the flow rate was about 100–150 gpm.

expressed as:

$$Q = K_d \cdot C \qquad\qquad (1)$$

where Q is the amount of ClO_4^- removed by the resin, and C is the feed ClO_4^- concentration. However, this relationship cannot be used to predict the amount of water treatment (or the amount of ClO_4^- sorption) at different feed concentrations because K_d generally is not a constant since sorption isotherms are usually nonlinear (Chapter 10).[2] Therefore, although the feed ClO_4^- concentration was five times higher than that reported in previous studies,[2] the volume of groundwater treated was still more than half of what was previously observed (i.e., 55,000 versus 100,000 BVs at the breakthrough).

Similarly, in a subsequent field demonstration experiment using groundwater with a much higher ClO_4^- concentration (~850–890 µg/L), we found that the resin bed was able to treat ~38,000 BVs of groundwater before breakthrough of ClO_4^- occurred (Figure 3). Apparently, even though the influent ClO_4^- concentration was more than three times higher than that in the first breakthrough experiment (Figure 2), the total volume of groundwater treated was reduced only ~30% (i.e., 38,000 versus 55,000 BVs at breakthrough). The total amount of ClO_4^- removed by the resin bed was nearly tripled to ~18.2 kg (Figure 3). At present, two treatment and regeneration cycles have been completed at this site.

Regeneration of Resin Bed

Regeneration of the spent resin bed using the $FeCl_3$-HCl displacement technique[11] was commenced after a breakthrough of ClO_4^- (>80% C/C_0) was observed (Figure 3). The spent resin bed was shipped off site and regenerated at Calgon Carbon Corporation in Pittsburgh, Pennsylvania. As has been discussed in Chapter 10, off-site (or centralized) regeneration and perchlorate destruction are preferred to minimize capital and operating costs, potential environmental impact, and handling of acidic regenerant solutions at the field site. A centralized resin regeneration and perchlorate destruction system is particularly advantageous when several treatment facilities are in operation simultaneously because the water treatment phase usually takes much longer than the time needed for regeneration and perchlorate destruction operations.

The spent resin canister was regenerated by counter-current flow, with regenerant entering through the lateral distributor at the bottom. The regenerant solution was composed of 1 M $FeCl_3$ and 4 M HCl,[9,11] and the solution was pumped through the resin bed at a rate of ~0.5 gpm (or ~0.004

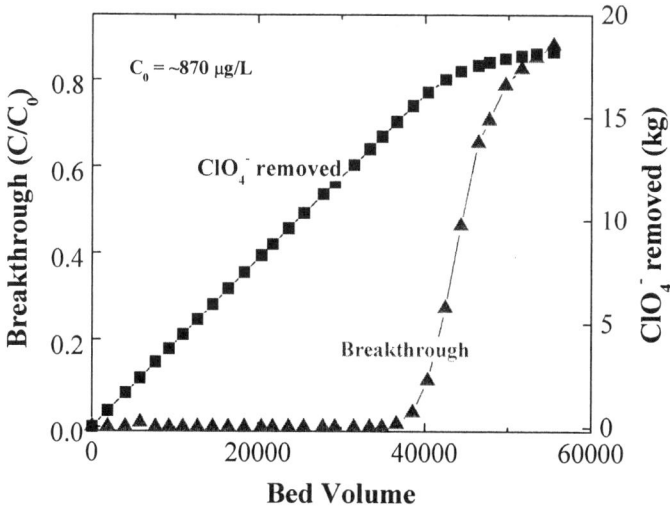

Figure 3. Perchlorate breakthrough and mass removal by the ion-exchange resin bed (~120 gal., Purolite A-530E) in a field groundwater treatment study in California (Site 1). The influent ClO_4^- concentration ranged from 850 to 890 µg/L, and the flow rate was about 150 gpm.

BV/min). Analyses of eluent samples indicated that perchlorate was effectively desorbed from the resin bed by the regenerant solution (Figure 4), as expected from previous laboratory and field studies.[9,11] In particular, most perchlorates appear to be eluted and highly concentrated within the first BV of the regenerant solution. The peak ClO_4^- concentration reached as high as ~350 mM (or ~35,000 mg/L) at ~3/4 BV and then declined rapidly to almost zero. In comparison with the elution profiles observed in small laboratory columns (10×22 mm), this rapid elution profile is indicative of an efficient regeneration process using the $FeCl_3$-HCl displacement technique. Previous laboratory studies have shown that about 3–5 BVs of the regenerant solution are needed to completely elute sorbed ClO_4^- at a flow rate of ~0.01–0.03 BV/min.[9,11] By using a larger column (5×30 cm), about 2 BVs of the regenerant solution were found necessary to completely desorb ClO_4^- from the resin bed.[9] These observations clearly indicate that better regeneration efficiency can be achieved through the use of larger resin beds and/or lower flow rates. This is because a larger resin bed and lower flow rate could eliminate potential wall effects and/or preferential flow paths within the resin bed. Additionally, a lower flow rate increases the contact equilibrium time between the regenerant solution and the resin, thereby increasing the regeneration efficiency.

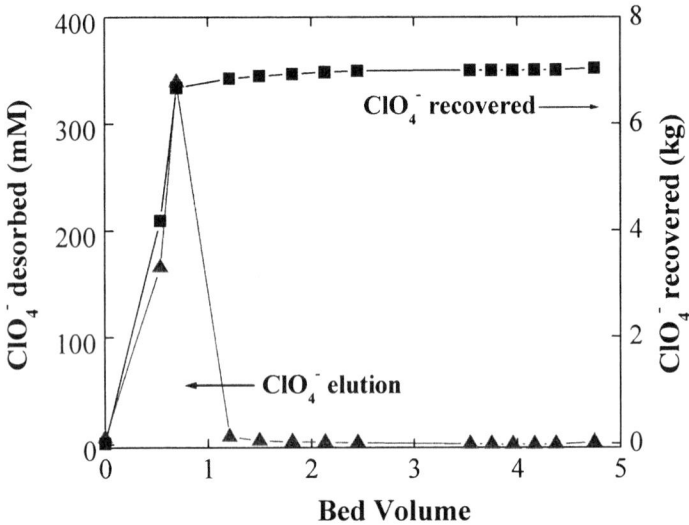

Figure 4. The first regeneration of the resin bed (~120 gal., Purolite A-530E) after water treatment at site 1 (see Figure 3). Regeneration was counterflow through a ½-in inlet at the bottom of the canister.

However, a mass balance analysis indicates that only about 7 kg of ClO_4^- was recovered during regeneration (Figure 4), which is less than half of the 18.2 kg of ClO_4^- sorbed by the resin bed during water treatment (Figure 3). This huge discrepancy was later explained by the fact that flow channeling had occurred during regeneration because the regenerant solution was pumped into the resin bed through the bottom distributor. Additionally, only about half of the 270-gallon resin canister was filled with resin, and resin beads may have floated to the top of the canister (due to a relatively high density of the regenerant solution), resulting in a significant dead space above the lateral distributors. Evidence of channeling includes the fact that no significant amount of ClO_4^- was found in the effluent solution after eluting with 1 BV of regenerant (Figure 4), but ClO_4^- concentrations as high as 11,300 mg/L were found in the drain solution at the end of regeneration operations. By analysis of the combined regenerant solutions, it was estimated that the drain solution contained about 6.5 kg of ClO_4^-. Therefore, a mass balance suggests that a total of 13.5 kg of ClO_4^- was recovered, and about 4.7 kg of ClO_4^- remained sorbed on the resin.

Previous laboratory studies[2,11] indicated that perchlorate still sorbed on the resin could be recovered in the next regeneration cycle if the regeneration process design was improved. Therefore, the partially regenerated resin bed was returned to service since the study objective was to evaluate the resin performance and regeneration efficiencies after repeated regeneration and treatment cycles. Residual ClO_4^- in the resin bed did cause a small bleed of ClO_4^- (~4% of the feed concentration, Figure 5). A sudden and early breakthrough of ClO_4^- was also observed when approximately 24,000 BVs of groundwater had been treated. However, this sudden breakthrough was attributed to an unexpected system shutdown and operating problems, which caused a sudden pressure drop across the system. As a result, short circuiting of the treatment system may have occurred, causing the early breakthrough of perchlorate. Nevertheless, the sorption of ClO_4^- continued and, at the end of this treatment cycle, an additional ~13 kg of ClO_4^- was removed by the resin bed. The total ClO_4^- loading was thus about 17.7 kg in the primary resin canister.

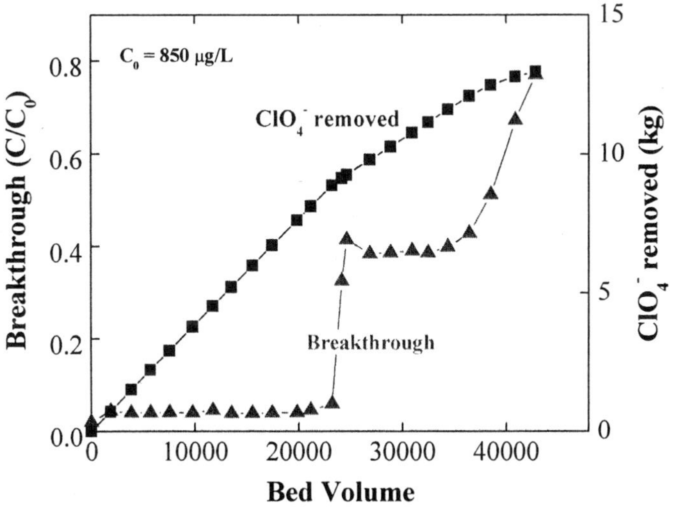

Figure 5. Perchlorate breakthrough and mass removal by the ion-exchange resin bed, which was regenerated after the first breakthrough (Figure 3). An early breakthrough at ~24,000 BVs was attributed to an operating problem during an unexpected system shutdown.

Based on the lesson learned from the first regeneration cycle, an improved and dedicated regeneration reactor was built at Calgon Carbon Corporation. The reactor consisted of a 2-ft diameter polyethylene tank to ensure an even flow thought the system. This improved design achieved plug flow by effectively preventing short circuiting, as evidenced by the rapid elution of ClO_4^- in the effluent solution (Figure 6). It is remarkable that the peak ClO_4^- effluent concentration reached as high as 1,130 mM (or equivalent to a 13% NH_4ClO_4 solution). No previous studies have reported such a high concentration of ClO_4^- in the regenerant solution. The perchlorate recovery estimated from these discrete sampling points was about 16 kg (Figure 6). However, a more accurate calculation of 17.2 kg recovered ClO_4^- was determined from analysis of the final composite regenerant solution in tanks. This number compared favorably with the ClO_4^- loading of 17.7 kg (before regeneration), yielding a recovery >97%. Given the margin of error in concentration and volume measurements, these results indicate a successful, complete regeneration of the resin bed. The performance of this regenerated resin bed is currently being re-evaluated for water treatment under the same field conditions.

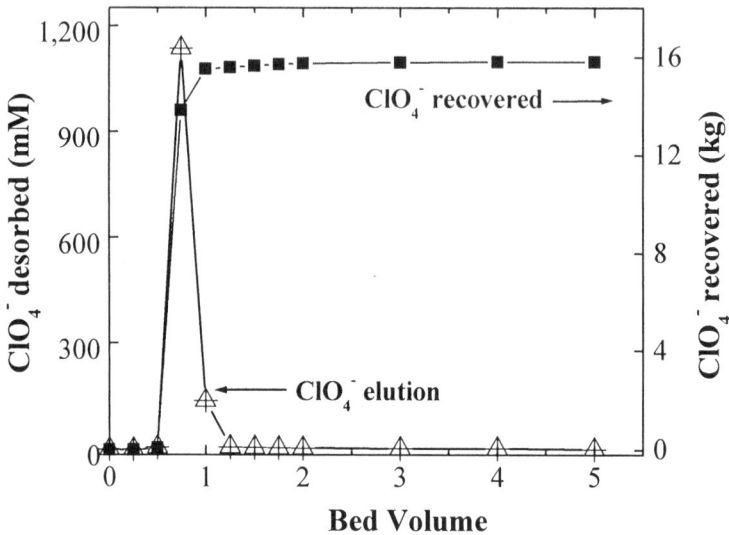

Figure 6. The second regeneration of the resin bed (~120 gal) after water treatment at Site 1 (Figure 4). Regeneration was performed in a dedicated regeneration canister, and the flow rate was 0.5 gpm.

Elution Profiles of Nitrate and Sulfate

The elution profiles of nitrate and sulfate also were monitored during the second regeneration of the resin bed shown in Figure 6. As described in Chapter 10, these competing anions were sorbed by the resin during water treatment because their concentrations in groundwater are orders of magnitude higher than that of ClO_4^-. However, because the bifunctional resin has a relatively low selectivity for nitrate and sulfate, these anions were expected to elute more quickly than ClO_4^- during the $FeCl_3$-HCl regeneration process. Indeed, results indicate that sulfate was eluted within the first half

BV, and nitrate was eluted after sulfate but before ClO_4^- (data not shown). This is analogous to chromatographic separation of ions based on their retention time (or selectivity) in ion-exchange columns. A more detailed elution experiment was performed in a small laboratory packed-resin column (10×22 mm), and results were shown in Figure 7. Sulfate was eluted first, as expected; its elution peak completely separated from ClO_4^- whereas the elution profile of nitrate overlapped with both sulfate and ClO_4^- (Figure 7). This elution sequence, i.e., sulfate → nitrate → perchlorate, is consistent with previous studies which showed that sulfate is the least strongly sorbed by styrenic Type-I anion-exchange resins whereas perchlorate is the most strongly sorbed.[13,14] Note that a slightly longer elution time or a higher BV of eluent was required to completely desorb ClO_4^-, sulfate and nitrate in the small laboratory columns than in the large reactor. This observation was attributed to a relatively fast flow (~0.02 BV/min) used in the laboratory elution experiment and the potential wall effects of small columns, as described earlier. Additionally, a shoulder was observed in the elution profile of nitrate (Figure 7), which was likely caused by the competitive desorption as sulfate was eluted off the column.

Separation of these anions in the spent regenerant solution is of great interest for subsequent waste segregation/minimization and perchlorate destruction in particular. For example, nitrate removal prior to thermal destruction of ClO_4^-

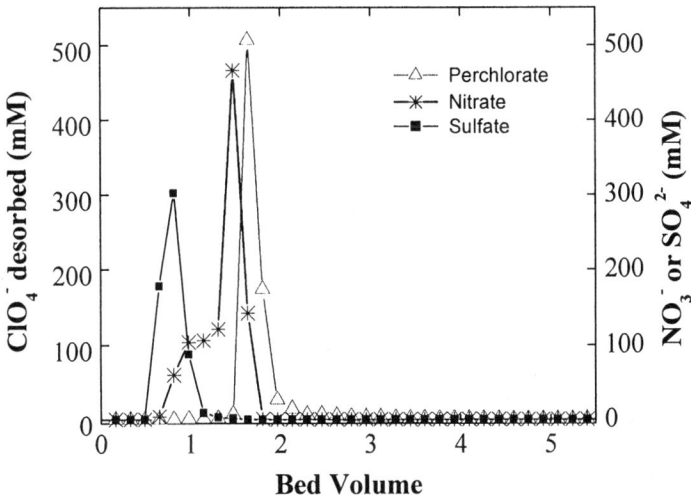

Figure 7. Elution profiles of sulfate, nitrate, and perchlorate during the regeneration of a small laboratory resin column (10×22 mm). The resin was obtained from the 120-gal canister, which was used for the first round of water treatment at Site 1 (see Figure 2).

(described in Chapter 10) is beneficial because less reducing agent (i.e., $FeCl_2$) is required. The isolation of ClO_4^- in the second half BV of the regenerant solution provides a highly concentrated, reduced volume of spent regenerant for further treatment. For example, perchlorate in the regenerant solution may be recovered as $KClO_4$ solids by precipitation or crystallization with KCl because of a relatively low solubility of $KClO_4$ (Chapter 10). A low volume of highly concentrated ClO_4^- solution gives higher yields of perchlorate recovery. These advantages could translate into significant cost savings and waste reduction for large-scale field applications.

CASE STUDY 2 – EDWARDS AFB SITE

Background

The second field demonstration site was located along the northern boundary of Edwards AFB, south of highway 58 between Mojave and Boron in southern California. The site had been used to support various research activities or programs associated with the development of solid fuel rocket technology beginning as early as 1960. Ammonium perchlorate was used in many of these facilities and resulted in groundwater contamination. Perchlorate has been detected in groundwater at concentrations up to 160,000 $\mu g/L$. In early 2001, a pilot-scale field demonstration experiment was initiated in an attempt to evaluate and demonstrate the efficacy of the ORNL-developed bifunctional resin, its regeneration, and perchlorate destruction technologies to remediate the contaminated groundwater. This pilot-scale experiment included one pairs of columns ($\sim 2 \times 12$ in), one packed with the bifunctional resin and the other packed with one of the best commercially-available monofunctional resins (Purolite A-520E) for comparison. Spent resins were regenerated and reevaluated for their efficiency in removing ClO_4^- from the contaminated groundwater. Results from the pilot-scale test have been reported elsewhere.[15,16] In brief, the bifunctional resin was found to be highly efficient in removing ClO_4^- despite the presence of relatively high concentrations of chloride (~ 400 mg/L) and sulfate (~ 180 mg/L) in groundwater. The perchlorate-loaded resin was successfully regenerated using the $FeCl_3$-HCl displacement technique, and nearly 100% of sorbed ClO_4^- was recovered after elution with as little as about 2 BVs of the regenerant solution.[9] This regeneration efficiency was a significant improvement (with two to five times less $FeCl_3$-HCl regenerant required) over that observed in laboratory studies using small columns (1-cm diam.).[11]

Based on results from the pilot-scale field test, a large-scale field demonstration was initiated in early 2003. Similar to the Case One study, the main goal was to evaluate the treatment efficiency and longevity of the bifunctional resin after repeated regeneration (using $FeCl_3$-HCl solution) and water treatment, and to evaluate the efficacy of using the ORNL-developed perchlorate-destruction technology.[12,17] The latter degrades ClO_4^- in spent regenerant solution, allowing regenerant recycling and resulting in negligible secondary waste production for the entire treatment process.

Treatment System Design and Operation

Groundwater at Edwards AFB (site 285) was pumped from four wells into an equalization tank and was then pumped into a two-stage filter system (with both 20- and 5-μm filters) followed by a sand filter to remove particulate materials (the sand filter was added after the first breakthrough was observed). The groundwater then passed at a flow rate of 20-35 gpm through the resin canisters to remove perchlorate. Four interchangeable resin canisters were used (Figure 8), each filled with approximately 100 gal of the bifunctional resin (WBR-112L) made to our specifications by Wandong Chemical Plant in China (also see Chapter 10). Three online canisters were designated as the lead (canister A), lag (canister B), and polishing canisters. The fourth canister was a reserve to facilitate changing canisters without shutting the system down. The treated groundwater was routed through two activated carbon canisters for removing organic contaminants before reinjection into the aquifer.

Figure 8. Perchlorate groundwater treatment system at the Edwards AFB site. The foreground shows four interchangeable resin canisters, three online canisters designated as the lead (canister A), lag (canister B), and polishing vessels, and one canister in reserve. The background shows an equalization tank and two carbon canisters for removing organic contaminants.

Each resin canister was also equipped with 8 lateral distributors at the bottom and top of the canisters for even distribution of groundwater (during the treatment phase) and/or the regenerant solution (during regeneration). Silicate gravel and sand were placed at the bottom of each canister to minimize channeling during the countercurrent, low flow regeneration process. During the water treatment phase, groundwater was pumped into the resin bed from top to bottom.

The groundwater geochemical properties are quite different from those found at the site 1 and can generally be characterized as having high ionic strength and alkalinity with relatively high concentrations of competing anions such as chloride (~400 mg/L) and sulfate (~180 mg/L) (Table 2). The feed ClO_4^- concentration varied between ~200 and 500 µg/L, and the effluent ClO_4^- concentrations after the lead, lag, and polishing resin canisters were monitored weekly during treatment. Samples were analyzed according to EPA Method 314.0 using an ion chromatograph equipped with Dionex AS16 analytical and AG16 guard columns.

Table 2. General geochemical properties of groundwater at Edwards Air Force Base in California.

Major anions and pH		Major cations	
Cl^- (mg/L)	400.4	Fe (mg/L)	0.01
NO_3^- (mg/L)	0.9	Ca (mg/L)	38.8
$SO_4^=$ (mg/L)	176.1	Mg (mg/L)	9.2
ClO_4^- (µg/L)	200–500	Na (mg/L)	383.3
pH	7.8	K (mg/L)	6.9

When breakthrough of ClO_4^- was observed in the lead canister, the system was taken offline, and the resin bed was regenerated using the $FeCl_3$-HCl elution technique. The lead canister was left in the lead position, even after regeneration, although the treatment system was designed to run continuously by switching resin canisters between the lead, lag, polishing, and reserve positions. The fixed setup was used to maximize the number of treatment and regeneration cycles within the limited demonstration time, allowing evaluation of the treatment efficiency and longevity of the bifunctional resin and ultimately, the cost effectiveness of this new highly-selective and regenerable ion-exchange system.

Unlike the Case One study, in which the second regeneration was performed offsite using a dedicated regeneration canister, the spent resin bed was regenerated directly *in situ* at the Base. Spent regenerant solution was segregated approximately every 2 BVs and stored in separate tanks because, on the basis of the pilot-scale test,[15,16] the desorbed ClO_4^- was mostly concentrated in the first 2 BVs of the regenerant solution, and the rest of the spent regenerant solution could be reused.[9,15] No discrete samples were taken during the regeneration process, and only composite samples in each regenerant tank were analyzed to obtain a mass balance between the amount of sorbed ClO_4^- and that recovered by the regeneration.

FIELD PERFORMANCE EVALUATION

The First Breakthrough and Regeneration

The treatment system was started on March 11, 2003 and operated at a flow rate of ~35 gpm. The system was designed to run at about 100 gpm, but was limited by the groundwater extraction rate at the site. The first breakthrough of ClO_4^- on the lead resin vessel was observed at about 20,000 BVs (Figure 9) or after treatment of about 2 million gallons of groundwater. This volume

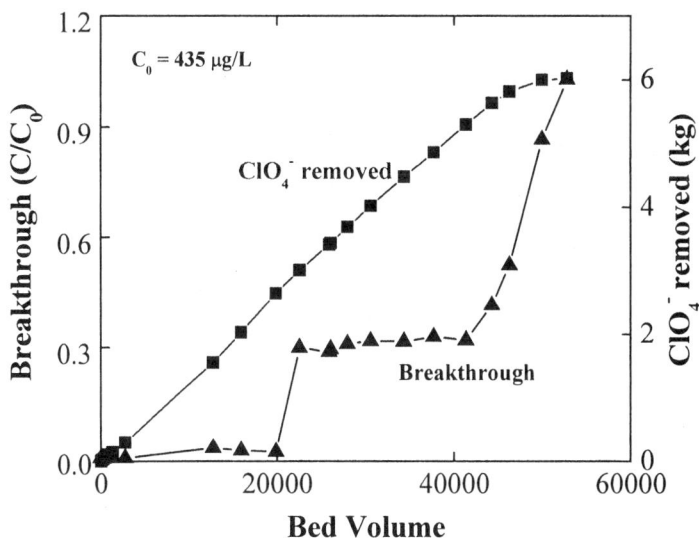

Figure 9.   Perchlorate breakthrough and mass removal by the lead ion-exchange resin bed (~100 gal) in a field groundwater treatment study at Edwards AFB in California. The average influent ClO_4^- concentration was ~435 µg/L.

was significantly less than anticipated based on resin performance in laboratory and pilot-scale field studies.[9,15] A significant pressure drop also was observed across the lead resin bed, suggesting possible buildup of mineral deposits and/or organic materials in the resin bed. Consequently, the system was shut down, and a backwash was performed in an attempt to remove any clay or mineral deposits in the resin bed. Indeed, analysis of the filters and the top layer of fouled resin (Figure 10) indicated the presence of a high load of suspended solids, primarily clays and iron oxides. The fouled resins were dark brown due to coatings of iron oxides and possibly manganese oxides (Figure 10, inset a). These coatings of iron and manganese oxides could have been formed by the oxidation of soluble Fe^{2+} and Mn^{2+} ions in the resin bed over time although the Fe^{2+} concentration in the groundwater was very low and Mn^{2+} was below the detection limit (Table 2).

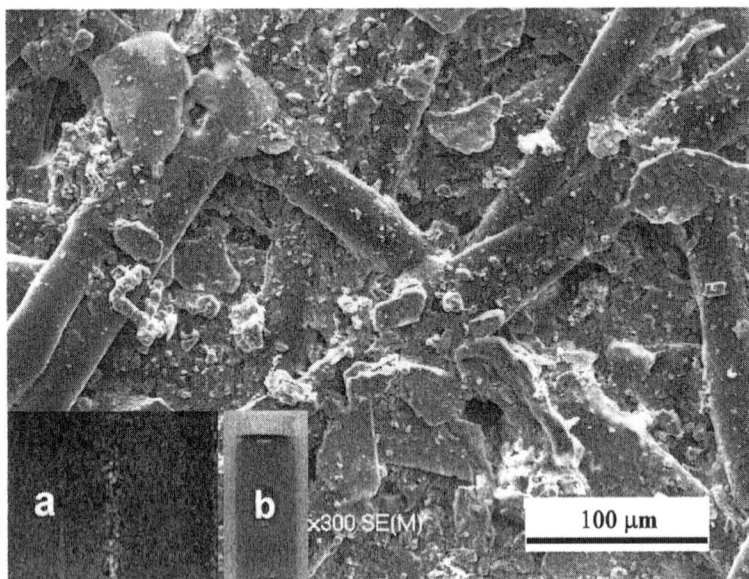

Figure 10. Scanning electron microscope image showing deposits of clay mineral particles on the 5-μm filter. Inset (a) shows the fouled resins at the top of the resin bed, and inset (b) shows suspended minerals and iron oxides after agitating the fouled resin in water.

After backwashing the resin bed, the system was brought back online. A full breakthrough of ClO_4^- was observed after treatment of an additional ~35,000 BVs of groundwater (Figure 9), and a total of about 6.1 kg of ClO_4^- (or ~7.2 kg NH_4ClO_4 equivalent) was retained by the resin bed in canister A.

Because the resin bed was severely fouled (Figure 10a), concerns were raised whether it could be regenerated and reused as the lead treatment canister. To address this concern, laboratory studies with small columns were undertaken to assess regeneration and performance of the fouled resin. Results (Figure 11a) indicate that the resin could be regenerated as expected. The elution profile of ClO_4^- was similar to that in previous laboratory studies using virgin resins (see Chapter 10).[11] The peak effluent ClO_4^- concentration reached as high as ~140 mM. More importantly, the breakthrough curve of the regenerated resin matched nicely with that of the virgin resin in accelerated laboratory flow-through experiments with a high flow rate (~17 BV/min) and a high influent ClO_4^- concentration (10 mg/L) (Figure 11b). It therefore appeared that the use of strong acid (4 M HCl) in the regenerant solution was helpful in cleaning up the resin bed, although it did not appear to remove much of the stain on the resin.

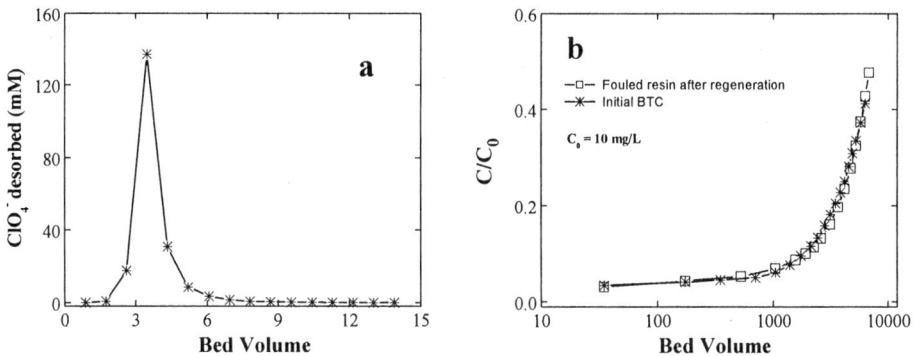

Figure 11. (a) Elution profile of ClO_4^- from regeneration of the fouled resin (WBR-112L) that had been used for groundwater treatment at Edwards AFB. The experiment was performed in a small laboratory column (10×22 mm), and the regenerant solution consisted of 1 M $FeCl_3$ and 4 M HCl. (b) Comparison of the breakthroughs of ClO_4^- in the regenerated versus the virgin resins in accelerated laboratory flow-through experiments (Chapter 10).

The field resin canister (A) was subsequently regenerated counter-currently using a mixture of 1M $FeCl_3$ and 4M HCl solution. The regeneration was operated at a flow rate of ~0.2 gpm, and the effluent regenerant solution was collected in portions of 190, 160, and 275 gal, respectively, in three polyethylene tanks. As expected, the desorbed ClO_4^- accumulated mostly in the first tank with a concentration of ~7,700 mg/L. The concentrations of ClO_4^- in the second and third tanks were 251 and 14 mg/L, respectively. A mass balance indicated that a total of 5.9 kg ClO_4^- was recovered, representing a 97% total recovery, by using the $FeCl_3$–HCl regeneration technique. Given analytical and field measurement errors, this high perchlorate recovery suggests a successful regeneration of the field resin bed. Therefore, the resin bed was returned to service as the lead treatment vessel after rinsing with potable water to remove residual ferric iron and hydrochloric acid in the resin bed.

Subsequent Perchlorate Breakthroughs

To minimize particulate matter and iron oxides entering the resin bed, a sandbag pre-filter was added. Figure 12 shows the second breakthrough and perchlorate mass removal by the resin. Results indicate that the regenerated resin bed treated more than 30,000 BVs of contaminated groundwater before breakthrough of ClO_4^- was observed. The total ClO_4^- removed was about 6.2 kg, which was nearly identical to the amount of ClO_4^- removed in the first run (Figure 9). These results demonstrate that resin performance is not adversely impacted by the new $FeCl_3$-HCl regeneration technique, and regeneration resulted in nearly complete recovery of exchange sites on the resin, as is consistent with previous laboratory studies.[2, 11]

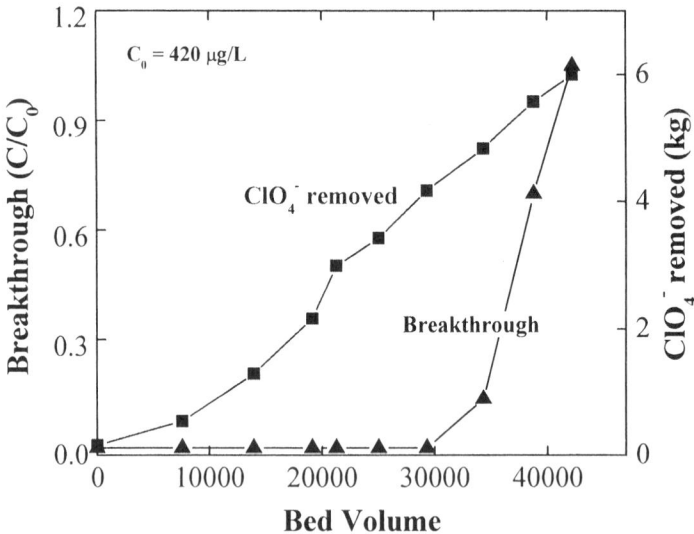

Figure 12. The second breakthrough and mass removal by the lead ion-exchange canister in a field groundwater treatment study at Edwards AFB. The average influent ClO_4^- concentration was ~420 μg/L.

However, in comparison with the amount of groundwater treated in the Case 1 study, i.e., ~55,000 and 38,000 BVs at influent ClO_4^- concentrations of ~250 and 870 μg/L, respectively, (Figures 1 and 2), a lower volume of groundwater was treated at the Edwards AFB site. These results could be attributed largely to the presence of competing ions at concentrations orders of magnitude higher than were present at Site 1 (Tables 1 and 2). For example, the chloride and sulfate concentrations were about 40 and 12 times higher, respectively, in the groundwater at Edwards AFB than at Site 1. These competing anions undoubtedly resulted in a decreased ClO_4^- selectivity or distribution coefficient (K_d) of the resin, as has been described in Chapter 10. Note that ClO_4^- removal is not directly proportional to the concentrations of competing anions and also depends on, among other factors, the anion exchange capacity and K_d of the resin, the type of competing ions in the influent, and the influent ClO_4^- concentration. The decrease in resin selectivity would have been even more pronounced if a non-selective polyacrylic anion-exchange resin had been used.

The dependence of perchlorate removal on feed ClO_4^- concentration is illustrated in Figure 13, which shows breakthrough curves and mass removal of ClO_4^- after the resin canister was regenerated the second and third times in the field. Over time, the ClO_4^- concentration in groundwater significantly decreased to an average of ~220 µg/L in the third run although other treatment conditions remained unchanged. As a result, no perchlorate breakthrough was observed until about 40,000 BVs of contaminated groundwater had been treated (Figure 13a). At full breakthrough, a total of ~4.1 kg of ClO_4^- had been removed by the resin bed. The reduction in ClO_4^- removal was expected because of decreased feed ClO_4^- concentration, although the treated groundwater volume increased substantially (from ~30,000 to 40,000 BVs). In the fourth run (after the resin bed was regenerated the third time), the groundwater flow was adjusted from various extraction wells (with varying ClO_4^- concentrations) in an attempt to raise the feed ClO_4^- concentration in the equalization tank. An average feed ClO_4^- concentration was maintained at ~320 µg/L until breakthrough of ClO_4^- occurred. As a result, the treated groundwater volume decreased to about 32,000 BVs at the breakthrough of ClO_4^- (Figure 13b). However, the feed

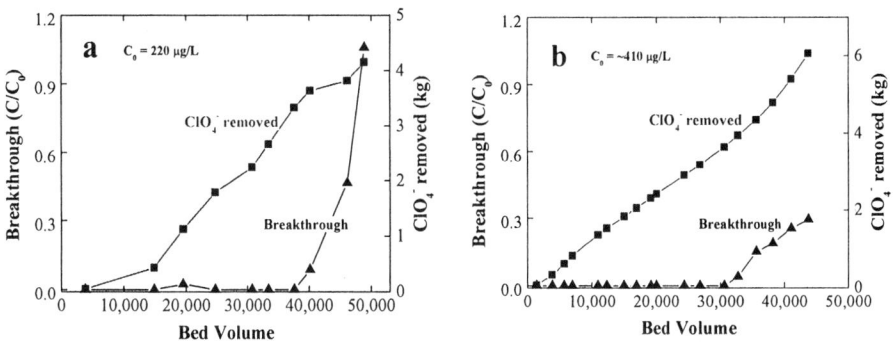

Figure 13. (a) The third and (b) the fourth breakthrough and mass removal by the lead ion-exchange canister (A) in a field groundwater treatment study at Edwards AFB.

ClO_4^- concentration increased substantially afterwards (from ~400 to >900 µg/L) because of the startup of soil flushing, which increased the ClO_4^- concentration in groundwater and resulted in a delayed full breakthrough (not observed at the completion of this manuscript). The total mass of ClO_4^- removed to date was about 6.1 kg and is almost identical to the amounts removed in the first and second treatment cycles (6.1 kg and 6.2 kg, respectively). Therefore, despite the resin bed was severely fouled in the first run, these results clearly demonstrate the efficacy and longevity of the bifunctional resin in removing ClO_4^- after repeated regeneration and treatment cycles over more than 2 years of field operations.

The Second and Third Regenerations

The second regeneration of the lead resin canister was initiated on July 27, 2004 after it reached full breakthrough (Figure 12). Regeneration conditions were similar to those described earlier for the first regeneration cycle. However, spent regenerant solution from the first regeneration in Tanks 2 and 3 was reused in the second regeneration cycle (plus some fresh regenerant solution) to minimize waste volume. Perchlorate concentrations in these tanks were relatively low (~251 and 14 mg/L, Table 3) and not expected to significantly impact regeneration efficiency.[2, 11] Table 3 lists the effluent volumes and ClO_4^- concentrations used for mass balance calculations. Results again indicate that perchlorate was eluted and highly concentrated roughly in the first two BVs of the regenerant solution; its concentrations in Tanks 1, 2, and 3 were about 7,600, 161, and 37 mg/L, respectively. The amount of ClO_4^- recovered was about 5.62 kg (after deduction of the residual ClO_4^- in the spent regenerant from the first regeneration). Therefore, the total ClO_4^- recovery was about 91% compared with the amount of ClO_4^- sorbed during water treatment (Figure 12).

The third regeneration of the lead resin canister was performed on January 7, 2005 after the third breakthrough was observed (Figure 13a). However, during the third breakthrough of resin canister A, the lag resin canister B also reached breakthrough so it was regenerated as well. Data from the regeneration of canister B are presented in Table 3 because they are pertinent to the third regeneration of the lead resin canister A. After regenerating canister B, spent regenerant solutions in all three tanks (715 gal) were unintentionally mixed and used for the third regeneration of canister A (after adding more $FeCl_3$ and HCl). This resulted in abnormal ClO_4^- concentrations of 2210, 3780, and 1580 mg/L in the effluent Tanks 1, 2 and 3, respectively (Table 3). The total amount of ClO_4^- recovered was 7.27 kg, but about 3.28 kg was carried over from the regeneration of canister B. Therefore, the actual mass of ClO_4^- recovered from the third regeneration of the lead resin canister was 3.99 kg, which is very close to the amount of ClO_4^- sorbed (~4.1 kg) on

Table 3. Perchlorate concentrations[*] and recoveries from three regeneration cycles of the lead resin canister (A) and one regeneration of the lag resin canister (B) at the Edwards AFB site.

Number of regeneration (canister A)	Tank 1		Tank 2		Tank 3		Total ClO_4^- recovered (kg)	ClO_4^- carryover (kg)	Actual ClO_4^- recovered (kg)	ClO_4^- sorbed on resin (kg)	ClO_4^- recovery (%)
	Volume (gal)	ClO_4^- (mg/L)	Volume (gal)	ClO_4^- (mg/L)	Volume (gal)	ClO_4^- (mg/L)					
1st regeneration	190	7,700	160	251	275	14	5.87	0	5.87	6.1	96
2nd regeneration	190	7,600	160	161	325	37	5.78	0.16	5.62	6.2	91
Canister B regen[*]	200	4,020	165	97	350	58	3.28	0.15	3.13	nd	nd
3rd regeneration	190	2,210	190	3,780	460	1,580	7.27	3.28	3.99	4.1	97

[*] Spent regenerant solutions in Tanks 2 and 3 after the first regeneration cycle were used in the second regeneration, whereas spent regenerant solutions in Tanks 1, 2, and 3 after regenerating the B canister were combined and used for the third regeneration of the lead canister A.

the resin bed during the third groundwater treatment cycle (Figure 13a). The recovery was thus nearly complete (~97%). These observations clearly demonstrate that a combination of the bifunctional resin and $FeCl_3$-HCl regeneration technologies could potentially offer an efficient and cost-effective means to remove ClO_4^- from contaminated water with greatly reduced secondary waste production.

Perchlorate Destruction and Regenerant Recycling

Perchlorate recovery and destruction allows recycling of spent regenerant solution, essentially eliminating secondary hazardous waste disposal issues. Chapter 10 details the theory of perchlorate recovery and destruction. Since eluted ClO_4^- is highly concentrated in the first BV of $FeCl_3$-HCl regenerant solution (and may be further reduced to one-half BV, Figure 6), it can readily be precipitated as pure solid phase $KClO_4$ or other salts. Alternatively, perchlorate can be destroyed by reducing it with ferrous ion and other reductants,[12] an approach used and demonstrated in the two case studies.

A 30 gal/d pilot-scale perchlorate destruction unit was custom built (Figure 17 in Chapter 10) for destroying ClO_4^- in spent regenerant solution. This capacity is usually sufficient for small volumes of spent regenerant solution (0.5–1 BV with highly concentrated ClO_4^-) for groundwater treatment systems of several hundred gallons per minute. During the demonstration, 30-40% ferrous chloride ($FeCl_2$) solution was used to reduce ClO_4^- in the regenerant solution. The amount of $FeCl_2$ needed depends on the spent regenerant ClO_4^- concentration. By chemical stoichiometry, eight moles of $FeCl_2$ is required to completely reduce one mole of ClO_4^- (Eq. 8 in Chapter 10), although excess $FeCl_2$ (20–30%) is usually used due to the presence of other oxidizing agents (such as nitrate and dissolved O_2) in the regenerant solution. Additionally, since Fe^{2+} is readily oxidized to Fe^{3+} in air, only fresh $FeCl_2$ solution should be used, and a N_2 gas purge is recommended during destruction operations.

Results to date indicate that perchlorate in spent regenerant solution can be effectively destroyed at an operating temperature of ~190–204°C (or 375–400°F) and a flow rate of ~25–32 gal/d (or a residence time of 45 min to 1 h) (Table 4). At the Edwards AFB site, the system was run overnight (~16 h) with influent ClO_4^- and $FeCl_2$ concentrations of 5280 mg/L and 9%, respectively. No ClO_4^- was detected in the effluent solution, giving a destruction efficiency of 100%. At a higher flow rate and influent ClO_4^- concentration but a lower operating temperature, perchlorate destruction efficiency was found to range from ~92 to 97% at Site 1 (Table 4). A $FeCl_2$ concentration of about 12% was used because of a relatively high influent ClO_4^- concentration. The results clearly demonstrate that perchlorate can be rapidly and effectively destroyed in spent regenerant solution. More importantly, while ClO_4^- is reduced, Fe^{2+} is oxidized to Fe^{3+} during the reaction, renewing the $FeCl_3$-HCl regenerant solution. Note that Fe^{3+} ion is usually depleted in the first BV of spent regenerant solution because it is sorbed by the resin (as $FeCl_4^-$) or used for the displacement of sorbed ClO_4^-. The destruction process replenishes Fe^{3+} without changing the regenerant properties, allowing reuse of the solution in many regeneration cycles and eliminating the need for disposal of perchlorate-containing hazardous regenerant wastes.

Table 4. Perchlorate destruction efficiency in spent regenerant solutions using a pilot-scale Perchlorate Destruction System (Figure 17 in Chapter 10) at varying temperatures and flow rates.

Site	Flow (gal/d)	Run time (h)	Temp. (°C)	ClO_4^- in (mg/L)	ClO_4^- out (mg/L)	% ClO_4^- destroyed
Edwards	22	16	204	5280	0	100
Site 1 [*]	25	12	190	7100	220	97
Site 1 [*]	32	12	190	7100	540	92

[*] Perchlorate destruction performed offsite at Calgon Carbon Corporation.

Two additional issues identified during the field test are noteworthy. The effluent solution at Edwards AFB appeared clear despite complete destruction of perchlorate, suggesting that iron was precipitated in the destruction system (or the flow-through coil). The relatively high operating temperature (204°C) may have caused precipitation of iron oxyhydroxides (perhaps magnetite) as has been reported in laboratory studies.[12] This could be problematic because iron oxides may plug the system and cause excessive system pressure. Therefore, a lower operating temperature (190°C) was used in the Case 1 study, and no iron precipitation was observed but effluent ClO_4^- concentration was slightly elevated. The presence of small quantities of ClO_4^- is not expected to significantly impact regeneration efficiency, however, as discussed earlier. Another issue related to the use of a laboratory HPLC pump for the destruction system (Prep 100, LabAlliance, State College, PA). Although the pump was rated at 36 gal/d at an operating pressure up to 4000 psi, it was not suited for field operations; its piston seals required frequent replacement (24 to 72 h), perhaps due to HCl and Fe^{3+}/Fe^{2+} ions in the regenerant solution. These issues are to be resolved by using a custom-built pump in the near future.

CONCLUSIONS

To date, the two treatment systems have been operating for about one year and more than two years at Site 1 and the Edwards AFB site, respectively; the former subjected to two treatment and two regeneration cycles, and the latter to four treatment and three regeneration cycles. At the Edwards AFB site, despite severe fouling of the resin bed during the first run, data suggest no significant deterioration of the resin to sorb ClO_4^- after repeated regenerations. The lead resin canister removed a total of ~6.1, 6.2, 4.1, and 6.1 kg of ClO_4^- from the 1st, 2nd, 3rd and 4th treatment cycles, respectively; the amounts of ClO_4^- recovered during the 1st, 2nd and 3rd regeneration were about 5.87, 5.62, and 3.99 kg, representing a total recovery of about 96, 91 and 97%, respectively. Given the margin of error associated with ClO_4^- analyses and volume measurements made in the field, we conclude that regeneration has proven successful in restoring ion-exchange sites and cleaning up the resin bed; nearly 100% ClO_4^- was desorbed and recovered by eluting with $FeCl_3$-HCl regenerant solution. Data also indicate that the desorbed ClO_4^- was highly concentrated in the first BV of regenerant solution, facilitating ClO_4^- recovery or destruction. A thermal destruction system was used and demonstrated high efficiency in degrading ClO_4^- in the spent regenerant solution. Thus the solution may be recycled, eliminating the need for disposal of hazardous secondary wastes containing ClO_4^-. The combination of the new selective ion-exchange, resin regeneration, and perchlorate-destruction technologies could lead to greatly enhanced treatment efficiency and longevity, and reduced operating and capital costs in comparison with single use, throwaway ion-exchange techniques. The new treatment system offers the following major advantages over conventional ion-exchange technologies:

- The use of highly selective ion-exchange resins allows long-term, unattended operations, up to one year when operated at an influent ClO_4^- concentration of <20 μg/L and 1 BV/min.

- High selectivity of the resin minimizes changes in water quality and chemistry since chemicals and nutrients are neither added nor removed.

- The fast ion-exchange kinetics permit high flow rates (thousands or tens of thousands gpm) with a relatively small treatment module.

- Spent resin is readily regenerated and recycled with the new $FeCl_3$-HCl regeneration technique; the eluted ClO_4^- is highly concentrated in the first BV of regenerant solution with a good mass balance and recovery.

- Eluted ClO_4^- can be readily recovered as pure perchlorate salts (see Chapter 10) or destroyed in a perchlorate destruction system, resulting in practically no secondary waste production from the entire treatment process.

- Even without using the recovery or destruction techniques, costs of one BV of $FeCl_2$ and HCl solution are still less by an order of magnitude than costs of using any of the throwaway resins on the market or by using brine-regenerable, non-selective Type-II or Type-I acrylic resins (Chapter 10). A total cost savings of >50% is possible.

- Perchlorate in the spent regenerant solution can be effectively destroyed and, more importantly, the regenerant can be recycled.

- The treatment process is environmentally safe, and no toxic chemicals are used. Offsite regeneration and perchlorate destruction may be used to further reduce capital and operating costs and environmental impact.

ACKNOWLEDGMENTS

The technical and field support operations provided by Y.K. Ku and H. Yan at ORNL, by C.C. Chiang at Calgon Carbon Corporation, and by A.J. Shepard, J. Oltion and T. Battey at Earth Tech are gratefully acknowledged. This work was supported in part by the Environmental Management Restoration Division at Edwards AFB. ORNL is managed by UT-Battelle, LLC, under contract DE-AC05-00OR22725 with the U.S. Department of Energy.

REFERENCES

1. Batista, J. R., McGarvey, F. X., Vieira, A. R. In *Perchlorate in the Environment*; Urbansky, E. T., Ed.; Kluwer/Plenum: New York, 2000, pp 135-145.

2. Gu, B., Brown, G. M., Alexandratos, S. D., Ober, R., Dale, J. A., Plant, S. In *Perchlorate in the Environment*; Urbansky, E. T., Ed.; Kluwer/Plenum: New York, 2000, pp 165-176.

3. Tripp, A. R., Clifford, D. A. In *Perchlorate in the Environment*; Urbansky, E. T., Ed.; Kluwer/Plenum: New York, 2000, pp 123-134.

4. Venkatesh, K. R. Removal of perchlorate and other contaminants from groundwater at JPL, a pilot study. *Final report to Jet Propulsion Laboratory, Pasadena, CA*, 1999:NAS7.000218, SSIC No. 009661.

5. Venkatesh, K. R., Klara, S. M., Jennings, D. L., Wagner, N. J. In *Perchlorate in the Environment*; Urbansky, E. T., Ed.; Kluwer/Plenum: New York, 2000, pp 147-153.

6. Calgon Carbon Corporation. Full-scale ISEP(R) groundwater treatment plant. 2000: http://www.perchloratenews.com/case-study4.html.

7. Final Project Report - Application of ion-exchange technology for perchlorate removal from San Gabriel Basin groundwater. *NAS7.000216, NASA-JPL, SSIC No. 9661,* 1999:Montgomery Watson, Pasadena, California.

8. Urbansky, E. T. Perchlorate chemistry: Implications for analysis and remediation. *Bioremed. J.,* 1998; *2*:81-95.

9. Gu, B., Ku, Y., Brown, G. M. Treatment of perchlorate-contaminated water using highly-selective, regenerable ion-exchange technology: a pilot-scale demonstration. *Remediation,* 2002; *12*:51-68.

10. Urbansky, E. T. Issues in managing the risks associated with perchlorate in drinking water. *J. Environ. Manag.,* 1999; *56*:79-95.

11. Gu, B., Brown, G. M., Maya, L., Lance, M. J., Moyer, B. A. Regeneration of perchlorate (ClO_4^-)-loaded anion exchange resins by novel tetrachloroferrate ($FeCl_4^-$) displacement technique. *Environ. Sci. Technol.,* 2001; *35*:3363-3368.

12. Gu, B., Dong, W., Brown, G. M., Cole, D. R. Complete degradation of perchlorate in ferric chloride and hydrochloric acid under controlled temperature and pressure. *Environ. Sci. Technol.,* 2003; *37*:2291-2295.

13. Gu, B., Ku, Y., Brown, G. Sorption and desorption of perchlorate and U(VI) by strong-base anion-exchange resins. *Environ. Sci. Technol.,* 2005; *39*:901-907.

14. Gu, B., Ku, Y., Jardine, P. M. Sorption and binary exchange of nitrate, sulfate, and uranium on an anion-exchange resin. *Environ Sci. Technol.,* 2004; *38*:3184-3188.

15. Gu, B., Ku, Y., Brown, G. M. Treatment of perchlorate-contaminated groundwater using highly-selective, regenerable anion-exchange resins at Edwards Air Force Base. 2002; *ORNL/TM-2002/53*:Oak Ridge National Laboratory, Oak Ridge, TN.

16. Gu, B., Ku, Y., Brown, G. M. Treatment of perchlorate-contaminated water using highly-selective, regenerable ion-exchange technology: a pilot-scale demonstration. *Fed. Facil. Environ. J.,* 2003; *14*:75-94.

17. Gu, B., Cole, D. R., Brown, G. M. Destruction of perchlorate in ferric chloride and hydrochloric acid solution with control of temperature, pressure, and chemical reagents. *US Patent 6,800,203,* 2004.

Chapter 12

The Microbiology of Perchlorate Reduction and its Bioremediative Application

John D. Coates[1] and Laurie A. Achenbach[2]

[1]Department of Plant and Microbial Biology, University of California, Berkeley, CA 94720
[2]Department of Microbiology, Southern Illinois University, Carbondale, IL 62901

INTRODUCTION

Microbial respiration of oxyanions of chlorine such as chlorate (ClO_3^-) and perchlorate (ClO_4^-) [(per)chlorate] under anaerobic conditions has been known for more than half a century.[1] The high reduction potential of (per)chlorate (ClO_4^-/Cl^- E^o = 1.287 V; ClO_3^-/Cl^- E^o = 1.03 V) makes them ideal electron acceptors for microbial metabolism.[2] In general, chlorine oxyanions in the environment result from anthropogenic sources including disinfectants, bleaching agents, herbicides,[3-5] and munitions.[6, 7] Ammonium perchlorate (NH_4ClO_4) represents approximately 90% of all perchlorate salts manufactured.[8] It is predominantly used by the munitions industry and the US Defense Department as an energetics booster or oxidant in solid rocket fuels.[6-10] Although a powerful oxidant, under most environmental conditions perchlorate is highly stable and non-reactive owing to the high energy of activation associated with its reduction.[9, 11] Because of the large molecular volume and single anionic charge, perchlorate also has a low affinity for cations and as a result, perchlorate salts, such as ammonium perchlorate, are generally highly soluble and completely dissociate into NH_4^+ and ClO_4^- in aqueous solutions. Furthermore, perchlorate does not sorb to any significant extent to soils or sediments and, in the absence of any biological interactions, its mobility and fate are largely influenced by the hydrology of the environment.[12]

Because of its unique chemical stability and high solubility, remediation efforts for perchlorate contamination have focused primarily on microbial processes [9] and many novel bioremediative technologies are currently being developed.[13] Enhanced in situ bioremediation (EISB) of groundwater is increasingly being used as the commercial solution for the remediation of perchlorate and of the 65 different case studies of perchlorate-treatment technologies outlined in a 2001 report published by Ground Water

Remediation Technologies Analysis Center, the majority (45 case studies) were either *in-situ* or *ex-situ* biological treatment technologies based on the unique ability of some microorganisms to reductively respire perchlorate completely to innocuous chloride in the absence of oxygen.[10] Given the large plume/source dimensions (i.e., width, depth) at many contaminated sites, EISB approaches often require extraction of impacted groundwater, amendment with soluble nutrients (electron donors), and reinjection of the nutrient-amended water into the aquifer to effectively mix and distribute the nutrients throughout the target treatment area. While a variety of delivery instrumentation has been used in EISB demonstrations (with varying levels of success), conventional injection wells are by far the most common tool used for nutrient delivery. Unfortunately, the repeated addition of electron donors such as acetate, ethanol, citrate, or more complex undefined substrates creates conditions within the injection well screen and the surrounding filter pack favoring the outgrowth of non-perchlorate reducing microbial communities. This results in ineffective treatment of the target contaminant, inefficient use of the added electron donor, plugging of the near-well aquifer matrix (biofouling), alteration of the physical-chemical nature of the aquifer matrix (mineral content, hydraulic conductivity, pH, redox etc.), and a further reduction in water quality through the direct or indirect release of undesirable end-products (Fe(II), HS^-, CH_4, mobilized heavy metals, etc.). The need to frequently rehabilitate nutrient delivery wells to overcome just one of these effects (biofouling) using conventional water well techniques (swab and purge) significantly diminishes the financial feasibility of EISB. As such, a better understanding of the microbiology involved in perchlorate reduction including the factors which control its bioremediative application and the issues associated with biofouling is warranted to allow the appropriate design of successful robust EISB remediation processes.

THE MICROBIOLOGY OF PERCHLORATE REDUCTION

The first studies published on the biological reduction of chlorine oxyanions indicated that microorganisms rapidly reduced chlorate that was applied as an herbicide for thistle control [1] and the application of this reductive metabolism was later proposed for the measurement of sewage and wastewater biological oxygen demand.[14, 15] Initial investigation of the microbiology of chlorate reduction suggested that it was mediated by nitrate-respiring organisms in the environment and chlorate uptake and reduction was simply a competitive reaction for the nitrate reductase system of these bacteria.[16-18] In support of this, many organisms were shown to be capable of the reduction of (per)chlorate including *Escherichia coli, Proteus mirabilis, Rhodobacter capsulatus* and *Rhodobacter sphaeroides.*[18, 19] Chlorite (ClO_2^-)

was generally produced as a toxic end product of this reduction, and there was no evidence that these organisms could couple growth to this metabolism. Furthermore, early studies demonstrated that membrane-bound respiratory nitrate reductases and assimilatory nitrate reductases could alternative reduce chlorate,[20] and selection for chlorate resistance has been used to obtain mutants that are unable to synthesize the molybdenum cofactor required for nitrate reduction.[21]

However, the assumption perchlorate reduction was mediated by a coincidental reaction of the nitrate reductase could not explain the presence of specialized enzymes such as the chlorate reductase C purified from *Proteus mirabilis,* which could only use chlorate as a substrate.[22] Now it is known that specialized organisms have evolved which can grow by the anaerobic reductive dissimilation of (per)chlorate into innocuous chloride.[23] More than fifty dissimilatory (per)chlorate-reducing bacteria are now in pure culture [24-35] and this number continues to increase.[23] These organisms have been isolated from a broad diversity of environments including both pristine and contaminated soils and sediments.[24-31] This was unexpected due to the assumed limited natural abundance of (per)chlorate. However, the diverse metabolic capabilities of (per)chlorate-reducing bacteria may explain their presence in environments where (per)chlorate is not found.

GENERAL CHARACTERISTICS OF DPRB

Phenotypic characterization revealed that the known dissimilatory (per)chlorate-reducing bacteria (DPRB) exhibit a broad range of metabolic capabilities including the oxidation of hydrogen,[31] simple organic acids and alcohols,[24-26,29,30] aromatic hydrocarbons,[32] hexoses,[29] reduced humic substances,[23,25,36] both soluble and insoluble ferrous iron,[24-26, 37-39] and hydrogen sulfide.[24, 25] No DPRB are known to utilize complex substrates such as methyl soyate, molasses, or various edible oils, compounds that are often utilized as electron donors for in-situ bioremediative technologies.

All of the known dissimilatory (per)chlorate-reducing bacteria are facultatively anaerobic or microaerophilic which is reasonable in light of the fact that molecular oxygen is produced as a transient intermediate of the microbial reduction of perchlorate.[24-26, 30, 31] Some, but not all, alternatively respire nitrate.[24,25] Generally, these organisms are assumed to use either chlorate or perchlorate as terminal electron acceptors [40] although this has only been demonstrated in a few isolated cases.[25, 28, 31] In contrast, there are several chlorate-reducing bacteria in pure culture, including the well-characterized *Ideonella dechloratans* [29] and *Pseudomonas chloritidismutans* strain AW-1,[41] which are incapable of the reductive respiration of perchlorate indicating that the dual metabolic capability cannot be assumed. In general, (per)chlorate-reducing bacteria are able to distinguish between light and heavy isotopes in the chlorine content of perchlorate.[42] Recent studies demonstrated that the perchlorate-reducing bacterium, *Azospira suillum,* preferentially utilizes perchlorate containing the lighter isotope (^{35}Cl), resulting in a significant fractionation (–15‰) of the isotopic content of the perchlorate as the organism grows in pure culture.[42] A subsequent study demonstrated similar isotopic fractionation of the chlorine content of perchlorate when DPRB were grown in natural sediments.[43] The results of these studies suggest that isotope-signature tracing can be successfully applied to monitor the microbial reduction and removal of perchlorate in environmental samples being treated for perchlorate contamination and distinguish this from abiotic reactions which may affect perchlorate concentration including dilution or absorption. However, it also implies that care must be taken when using isotope fingerprinting to identify perchlorate sources as any intrinsic microbial reduction occurring in the environment may alter the isotope fingerprint obtained resulting in false identifications.

PHYLOGENY OF DPRB

(Per)chlorate reducing bacteria are phylogenetically diverse[24,26,31] with members in the alpha (α), beta (β), gamma (γ), and epsilon (ϵ) subclasses of the Proteobacteria (Fig. 1).[23,24,26,31,44] As such, the metabolic capability of (per)chlorate reduction is widespread throughout the Proteobacteria, which has some interesting evolutionary implications due to the relatively short time in which (per)chlorate reduction could have evolved and the assumed limited geographical distribution of natural sources of these compounds.[23,45,46] Several of the known (per)chlorate-reducing isolates are representatives of previously defined genera (*Pseudomonas, Magnetospirillum, Wolinella*)[24,31] not recognized for the capability of (per)chlorate respiration. However, the majority of the known (per)chlorate-reducing bacteria are closely related to each other and to the bacterial species *Rhodocyclus tenuis* and *Ferribacterium limneticum* in the class Betaproteobacteria. In general, the known close relatives to the (per)chlorate-reducing isolates do not grow by (per)chlorate respiration regardless of the similarity of their 16S rDNA sequence, thus making predictions of metabolic functionality based on 16S rDNA sequence analysis futile.[47] As such, the potential for this metabolism in any given environment must be identified through the application of functional gene or biochemical probes which target the genes or proteins specifically involved in reductive pathway used by the microorganisms for (per)chlorate metabolism.[23]

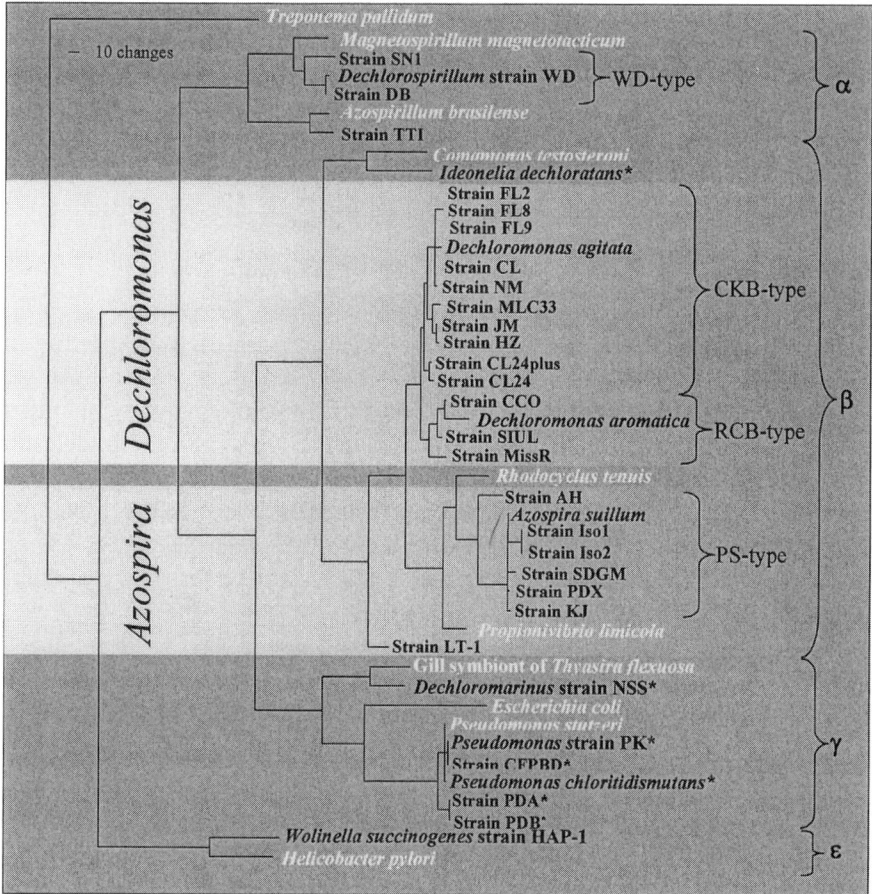

Figure 1. (Per)chlorate-reducing bacteria (in bold) and chlorate-reducing bacteria (marked with an asterix) bacteria are found in four (alpha, beta, gamma and epsilon) of the five subclasses of the Proteobacteria.   The (per)chlorate-reducing populations in the environment are predominately members of the *Dechloromonas* and *Azospira* species of the beta subclass of the Proteobacteria. The *Dechloromonas* species can be subdivided into the CKB-type and the RCB-type based on signature nucleotide sequences within the small subunit of the ribosomal RNA gene (16S rDNA). These, and similar signature nucleotide sequences within the 16S rDNA of the *Azospira* species and the *Dechlorospirillum* species in the alpha Proteobacteria, have allowed the development of specific molecular probes for the rapid screening of environmental samples for the presence of these microorganisms. The close non-(per)chlorate-reducing relatives are shown in white.

The (per)chlorate reducers of the beta subclass of the proteobacteria represent two novel genera with monophyletic origin, the *Dechloromonas* species and the *Azospira* (formerly *Dechlorosoma*) species (Fig. 1).[44, 48] The genus *Dechloromonas* can be further subdivided into the RCB-type and CKB-type based on signature nucleotide sequences within the 16S rRNA gene sequence (Fig. 1). Members of both the *Dechloromonas* and *Azospira* genera are ubiquitous [24] and have been identified and isolated from nearly all environments screened including pristine and contaminated field samples,[24,46,49,50] and even in soil and lake samples collected from Antarctica.[46] As such, these two groups are considered to represent the dominant (per)chlorate-reducing bacteria in the environment.[24,44,51] A third group, the *Dechlorospirillum* species[2,52] which are closely related to the magnetotatic *Magnetospirillum* species in the alpha subgroup of the Proteobacteria, are also commonly found in contaminated soils and bioreactors treating groundwater contaminated with perchlorate (Fig. 1).[23] The *Magnetospirillum* genus have been described based on its ability to form magnetosomes, an intracellular form of magnetite, when growing microaerophilically on iron-based media which confers on these microorganisms a unique magnetotactic characteristic. The best described of the *Dechlorospirillum* species is *D. anomalous* strain WD which shows almost 97% 16S rDNA sequence identity to *Magnetospirillum gryphiswaldense.*[26] Like *M. gryphiswaldense, D. anomalous* strain WD is a microaerophile, however, in contrast to *M. gryphiswaldense, D. anomalous* does not produce magnetosomes.[26]

ENVIRONMENTAL FACTORS CONTROLLING DPRB ACTIVITY

Many environmental factors have been shown to affect microbial (per)chlorate reduction, including trace elements, pH, salt concentration, and presence of other electron acceptors. Pure-culture studies have demonstrated that members of the *Dechloromonas* and *Azospira* genera can grow over a broad range of environmental conditions. These microorganisms are generally not nutritionally fastidious, however, molybdenum is a required trace element for perchlorate reduction.[53] Although most pure-culture studies have been performed in media supplemented with a defined or undefined vitamin source, it has been demonstrated for at least one microorganism, *Dechloromonas agitata,* that vitamin supplementation was unnecessary for growth in minimal media.[25] Several lines of evidence suggest the optimal pH for perchlorate reduction occurs around neutral pH. The *Dechloromonas* and *Azospira* species generally grow optimally at pH values near neutrality in freshwater environments.[24-26] Even so, recent field studies have shown that related deep-branching members of the *Dechloromonas* and *Azospira* genera often predominate in sites of adverse pH or salinity with some species being capable of growth and perchlorate respiration at pH values as low as pH 5 [54] supporting the environmental dominance of these species.

To date, no microorganism isolated has been demonstrated to reduce perchlorate in salinities greater than 2%. This presents a problem for the biological treatment of waste brine concentrated with perchlorate collected by ion-exchange processes. An enrichment culture from the Great Salt Lake was able to carry out perchlorate reduction in salt brines as concentrated as 11% NaCl [55] but perchlorate removal efficiency was not reported while another enrichment culture developed from marine sediments was reported to reduce 70 to 90 mg.L^{-1} perchlorate at 6% NaCl within 24 hours.[56] One putative perchlorate reducer, *Citrobacter* sp. strain IsoCock1, was reported to partially reduce perchlorate in salt concentrations as high as 7.5%, however, neither growth coupled to perchlorate reduction nor complete reduction of perchlorate to chloride was demonstrated for the microorganism.[35] By contrast, one chlorate-reducing microorganism, *Dechloromarinus chlorophilus,* isolated from a marine sediment, can grow and reduce chlorate, but not perchlorate, in salinities greater than 5% [57] suggesting that the metabolism is not limited by the chloride content.

Both oxygen and nitrate can be inhibitors of perchlorate reduction.[23,53] An investigation of the environmental factors controlling reduction of perchlorate by the dissimilatory perchlorate-reducer *Azospira suillum* demonstrated that perchlorate reduction only occurred under anaerobic conditions in the presence of perchlorate.[53] Anaerobic conditions alone were not enough to induce expression of an active form of the chlorite dismutase, a central enzyme involved in (per)chlorate metabolism by DPRB,[23,24,58,59] and its activity was only observed under anaerobic conditions in the presence of perchlorate.[53] Furthermore, molecular studies focused on the identification of the gene encoding the perchlorate reductase demonstrated that its expression was down regulated by the presence of atmospheric oxygen.[45, 60] Phenotypic studies demonstrated that dissolved oxygen concentrations less than 2 mg L^{-1} were enough to completely inhibit perchlorate reduction by *A. suillum.*[53]

Similar to oxygen, nitrate also negatively regulated the production of the active enzymes involved in perchlorate metabolism and inhibited dissimilatory perchlorate reduction by *A. suillum.*[53] If present, nitrate was preferentially used even if the cultures had previously been grown on perchlorate. However, such preferential use is not predictable as several notable exceptions are known to exist.[34,53] For example, nitrate had no significant inhibitory effect on perchlorate reduction by the perchlorate-reducer *Dechloromonas agitata* strain CKB, which is the only perchlorate-reducer described that is incapable of growth by dissimilatory nitrate reduction.[25,51] Interestingly, although *D. agitata* does not grow by nitrate reduction, during perchlorate reduction the nitrate in the culture medium was concomitantly reduced to nitrite, which accumulated in the culture broth suggesting that the nitrate is co-reduced by the (per)chlorate reductase.[53] Previous studies performed on the DPRB strain perc1ace and also indicated that nitrate was concomitantly reduced with perchlorate by this DPRB.[34] However, in contrast to *D. agitata*, strain perc1ace could grow by nitrate reduction and nitrite did not accumulate in the culture broth.[34] Similarly, perchlorate reduction by *Citrobacter* strains was shown to readily occur in the presence of nitrate [61] further indicating that the ability of DPRB to reduce (per)chlorate in the presence of nitrate appears to be species specific.

THE MICROBIOLOGY OF BIOFOULING DURING PERCHLORATE BIOREMEDIATION

As outlined above DPRB are ubiquitous and capable of growth under a wide range of environmental conditions utilizing a diverse range of electron donors.[23,24] As such, bioaugmentation is not required for an in situ bioremediative strategy for most environments. Current bioremediation strategies for groundwater are based on stimulating the activity of indigenous DPRB through injection of a suitable electron donor such as acetate, ethanol, citrate or a more complex substrate into the subsurface. Because microbial perchlorate reduction is generally inhibited by the presence of O_2 and to some extents nitrate,[53,62] excess electron donor must be added to biologically remove these components. Although a promising bioremediative strategy, electron donor amendments through one or more injection wells into the subsurface often additionally stimulate other undesirable non-perchlorate-reducing microorganisms within the injection well screen and the surrounding filter pack. Over the long-term this results in ineffective treatment of the target contaminant, inefficient waste of the added electron donor, plugging of the near-well aquifer matrix (biofouling), alteration of the physical-chemical nature of the aquifer matrix (mineral content, hydraulic conductivity, pH, etc.), and a further reduction in water quality through the direct or indirect release of undesirable end-products (Fe(II), HS^-, CH_4, mobilized heavy metals, etc.). There are two primary reasons for this:

i) Because of their metabolic diversity many alternative electron donors can be utilized for stimulating the activity of indigenous DPRB, however, the stimulatory effects of the selected electron donor may be either direct or indirect. Indirect stimulation occurs when the added electron donor must first be biotransformed through the activity of indigenous non-perchlorate-reducing bacteria into more simple substrates that can be utilized by the indigenous DPRB. Substrates that must first be biotransformed often result in the rapid outgrowth of fermentative microorganisms that are not directly involved in perchlorate reduction. This is particularly true with complex substrates such as methyl soyate, molasses, or various edible oils, but may also occur with even relatively simple compounds such as citrate which must first be fermented into acetate, propionate, and lactate by organisms like *Lactobacilli* species before microbial perchlorate reduction is stimulated. This is because, although metabolically diverse, there are no DPRB known which can directly couple the degradation of these compounds to perchlorate reduction.[23] Even the ubiquitous and metabolically versatile *Dechloromonas* and *Azospira* genera are only capable of biodegrading low molecular weight monocarboxylic and dicarboxylic acids, simple alcohols, and monoaromatics.[23]

ii) In the natural environment microbial communities form structured populations based primarily upon the thermodynamic preferential use of specific electron acceptors and the affinity of the respective microbial trophic groups for the available electron donors.[63,64] Thermodynamic considerations indicate that microbial perchlorate reduction is energetically similar to nitrate reduction and more favorable than Fe(III) reduction, sulfate reduction, or methanogenesis and should therefore occur first.[2,23] Thus, the addition of excess amounts of a labile electron donor results in the formation of a series of redox zones which radiate out from the injection well as available alternative terminal electron acceptors along the ground water flow path are depleted and new ones become available (Fig. 2).[63-67] In any given location, as one electron acceptor is depleted, the microbial community will evolve to take advantage of the next most thermodynamically favorable electron acceptor that is available.[63, 64, 67] The particular electron acceptor being used at any specific location is a function of the length of time the location is exposed to the electron donor, and the respective concentrations of the electron donor and the individual electron acceptors. Thus, as oxygen, and nitrate deplete in the near-well environment, perchlorate reduction becomes active. As the perchlorate is depleted in the same environment the terminal electron accepting process again changes to the next most thermodynamically favorable available electron acceptor, usually Mn(IV) or Fe(III). As the near well environment is exposed to the highest electron donor concentration for the longest time, the available electron acceptors in this environment will ultimately be completely depleted resulting in the creation of an active zone of methanogenesis (Fig 2). This process continues at each point as the electron donor radiates out from the injection well until the electron donor has been completely consumed (i.e. it zone of impact) and results in the establishment of substantial communities of non-perchlorate-reducing respiratory microbial populations through which the injected electron donor has to travel prior to its consumption coupled to perchlorate reduction (Fig. 2). As these sequential terminal electron accepting redox zones are established the zone of onset of perchlorate reduction is pushed further and further away from the injection well until it is ultimately outside

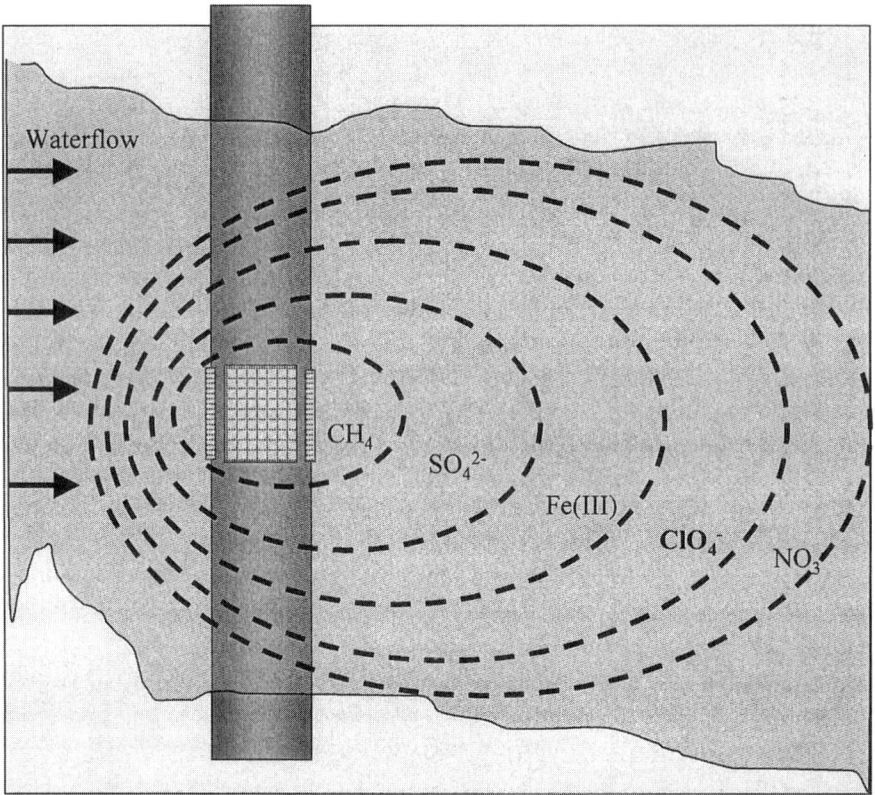

Figure 2. Schematic outlining the electron accepting redox zones created in a perchlorate-contaminated aquifer system radiating out from the screened area of an injection well during extended injection of an electron donor. The shape and size of the individual redox zones is determined by the in-situ concentrations of the individual electron acceptors and the aquifer flow rate.

of the zone of impact of the added electron donor (i.e. the electron donor is completely utilized before it has traveled to the zone of active perchlorate reduction). At this point, further electron donor addition will no longer have a significant effect on perchlorate concentrations downstream of the injection well and will simply maintain the non-perchlorate-reducing extraneous microbial communities.

As such, the long-term electron donor amendments into the subsurface will negatively impact the environment directly diminishing the effectiveness of the bioremediative strategy. Typical effects include:

(i)   Stimulation of substantial non-perchlorate-reducing microbial communities in the near-well environment resulting in a significant loss of the injected electron donor to metabolisms other than perchlorate reduction decreasing the distribution and zone of impact of the injected materials.

(ii) Biofouling or loss of hydraulic conductivity in the near-well environment as a result of the stimulated large microbial communities causing a significant reduction in the injection rate and further decreasing the distribution and zone of impact of the injected materials.

(iii) Increased activity of non-perchlorate reducing microbial communities such as Fe(III) reducers, sulfate reducers, and methanogens in the near-well environment causing additional reduction in water quality through an increase in soluble Mn(II) and Fe(II), a release of adsorbed metals and phosphates, and a production of sulfides and methane gas.

(iv) Solubilization and mobilization of normally immobile toxic metals (e.g. copper, zinc, or chromium) through direct complexation with added electron donors such as citrate or oxalate.

The extent of each of these effects is a function of the nature of the electron donor selected and its chemical and biological reactivity. As such the biogeochemical characteristics of an ideal electron donor can be identified.

(i) An ideal electron donor should be non-fermentable. The electron donor should be directly utilizable by the indigenous DPRB discounting the establishment of a sizable fermentative microbial population and thus reduce biofouling.

(ii) An ideal electron donor should be biocidal at elevated concentrations. The electron donor itself should inhibit the growth of all microorganisms at elevated concentrations thus allowing it to diffuse out from the injection well until it is diluted to a concentration where it becomes suitable for perchlorate reduction. This will reduce near-well biofouling and increase the zone of impact.

(iii) An ideal electron donor should, if possible, limit its stimulatory effects to perchlorate reduction. It should be non-biodegraded by Mn(VI)-reducing, Fe(III)-reducing, sulfate-reducing or methanogenic bacteria which divert the valuable electron donor capacity away from perchlorate reduction, enhance biofouling, and result in a decrease in water quality through the production of undesirable endproducts..

(iv) An ideal electron donor should not impact the geochemistry of the environment. Many compounds such as citrate or oxalate readily complex and solubilize insoluble metals, mobilizing these through the ground water until the organic is biodegraded at which point the metal re-precipitates back out of solution.[68] Such mobilization and re-precipitation may result in localized mineral formation causing irreversible reduction in porosity.

Although no one potential electron donor matches all of the requirements outlined, several of these requirements can be satisfied through the judicious application of individual or mixtures of non-fermentable electron donors. The use of complex substrates such as molasses, or edible vegetable oils should be carefully considered. Although such compounds will initially stimulate successful microbial removal of perchlorate in-situ, their log-term application will result in several undesirable biogeochemical and biofouling effects.

CONCLUSIONS

The field of microbial perchlorate reduction has clearly advanced significantly in a very short period from a poorly understood metabolism to a burgeoning scientific field of discovery. As outlined above, there is now a much greater appreciation of the microbiology involved and the application of the knowledge to the successful treatment of contaminated environments. Overall, the future is promising even though research in this field is still in its infancy. Nothing is known of the evolutionary root of this metabolism. From a biogeochemical perspective, a better understanding of how perchlorate is formed in the natural environment and what geochemical conditions are required for its formation might give some insight into plotting the metabolism against a realistic evolutionary timeline. From a microbial perspective, it will be important to look for this metabolism in more extreme environments such as hypersaline or hyperthermophilic environments to obtain DPRB isolates across a broader phylogeny to establish a broad-base molecular chronometer. With the development of this field comes a better understanding of the ideal electron donors available and the individual factors which truly control the activity of these organisms in-situ allowing for the design of more effective and robust enhanced in situ bioremediation technologies.

ACKNOWLEDGEMENTS

Research on the microbial reduction of (per)chlorate in the laboratories of JDC and LAA is supported by grants from the US Department of Defense SERDP program.

REFERENCES

1. Aslander, A. 1928. Experiments on the eradication of Canada Thistle, *Cirsium arvense*, with chlorates and other herbicides. J Agric Res. 36:915.

2. Coates, J. D., U. Michaelidou, S. M. O'Connor, R. A. Bruce, and L. A. Achenbach 2000. The diverse microbiology of (per)chlorate reduction., p. 257-270. *In* E. D. Urbansky (ed.), Perchlorate in the Environment. Kluwer Academic/ Plenum, New York.

3. Germgard, U., A. Teder, and D. Tormund 1981. Chlorate formation during chlorine dioxide bleaching of softwood kraft pulp. Paperi ja Puu. 3:127-133.

4. Rosemarin, A., K. Lehtinen, and M. Notini 1990. Effects of treated and untreated softwood pulp mill effluents on Baltic sea algae and invertebrates in model ecosystems. Nord. Pulp and Paper Res. J. 2:83-87.

5. Agaev, R., V. Danilov, V. Khachaturov, B. Kasymov, and B. Tishabaev 1986. The toxicity to warm-blooded animals and fish of new defoliants based on sodium and magnesium chlorates. Uzb. Biol. Zh. 1:40-43.

6. Urbanski, T. 1984. Salts of nitric acid and of oxy-acids of chlorine, p. 444-461. *In* T. Urbanski (ed.), Chemistry and Technology of Explosives, vol. 4. Pergamon Press, Elmsford, N.Y.

7. Urbanski, T. 1984. Composite propellants, p. 602-620. *In* T. Urbanski (ed.), Chemistry and Technology of Explosives, vol. 4. Pergamon Press, New York.

8. Motzer, W. E. 2001. Perchlorate: problems, detection, and solutions. Environ For. 2:301-311.

9. Urbansky, E. T. 1998. Perchlorate chemistry: implications for analysis and remediation. Bioremed J. 2:81-95.

10. Roote, D. S. 2001. Technology status report perchlorate treatment technologies first edition. Technology status report DAAE30-98-C-1050. Ground-Water Remediation Technologies Analysis Center.

11. Urbansky, E. T. 2002. Perchlorate as an environmental contaminant. Environ Sci Pollut Res. 9:187-192.

12. Urbansky, E. T., and S. K. Brown 2003. Perchlorate retention and mobility in soils. J. Environ Monit. 5:455-462.

13. Xu, J., Y. Song, B. Min, L. Steinberg, and B. E. Logan 2003. Microbial degradation of perchlorate: principles and applications. Environ Eng Sci. 20:405-422.

14. Bryan, E. H., and G. A. Rohlich 1954. Biological reduction of sodium chlorate as applied to measurement of sewage BOD. Sew Ind Waste. 26:1315-1324.

15. Bryan, E. H. 1966. Application of the chlorate BOD procedure to routine measurement of wastewater strength. J Wat Pol Con Fed. 38:1350-1362.

16. Hackenthal, E., W. Mannheim, R. Hackenthal, and R. Becher 1964. Die Reduktion Von Perchlorat Durch Bakterien. I.* Untersucungen An Intaken Zellen. Biochem Pharm. 13:195-206.

17. Hackenthal, E. 1965. Die reduktion von perchlorat durch bacterien - II. Die identitat der nitratreduktase und des perchlorat reduzierenden enzyms aus *B. cereus*. Biochem. Pharm. 14:1313-1324.

18. de Groot, G. N., and A. H. Stouthamer 1969. Regulation of reductase formation in *Proteus mirabilis*. I. Formation of reductases and enzymes of the formic hydrogenlyase complex in the wild type and in chlorate resistant mutants. Arch Microbiol. 66:220-233.

19. Roldan, M. D., F. Reyes, C. Moreno-Vivian, and F. Castillo 1994. Chlorate and Nitrate reduction in the phototrophic bacteria *Rhodobacter capsulatus* and *Rhodobacter sphaeroides*. Cur Microbiol. 29:241-245.

20. Stewart, V. 1988. Nitrate respiration in relation to facultative metabolism in enterobacteria. Microbiol. Rev. 52:190-232.

21. Neidhardt, F. C., R. Curtiss, J. Ingraham, E. Lin, K. Brooks Low, B. Magasanik, W. Rfznikopp, M. Riley, M. Schaechter, and H. E. Umbarger (eds.) 1996 *Escherichia coli* and *Salmonella* - Cellular and Molecular Biology. ASM Press, Washington DC.

22. Oltmann, L. F., W. N. M. Reijnders, and A. H. Stouthamer 1976. Characterization of purified nitrate reductase a and chlorate reductase c from *Proteus mirabilis*. Archi Microbiol. 111:25-35.

23. Coates, J. D., and L. A. Achenbach 2004. Microbial perchlorate reduction: rocket fuelled metabolism. Nat Rev Microbiol. 2:569-580.

24. Coates, J. D., U. Michaelidou, R. A. Bruce, S. M. O'Connor, J. N. Crespi, and L. A. Achenbach 1999. The ubiquity and diversity of dissimilatory (per)chlorate-reducing bacteria. Appl Environ Microbiol. 65:5234-5241.

25. Bruce, R. A., L. A. Achenbach, and J. D. Coates 1999. Reduction of (per)chlorate by a novel organism isolated from a paper mill waste. Environ Microbiol. 1:319-331.

26. Michaelidou, U., L. A. Achenbach, and J. D. Coates 2000. Isolation and characterization of two novel (per)chlorate -reducing bacteria from swine waste lagoons., p. 271 - 283. *In* E. D. Urbansky (ed.), Perchlorate in the Environment. Kluwer Academic/ Plenum, New York.

27. Romanenko, V. I., V. N. Korenkov, and S. I. Kuznetsov 1976. Bacterial decomposition of ammonium perchlorate. Mikrobiologiya. 45:204-209.

28. Stepanyuk, V., G. Smirnova, T. Klyushnikova, N. Kanyuk, L. Panchenko, T. Nogina, and V. Prima 1992. New species of the *Acinetobacter* genus *Acinetobacter thermotoleranticus* sp. nov. Mikrobiologiya. 61:347-356.

29. Malmqvist, A., T. Welander, E. Moore, A. Ternstrom, G. Molin, and I.-M. Stenstrom 1994. *Ideonella dechloratans* gen. nov., sp. nov., a new bacterium capable of growing anaerobically with chlorate as an electron acceptor. Sys. Appl. Microbiol. 17:58-64.

30. Rikken, G., A. Kroon, and C. van Ginkel 1996. Transformation of (per)chlorate into chloride by a newly isolated bacterium: reduction and dismutation. Appl. Microbiol. Biotechnol. 45:420-426.

31. Wallace, W., T. Ward, A. Breen, and H. Attaway 1996. Identification of an anaerobic bacterium which reduces perchlorate and chlorate as *Wolinella succinogenes*. J Ind Microbiol. 16:68-72.

32. Coates, J. D., R. Chakraborty, J. G. Lack, S. M. O'Connor, K. A. Cole, K. S. Bender, and L. A. Achenbach 2001. Anaerobic benzene oxidation coupled to nitrate reduction in pure culture by two strains of *Dechloromonas*. Nature. 411:1039-1043.

33. Zhang, H. S., M. A. Bruns, and B. E. Logan 2002. Chemolithoautotrophic perchlorate reduction by a novel hydrogen-oxidizing bacterium. Environ Microbiol. 4:570-576.

34. Herman, D. C., and W. T. Frankenberger, Jr. 1999. Bacterial reduction of perchlorate and nitrate in water. J Environ Qual. 28:1018-1024.

35. Okeke, B. C., T. Giblin, and W. T. Frankenberger 2002. Reduction of perchlorate and nitrate by salt tolerant bacteria. Environ Pollut. 118:357-363.

36. Coates, J. D., K. A. Cole, R. Chakraborty, S. M. O'Connor, and L. A. Achenbach 2002. The diversity and ubiquity of bacteria utilizing humic substances as an electron donor for anaerobic respiration. Appl Environ Microbiol. 68:2445-2452.

37. Chaudhuri, S. K., J. G. Lack, and J. D. Coates 2001. Biogenic magnetite formation through anaerobic biooxidation of Fe(II). Appl Environ Microbiol. 67:2844-2848.

38. Lack, J. G., S. K. Chaudhuri, S. D. Kelly, K. M. Kemner, S. M. O'Connor, and J. D. Coates 2002. Immobilization of radionuclides and heavy metals through anaerobic biooxidation of Fe(II). Appl Environ Microbiol. 68:2704-2710.

39. Lack, J. G., S. K. Chaudhuri, R. Chakraborty, L. A. Achenbach, and J. D. Coates 2002. Anaerobic biooxidation of Fe(II) by *Dechlorosoma suillum*. Microb Ecol. 43:424-431.

40. Logan, B. 1998. A review of chlorate- and perchlorate-respiring microorganisms. Bioremed J. 2:69-79.

41. Wolterink, A. F. W. M., A. B. Jonker, S. W. M. Kengen, and A. J. M. Stams 2002. *Pseudomonas chloritidismutans* sp. nov., a non-denitrifying, chlorate-reducing bacterium. Int J Syst Evol Microbiol. 52:2183-2190.

42. Coleman, M. L., M. Ader, S. Chaudhuri, and J. D. Coates 2003. Microbial isotopic fractionation of perchlorate chlorine. Appl Environ Microbiol. 69:4997-5000.

43. Sturchio, N. C., P. B. Hatzinger, M. Arkins, C. S u h, and L. Heraty 2003. Chlorine isotope fractionation during microbial reduction of perchlorate. Environ Sci Technol. 37:3859-3863.

44. Achenbach, L. A., R. A. Bruce, U. Michaelidou, and J. D. Coates 2001. *Dechloromonas agitata* N.N. gen., sp. nov. and *Dechlorosoma suillum* N.N. gen., sp. nov. two novel environmentally dominant (per)chlorate-reducing bacteria and their phylogenetic position. Int J Syst Evol Microbiol. 51:527- 533.

45. Bender, K. S., C. Shang, R. Chakraborty, S. M. Belchik, J. D. Coates, and L. A. Achenbach 2005. Identification, characterization, and classification of genes encoding perchlorate reductase. J Bacteriol. 187: 5090-5096

46. Bender, K. S. 2003. The genetics of (per)chlorate reduction. PhD. Thesis Southern Illinois University.

47. Achenbach, L. A., and J. D. Coates 2000. Disparity between bacterial phylogeny and physiology. ASM News. 66:714-716.

48. Tan, Z., and B. Reinhold-Hurek 2003. *Dechlorosoma suillum* Achenbach et al. 2001 is a later subjective synonym of *Azospira oryzae* Reinhold-Hurek and Hurek 2000. Int J Sys Evol Microbiol. 53:1139-1142.

49. Coates, J. D., R. A. Bruce, J. A. Patrick, and L. A. Achenbach 1999. Hydrocarbon bioremediative potential of (per)chlorate-reducing bacteria. Bioremed J. 3:323-334.

50. Logan, B. E., H. Zhang, P. Mulvaney, M. G. Milner, I. M. Head, and R. F. Unz 2001. Kinetics of perchlorate- and chlorate-respiring bacteria. Appl Environ.Microbiol. 67:2499-2506.

51. Coates, J. D. 2005. Bacteria that respire oxyanions of chlorine. *In* D. Brenner, N. Krieg, J. Staley, and G. Garrity (eds), *Bergey's Manual of Sytematic Bacteriology*, vol. 2. Springer-Verlag, New York, NY.

52. Michaelidou, U. 2005. An investigation of the environmental significance of microbial (per)chlorate reduction. MSc. Thesis Southern Illinois University.

53. Chaudhuri, S. K., S. M. O'Connor, R. L. Gustavson, L. A. Achenbach, and J. D. Coates 2002. Environmental factors that control microbial perchlorate reduction. Appl Environ Microbiol. 68: 4425-4430.

54. Pollock, J. 2003. Diversity of perchlorate-reducing bacteria in relation to environmental factors. MSc. Thesis Southern Illinois University.

55. Logan, B. E., J. Wu, and R. F. Unz 2001. Biological perchlorate reduction in high-salinity solutions. Wat Res. 35:3034-3038.

56. Cang, Y., D. J. Roberts, and D. A. Clifford 2004. Development of cultures capable of reducing perchlorate and nitrate in high salt solutions. Wat Res. 38:3322-3330.

57. Bruce, R. A. 1999. The microbiology and bioremediative potential of (per)chlorate-reducing bacteria MSc. Thesis Southern Illinois University.

58. van Ginkel, C., G. Rikken, A. Kroon, and S. Kengen 1996. Purification and characterization of chlorite dismutase: a novel oxygen-generating enzyme. Arch. Microbiol. 166:321-326.

59. Bender, K. S., S. M. O'Connor, R. Chakraborty, J. D. Coates, and L. A. Achenbach 2002. The chlorite dismutase gene of *Dechloromonas agitata* strain CKB: Sequencing, transcriptional analysis and its use as a metabolic probe. Appl Environ Microbiol. 68:4820-4826.

60. Bender, K. S. 2003. The genetics of (per)chlorate reduction. PhD. Thesis Southern Illinois University.

61. Bardiya, N., and J. H. Bae 2004. Role of *Citrobacter amalonaticus* and *Citrobacter farmeri* in dissimilatory perchlorate reduction. J Bas Microbiol. 44:88-97.

62. O'Connor, S. M., and J. D. Coates 2002. A universal immuno-probe for (per)chlorate-reducing bacteria. Appl. Environ. Microbiol. 68:3108-3113.

63. Lovley, D. R., and F. H. Chapelle 1995. Deep subsurface microbial processes. Rev Geophys. 33:365-381.

64. Coates, J. D., and L. A. Achenbach 2001. The Biogeochemistry of Aquifer Systems, p. 719-727. *In* C. J. Hurst, G. R. Knudsen, M. J. McInerney, L. D. Stetzenbach, and M. W. Walter (eds), Manual of Environmental Microbiology, 2nd ed. ASM Press, Washington, DC.

65. Lovley, D. R., and S. Goodwin 1988. Hydrogen concentrations as an indicator of the predominant terminal electron accepting reactions in aquatic sediments. Geochim. Cosmochim. Acta. 52:2993-3003.

66. Champ, D. R., J. Gulens, and R. E. Jackson 1979. Oxidation-reduction sequences in ground water flow systems. Can. J. Earth Sci. 16:12-23.

67. Christensen, T. H., P. L. Bjerg, S. A. Banwart, R. Jakobsen, G. Heron, and H. Albrechtsen 2000. Characterization of redox conditions in groundwater contaminant plumes. J. Contam. Hydrol. 45:165-241.

68. Ehrlich, H. L. 1990. Geomicrobiology, Third edition, Revised and Expanded ed. Marcel dekker, Inc., New York.

Chapter 13

The Biochemistry and Genetics of Microbial Perchlorate Reduction

Laurie A. Achenbach,[1] Kelly S. Bender,[2] Yvonne Sun,[3] and John D. Coates[3]

[1]*Southern Illinois University, Carbondale, IL 62901;* [2]*BioInsite, LLC, Carbondale, IL 62903;* [3]*University of California-Berkeley, Berkeley, CA, 94720*

INTRODUCTION

While regarded as a relatively new form of microbial metabolism, our understanding of the perchlorate reduction pathway has progressed dramatically in the last five years. Genetic analyses have offered the first glimpse into the enzyme encoding genes integral to perchlorate metabolism as well as putative regulation targets. This genomic analysis has also offered tools for detecting and monitoring dissimilatory perchlorate-reducing bacteria for bioremediative purposes. Mimicking the physiological diversity of these microorganisms, a surprising realm of genetic diversity has been discovered in the perchlorate reduction loci of dissimilatory perchlorate-reducing bacteria (DPRB). The perchlorate reduction pathway consists of two central enzymes: perchlorate reductase and chlorite dismutase (Fig. 1). The first enzymatic step of the pathway, perchlorate reduction to chlorite, is performed by perchlorate reductase. The toxic chlorite formed from this reduction is subsequently converted to chloride and oxygen by chlorite dismutase. Here we review the identification, analysis, and phylogenetic conservation of the genes encoding perchlorate reductase and chlorite

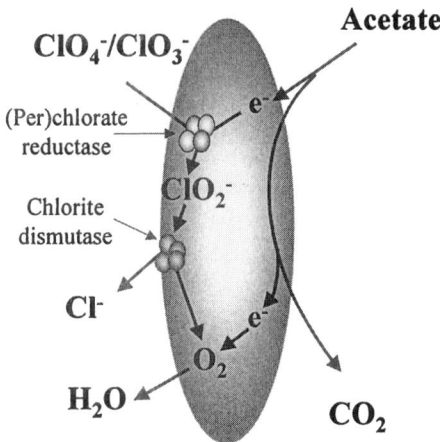

Figure 1. Model of the pathway involved in the respiratory reduction of (per)chlorate by (per)chlorate-reducing bacteria.

dismutase based on the timeline of the respective discoveries, as well as their potential application in environmental biotechnology.

CHLORITE DISMUTASE GENE

The dismutation of chlorite into chloride and O_2 represents a central step in the reductive pathway of perchlorate and chlorate [collectively (per)chlorate] that is common to all dissimilatory perchlorate-reducing bacteria (DPRB) and is mediated by a single enzyme, chlorite dismutase (CD). Following purification of the CD enzyme,[1,2] the chlorite dismutase gene (cld) was isolated and characterized from both the (per)chlorate-reducing microorganism *Dechloromonas agitata* strain CKB[3] and the chlorate-reducing microorganism *Ideonella dechloratans*.[4] In the case of *D. agitata,* sequence analysis identified an open-reading frame of 834 base pairs that encodes a mature protein with an N-terminal sequence identical to that of the CD protein previously purified from *D. agitata*. Further information regarding promoter structure of the *D. agitata cld* gene was obtained using primer extension analysis. This molecular technique identified a single transcription start site located 8 bp downstream of a putative AT-rich -10 promoter region (5'-AAATTT-3'), indicating that a single promoter regulates expression of this gene.[3] While brief, this analysis provided the first glimpse into a putative promoter sequence involved in the perchlorate reduction pathway, providing a target for future regulation studies.

Chlorite Dismutase Protein

Analysis of the predicted protein product of the *D. agitata* chlorite dismutase gene indicated a 277 amino acid protein, including a leader peptide of 26 amino acids targeting the protein to the cell membrane.[3] This signal peptide had a characteristic positive-charge at the N terminus, a hydrophobic helix, and a cleavage site motif of SQA-QQA. While the leader sequence was not present in the N-terminal sequence of the purified CD enzyme, the identification of a ribosome binding site located immediately upstream of the potential leader peptide supports the hypothesis that the leader peptide of the *D. agitata cld* gene is translated and later removed to produce the mature CD protein.[3] Previous biochemical studies also demonstrated CD activity was associated with both the cell membrane and soluble fractions of a *D. agitata* lysed-cell preparation,[5] further suggesting that the CD is loosely bound to the membrane or present in the periplasm. In *I. dechloratans*, the CD enzyme activity was located primarily in the periplasmic extract.[6] Furthermore, a CD specific immunoprobe readily bound to whole cells of both of these organisms suggesting that the antigenic portion of the CD was present on the outer membrane.[7] All of these studies lend further support to the membrane-bound nature of the chlorite dismutase enzyme.

Recently, a protein with remote homology (15% identity) to CD was described.[8] This protein, TT1485 from *Thermus thermophilus* HB8, produced oxygen with the substrates chlorite and H_2O_2, but not with hypochlorite and chlorate. However, the catalytic activity of reconstituted TT1485 for chlorite was much lower than that of CD and the authors further concluded that TT1485 exerts a function other than chlorite degradation. TT1485 may act as a heme peroxidase in a downstream pathway of superoxide dismutation to protect the cell against the harmful effects of reactive oxygen species.[8] As part of that work, Ebihara and coworkers (2005) built a structural model of the heme-binding site of CD and indicated the key residues involved in the active site of CD. Due to the unique conversion of chlorite to chloride and oxygen, this enzyme model may provide a tool for future CD kinetic studies.

Transcriptional Regulation of the *cld* Gene

To understand the genetic regulation of the *cld* gene, and thus factors involved in regulating perchlorate metabolism, RNA-based studies were performed. Northern blot analysis with purified *D. agitata* RNA indicated that the chlorite dismutase gene is basally expressed under aerobic conditions but, as indicated by the previous physiological observations, is transcriptionally up-regulated when cells are grown under perchlorate-reducing conditions.[3] In contrast, CD expression is constitutive in the chlorate reducing microorganisms *Pseudomonas* strain PDA[9] and *Pseudomonas* strain PK (JD Coates and LA Achenbach, unpublished). It was also apparent from Northern blot analyses that the chlorite dismutase gene is not transcribed as part of an operon as in each case only a single band representing *cld* mRNA was detected.[3] Thus, despite the absolute requirement for chlorite dismutation in dissimilatory perchlorate reduction, the *cld* and perchlorate reductase genes appear to be separately regulated.

Further transcriptional attributes were discovered through examination of the *Dechloromonas aromatica* genome sequence (http://www.jgi.doe.gov), upstream of the *cld* gene (Y. Sun, JD Coates, and LA Achenbach, unpublished) (Fig. 2). This analysis revealed the presence of two heptamer sequences that match the consensus for the NarL regulator protein binding site.[10] Similar to the two central NarL heptamers of the *fdnG* gene that are essential for regulation by NarL,[11] these heptamer sequences were separated by 9 bp and arranged in an inverted repeat. The NarL protein is known to repress transcription of NarL-regulated genes (*e.g.* fumarate reductase and DMSO reductase) in high nitrate concentrations but to activate transcription

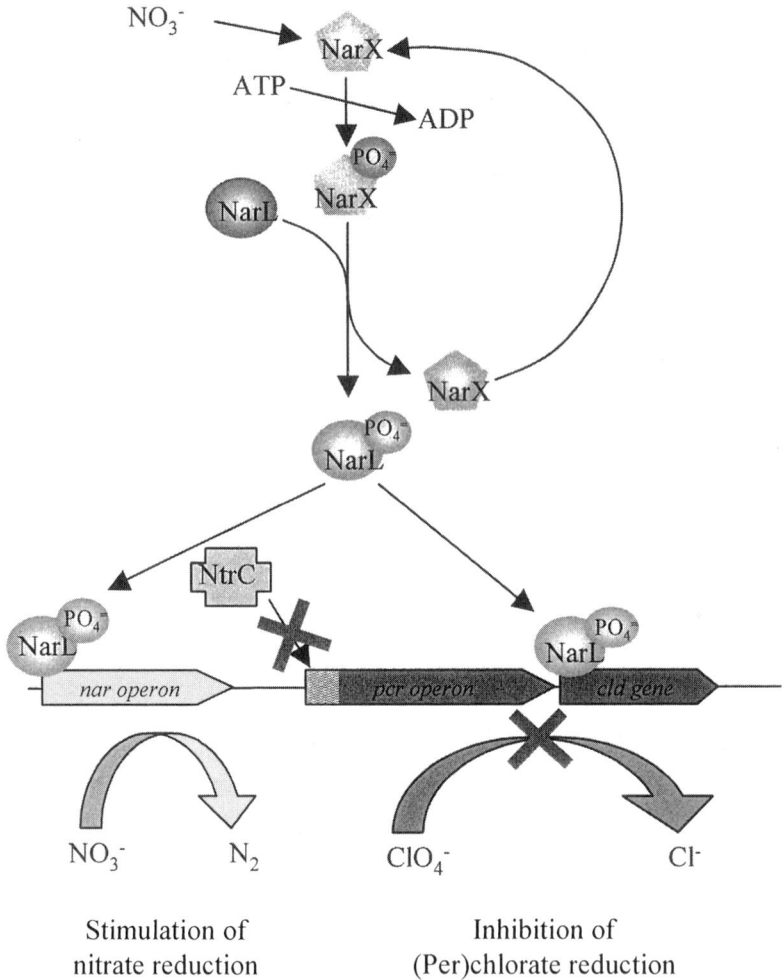

Figure 2. A proposed model for the transcriptional up-regulation of the nitrate reductase operon (*nar*) and suppression of perchlorate reduction through a blocking of the promoter region of the perchlorate reductase operon (*pcr*) by the NtrC response regulator and suppression of the chlorite dismutase gene (*cld*) gene expression by the NarL response regulator after its activation through a phosphorylation event from NarX which is initiated by nitrate.

in response to nitrite. Whether the *cld* gene is transcriptionally regulated by the NarL protein or is under the control of another regulatory protein that recognizes these heptamer sites remains unknown. However, given the additional observation of a 17-bp sigma(54) recognition[12] site located 113 bp upstream of the perchlorate reductase operon *pcr*, it is possible that the genes for perchlorate reduction in *D. aromatica* are regulated by nitrate concentration. One possible scenario is that under conditions of high nitrate concentrations, the sigma(54)-dependent promoter of the perchlorate reductase operon is inactive and the NarL protein represses chlorite dismutase gene expression. As the nitrate is depleted, the response regulator NtrC is phosphorylated and binds upstream of the *pcr* operon allowing transcription by sigma(54) RNA polymerase. Under these conditions, the NarL protein, known to activate gene expression in response to nitrite, will then upregulate expression of *cld*. It is interesting to note that, in support of this hypothesis, nitrate levels must be depleted before perchlorate reduction is initiated when *Dechloromonas* species are grown in media containing perchlorate and nitrate.[13]

While the *D. aromatica* genome implicates nitrate as a key factor in the transcriptional regulation of perchlorate metabolism, the effect of nitrate as a terminal electron acceptor on perchlorate reduction appears to be strain dependent. For some strains, nitrate has no effect on the reduction of perchlorate,[13, 14] while in other strains nitrate is preferentially reduced before perchlorate.[13] *Dechloromonas agitata* concomitantly reduces nitrate to nitrite when perchlorate is present, suggesting reduction of the two electron acceptors by the same enzyme.[2] However, common to all DPRB strains is the effect of oxygen which inhibits perchlorate reduction in all kinetic studies performed to date.[13, 15-17] Nothing is known about the role other terminal electron acceptors may play in regulating the perchlorate reduction pathway.

HYBRIDIZATION ANALYSIS WITH A *CLD* GENE PROBE

Microbial dissimilatory (per)chlorate reduction is a phylogenetically diverse metabolism and DPRB isolates have been placed in four of the five subclasses of the Proteobacteria.[5] Furthermore, 16S rDNA sequence analysis indicates that several of these organisms show less than 0.5 % divergence in their 16S rDNA sequence with non-perchlorate-reducing relatives.[18,19] As such, DPRB molecular probes based on signature nucleotide sequences within the 16S rDNA sequence are of limited use. To determine the utility of the *D. agitata cld* gene as a metabolic probe, Bender *et al.* screened the genomic DNAs of several phylogenetically-distinct DPRB representing the alpha, beta, and gamma subclasses of the Proteobacteria in a slot-blot assay.[3] In this study the *D. agitata cld* probe hybridized to all DPRB tested and failed to hybridize to any of the non-DPRB, despite their close phylogenetic relationships. Interestingly, a weak hybridization signal was obtained from *Magnetospirillum magnetotacticum*, a bacterium closely related to the DPRB *Dechlorospirillum anomolous* strain WD.[5, 18, 20] This was unexpected as previous studies in our lab demonstrated that several *Magnetospirillum* species including *M. magnetotacticum* did not grow by dissimilatory perchlorate reduction.[20] A subsequent search of the *M. magnetotacticum* genome revealed the presence of a putative chlorite dismutase gene. The physiological ramifications of this discovery are as yet unknown. Despite the remote possibility of detecting a *cld* gene in non-perchlorate-reducing strains, this study indicated the utility of the *cld* gene as a molecular beacon for perchlorate reduction.

Nested *cld* Gene Primers

While hybridization studies are informative, environmental detection of specific genes using the polymerase chain reaction (PCR) is a more sensitive technique that can be applied to smaller amounts of DNA. To detect DPRB in the environment, two degenerate primer sets targeting the chlorite dismutase gene were developed.[21] In this approach, nested PCR amplification was used to increase the sensitivity of the molecular detection method. Screening of environmental samples indicated that all products amplified by this method were *cld* gene sequences. These sequences were obtained from pristine sites as well as contaminated sites from which DPRB were isolated. More than one *cld* phylotype was also identified from some environmental samples, indicating the presence of more than one DPRB strain at those sites. Subsequent sequence analysis of these *cld* products indicated differences of only one or two nucleotides with the predicted protein products reflecting these changes. Due to the observation of different *cld* gene sequences from the same environmental sample, DGGE (denaturing gradient gel electrophoresis) may be a useful tool in determining the number of and prevalent phylotypes in a given sample.[22] Since DGGE could also be used to address the effect of ecological changes on the diversity of *cld* sequences present, the nested *cld* primer sets can be used to analyze and monitor DPRB populations in the environment. The use of these primer sets represents a direct and sensitive molecular method for the qualitative detection of perchlorate-reducing bacteria in the environment, thus offering another tool for monitoring natural attenuation.

Reverse Transcriptase-PCR

While the nested PCR approach is efficient at detecting *cld* genes in the environment, traditional PCR cannot be used to determine the relative abundance or activity of DPRB in a given site. Based on the lack of perchlorate in most environments and the ability of DPRB to use alternate metabolisms, there is some question that the organisms detected using these primers are actively reducing perchlorate. We have addressed this question using reverse-transcription PCR to detect *cld* transcripts from different environments. In this procedure, total RNA is isolated from an environmental sample and used in a reverse transcription reaction with the *cld*-specific primers to produce cDNA. The cDNA is then PCR amplified using an internal set of *cld* primers to produce a pool of environmental *cld* gene fragments. Using this approach and a similar one for perchlorate reductase, we have demonstrated active transcription of both the chlorite dismutase and perchlorate reductase genes in perchlorate-contaminated environments, indicating active microbial perchlorate reduction *in situ* (LA Achenbach, S. Taft, and JD Coates, unpublished).

These same strategies can also be used to monitor the sustainability of natural attenuation over long periods of time. For example, it is known that physiological changes in bacteria in response to the environment can affect the long-term bioremediative potential.[23] Thus, the *cld* primer sets can be used in an RNA approach to assess the long-term attenuation potential of a bacterial community. Quantitative PCR using this primer set could also determine if an increase in catabolic gene copy number occurs after exogenously supplying a growth amendment. An increase in gene copy number would imply that the perchlorate-reducing potential of the site had been enhanced and that stimulation of these bacteria may lead to the natural attenuation of perchlorate. Thus, quantitative PCR using a metabolic primer set could aid in monitoring the effectiveness of a bioremediative strategy.

PERCHLORATE REDUCTASE OPERON

The perchlorate reductase operon (*pcr*) has recently been identified in the genomes of two perchlorate-reducing bacteria, *D. agitata* and *D. aromatica*.[24] There are four genes in the transcriptional unit, *pcrABCD*, encoding two structural subunits (*pcrA* and *pcrB*), a cytochrome (*pcrC*), and a molybdenum chaperone subunit (*pcrD*). Amino acid sequence analysis of the products encoded by the *pcr* operon indicated similarities to subunits of microbial nitrate reductase, selenate reductase, dimethyl sulfide dehydrogenase, ethylbenzene dehydrogenase, and chlorate reductase, all members of the type II DMSO reductase family. While the *ser*, *ddh*, and *clr* operons all shared the same gene order *ABDC*, the *pcr* operon mimics the *ebd* operon arrangement of *ABCD*.[24]

Functional characterization of the *pcrA* gene was provided by a recently developed gene knockout system for *D. aromatica*[25] in which insertional inactivation of the *pcrA* gene with a tetracycline resistance cassette abolished both perchlorate and chlorate reduction.[24] As expected, *pcrA* mutant *D. aromatica* cells were able to grow aerobically as well as anaerobically via nitrate reduction, indicating separate metabolic pathways for perchlorate and nitrate.[24] Transcriptional analysis indicated that the *pcr* gene cluster is expressed in anaerobic perchlorate and chlorate grown cultures of *D. agitata*.[24] However, aerobic cultures with either perchlorate, chlorate, or nitrate as the electron acceptor displayed no induction of *pcr* transcription, indicating the ability of oxygen to completely inhibit expression of the perchlorate reductase operon.[24] Conversely, the *D. agitata cld* gene exhibits basal expression under aerobic conditions (see above), thus further implicating separate regulation of CD and PR production.

As discussed above, the chlorite dismutase gene is present in all (per)chlorate reducers and, as such, detection of this gene is unable to distinguish between perchlorate-reducing bacteria and those that can only reduce chlorate. To address this issue, two sets of degenerate primers specific to the *pcrA* gene were recently designed and used in a nested PCR approach (L. Achenbach, unpublished). These primers have been used to specifically detect DPRB in the environment and, when used in an RNA-based assay, to assess the production of perchlorate reductase in contaminated soils. These *pcrA* primers could also be used to assess *pcrA* gene numbers by quantitative PCR as a means of tracking the concentration and distribution of DPRB *in situ* during the course of a bioremediative strategy.

PERCHLORATE REDUCTASE PROTEIN SUBUNITS

The structural α-subunit of perchlorate reductase encoded by *pcrA* is an electron transfer protein of approximately 100 kDa possessing a pterin molybdenum cofactor and iron-sulfur centers.[24, 26] Analysis of the *pcrA* gene product from *D. agitata* and *D. aromatica* identified a 927 amino acid protein containing a molybdopterin-binding domain and a twin-arginine signal motif. Previous studies have suggested that the twin-arginine motif tags proteins involved in electron transfer reactions for transport to the periplasm via the Tat pathway.[27] Thus the signal motif indicates that the α-subunit has prosthetic groups that are formed in the cytoplasm prior to secretion for transport to the periplasm via a Sec-independent transport pathway.

The structural β-subunit, produced by the *pcrB* gene, is a 333 amino acid polypeptide of approximately 40 kDa that has four cysteine-rich clusters for Fe-S center binding which are likely responsible for electron transfer to the molybdopterin-containing α-subunit of perchlorate reductase.[24, 26] Since no signal sequence was detected in the PcrB protein, the β-subunit of perchlorate reductase is likely translocated with the α-subunit in a manner similar to that proposed for dimethyl sulfide dehydrogenase and selenate and chlorate reductase enzymes.[24]

The γ-subunit of perchlorate reductase, encoded by *pcrC*, is a *c*-type cytochrome of approximately 25 kDa responsible for connecting the reductase to the membrane.[24] This subunit was believed to have been lost during purification of the enzyme from GR-1.[26] This is further supported by previous physiological studies which indicated a functional role of type-c cytochrome(s) as electron carriers in the reduction of chlorate and perchlorate by DPRB.[26] Although the *pcrC* translation product is not a membrane bound protein and therefore cannot directly link the PcrAB complex to the membrane, Bender and co-workers (2005) suggested an enzyme model in which a NirT-type cytochrome links the periplasmic PcrABC perchlorate reductase complex to the membrane quinol pool (Fig. 3). Although PcrC shares the unique tetra-heme organization of the c^{554} cytochrome from *Nitrospira europaea*, amino acid sequence analysis revealed a high amount of sequence diversity in the γ-subunits of type II DMSO reductase family members.[24]

The δ-subunit of perchlorate reductase, PcrD, is thought to be a system-specific molybdenum chaperone protein of approximately 25 kDa,[24] a hypothesis that is supported by the absolute requirement of molybdenum for active perchlorate reduction.[13] Like other type II DMSO reductase family δ-

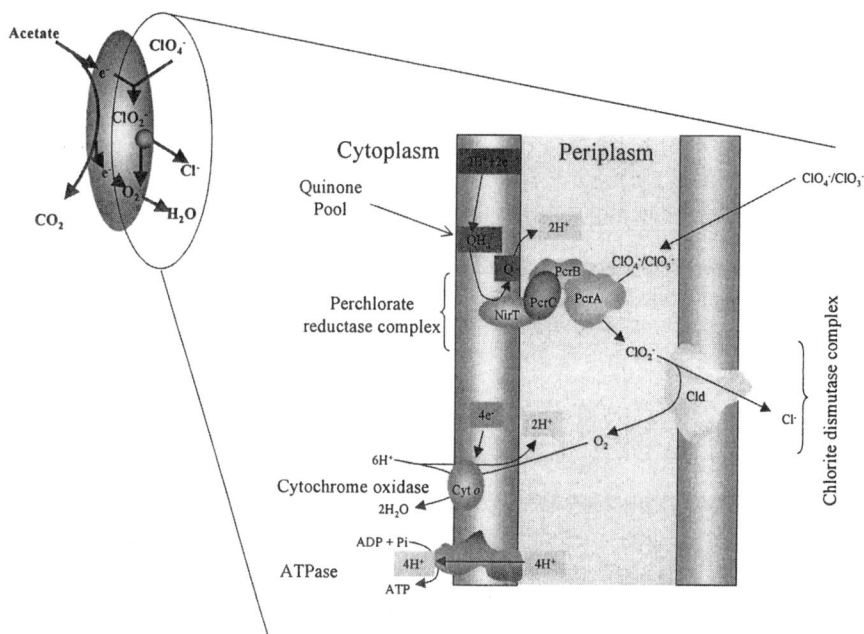

Figure 3. Predicted model of electron transfer during (per)chlorate reduction. Electrons from a quinone pool (Q) are transferred from the membrane via a NirT-type cytochrome to the (per)chlorate reductase (PcrABC). While the PcrD is absent from the functional enzyme, the protein is predicted to be involved in enzyme assembly. The chlorite thus produced is subsequently dismutated by the chlorite dismutase (Cld) into chloride and oxygen which is further respired to H_2O by the organism involving a cytochrome oxidase (Cyt *o*). Energy is conserved by the generation of a proton motive force through the translocation of protons across the membrane which drives ATP formation using a ATPase.

subunits, PcrD is believed to be involved in the assembly of the mature perchlorate reductase prior to periplasmic translocation but not to be part of the active enzyme.

GENE ORGANIZATION OF *cld* AND *pcr* GENES

Recently, it was discovered that the organization and transcriptional direction of the genes involved in perchlorate reduction differ from strain to strain independent of phylogenetic similarity, with the *cld* being the most capricious at the organization level[28] (Fig. 4). For example, in *D. agitata*, the *cld* gene was found to be located upstream and transcribed in the same direction as the *pcr* operon while in *D. aromatica*, the *cld* gene was downstream of the *pcr* operon. Similarly, the *cld* gene in the chlorate reducers *Pseudomonas* strain PK and *I. dechloratans* was located upstream of the chlorate reductase genes but transcribed in the opposite direction.[28, 29] Also found in the two chlorate reducers was the presence of a transposase gene downstream of the *cld* gene.[28, 29] Both transposases were transcribed in

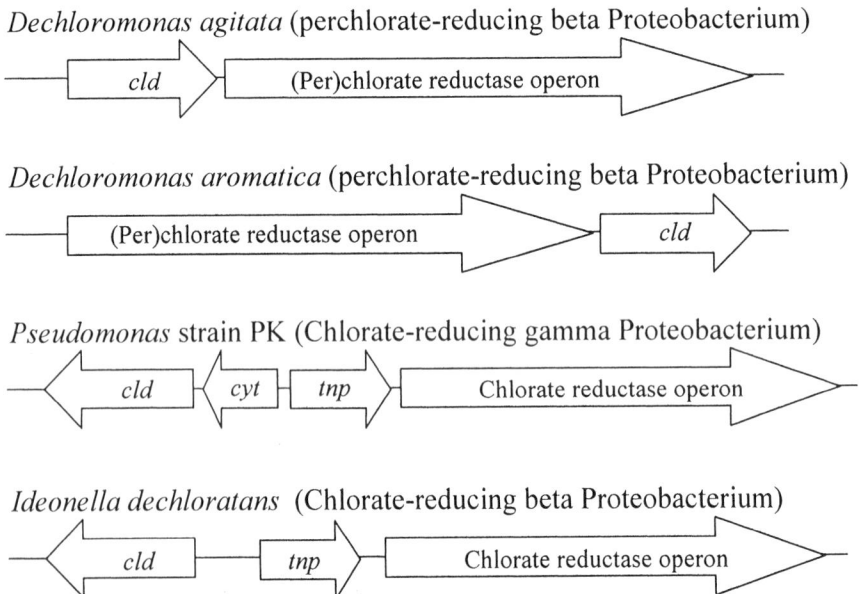

Dechloromonas agitata (perchlorate-reducing beta Proteobacterium)

Dechloromonas aromatica (perchlorate-reducing beta Proteobacterium)

Pseudomonas strain PK (Chlorate-reducing gamma Proteobacterium)

Ideonella dechloratans (Chlorate-reducing beta Proteobacterium)

Figure 4. Genomic organization of the chlorite dismutase gene *cld*, perchlorate reductase operon, and chlorate reductase operon in three (per)chlorate-reducing isolates. A gene encoding a *c*-type cytochrome (*cyt*) lies between *cld* and a transposase gene (*tnp*) in *Pseudomonas* strain PK. A different transposase gene is located upstream of *cld* in *I. dechloratans*. Arrows indicate gene location and direction of transcription. Reprinted with permission.

the opposite direction from the *cld* gene and possessed little sequence identity. The varied arrangements of the genes involved in perchlorate and chlorate reduction could be based on separate pathway evolution or could be an artifact of horizontal gene transfer.

Phylogeny of the *pcrA* Gene

Enzymes within the prokaryotic type II DMSO reductase family, including perchlorate reductase, reside in the periplasm and share a common pterin molybdenum cofactor known as bis(molybdopterin guanine dinucleotide)Mo.[30] Phylogenetic analysis of the α-subunit protein sequences from known microbial DMSO enzymes including perchlorate reductase revealed that type I, type II, and type III DMSO enzymes form separate clades within the resulting tree.[24] In this analysis, the perchlorate reductases from *D. agitata* and *D. aromatica* formed a clade with other type II enzymes such as ethylbenzene dehydrogenase, dimethyl sulfide dehydrogenase, selenate reductase, nitrate reductase, and chlorate reductase, all heterotrimeric structures that contain conserved cysteine residues for Fe-S binding within the β-subunit. All of these enzymes also contain a type II DMSO signature motif $(HX_3CX_2CX_{(n)}C)$ for binding one 4Fe-4S center in domain I and a conserved Asp residue for Mo ion binding (Bender, 2005 and references therein). Interestingly, this analysis also revealed that perchlorate reductase is more closely related to nitrate reductase than to chlorate reductase and that the distance between the perchlorate and chlorate reductase is indicative of distinct enzymes,[24] a conclusion that was supported by molecular probe experiments in which a perchlorate reductase probe hybridized to organisms capable of perchlorate reduction, but not to those capable of only chlorate reduction.[24]

Phylogeny of the *cld* Gene

From the development of the degenerate *cld* primer set, the first library of *cld* gene sequences was generated.[21] This sequence information was used to determine if the *cld* gene phylogeny tracked that of the 16S rDNA gene and possibly gain some insight into the evolution of perchlorate reduction. Comparison of the *cld* and 16S rDNA unrooted phylogenetic trees resulted in incongruent topologies. While members of the α-proteobacteria formed a distinct cluster on both trees, the *cld* gene sequences from some members of the β-proteobacteria clustered with those from the γ-proteobacteria. This aberration indicated that *cld* gene sequences derived from members of the same genus are not monophyletic.[21] In addition, extremely short branch lengths generated on the tree reflect the high sequence similarity of these genes and indicate possible transfer of the *cld* gene among members of the β-

and γ-proteobacteria. These incongruent tree topologies suggest a role for horizontal gene transfer in the evolution of the perchlorate reduction pathway. Since DPRB can grow by alternate metabolisms, the *cld* gene may not be subject to intense selective pressure. As such, mutation may occur until the gene sequence becomes functional with respect to the codon usage and regulation of the host. However, more extensive data is needed on the codon biases and G+C content of housekeeping genes in other DPRB isolates before further conclusions can be drawn. While one can only speculate on the possible mechanism of transfer, transposition is a likely candidate due to the transposase genes identified directly upstream of the *cld* gene in *Pseudomonas* sp. strain PK[28] and *Ideonella dechloratans*[29]. While phylogenetic comparisons of the *cld* gene and the 16S rDNA gene indicate that horizontal transfer is involved in the evolution of (per)chlorate reduction, an interesting question still remains regarding the progenitor of this metabolism and the selective advantage for retaining the metabolic machinery given that perchlorate has only been widespread in the environment, primarily as a result of anthropogenic sources, in the last 50 years and that many DPRB are found in pristine areas.

CONCLUSION

The identification and analysis of the genes encoding perchlorate reductase and chlorite dismutase has provided not only a building block for pathway understanding, but has also provided a tool for bioremediative and phylogenetic studies. On-going genome sequencing will further facilitate transcriptional profiling under perchlorate-reducing conditions via microarray analyses. This analysis will give a more inclusive look into transcriptional expression patterns associated with the perchlorate metabolism. While further advancements in the genetic analysis of perchlorate-reducing bacteria continue, the recent development of a genetic system in *D. aromatica* will provide an invaluable tool for corroborating microarray results and solidifying hypotheses regarding microbial perchlorate metabolism.

ACKNOWLEDGEMENTS

Research on the microbial reduction of (per)chlorate in the laboratories of LAA and JDC is supported by grants from the US Department of Defense SERDP program.

REFERENCES

1. van Ginkel, C.G., Rikken, G.B., Kroon, A.G.M., and Kengen, S.W.M. Purification and characterization of the chlorite dismutase: a novel oxygen-generating enzyme. Arch. Microbiol. 1996; 166:321-326.

2. Bruce, R.A., Achenbach, L.A., and Coates, J.D. Reduction of (per)chlorate by a novel organism isolated from paper mill waste. Environ. Microbiol. 1999; 1:319-329.

3. Bender, K.S., O'Connor, S.M., Chakraborty, R., Coates, J.D., and Achenbach, L.A. Sequencing and transcriptional analysis of the chlorite dismutase gene of *Dechloromonas agitata* and its use as a metabolic probe. Appl. Environ. Microbiol. 2002; 68:4820-4826.

4. Danielsson-Thorell, H., Karlsson, J., Portelius, E., and Nilsson, T. Cloning, characterisation, and expression of a novel gene encoding chlorite dismutase from *Ideonella dechloratans*. Biochim. Biophys. Acta 2002; 1577:445-451.

5. Coates, J.D., Michaelidou, U., Bruce, R.A., O'Connor, S.M., Crespi, J.N., and Achenbach, L.A. Ubiquity and diversity of dissimilatory (per)chlorate-reducing bacteria. Appl. Environ. Microbiol. 1999; 65:5234-5241.

6. Stenklo, K., Danielsson-Thorell, H., Bergius, H., Aasa, R., and Nilsson, T. Chlorite dismutase from *Ideonella dechloratans*. J. Biol. Inorg. Chem. 2001; 6:601-607.

7. O'Connor, S.M. and Coates, J.D. Universal immunoprobe for (per)chlorate-reducing bacteria. Appl. Environ. Microbiol. 2002; 68:3108-3113.

8. Ebihara, A., Okamoto, A., Kousumi, Y., Yamamoto, H., Masui, R., Ueyama, N., Yokoyama, S., and Kuramits, S. Structure-based functional identification of a novel heme-binding protein from *Thermus thermophilus* HB8. J. Struct. Funct. Genomics 2005; 6:21-32.

9. Xu, J., Trimble, J.J., Steinberg, L., and Logan, B.E. Chlorate and nitrate reduction pathways are separately induced in the perchlorate-respiring bacterium *Dechlorosoma* sp. KJ and the chlorate-respiring bacterium *Pseudomonas* sp. PDA. Water Res. 2004; 38:673-680.

10. Dong, X.R., Li, S.F., and DeMoss, J.A. Upstream sequence elements required for NarL-mediated activation of transcription from the *narGHJI* promoter of *Escherichia coli*. J. Biol. Chem. 1992; 267:14122-14128.

11. Darwin, A.J., Li, J., and Stewart, V. Analysis of nitrate regulatory protein NarL-binding sites in the *fdnG* and *narG* operon control regions of *Escherichia coli* K-12. Mol. Microbiol. 1996; 20:621-632.

12. Barrios, H., Valderrama, B., and Morett, E. Compilation and analysis of [sigma][54]-dependent promoter sequences. Nucleic Acids Res. 1999; 27:4305-4313.

13. Chaudhuri, S.K., O'Connor, S.M., Gustavson, R.L., Achenbach, L.A., and Coates, J.D. Environmental factors that control microbial perchlorate reduction. Appl. Environ. Microbiol. 2002; 68:4425-4430.

14. Wallace, W., Ward, T., Breen, A., and Attaway, H. Identification of an anaerobic bacterium which reduces perchlorate and chlorate as *Wolinella succinogenes*. J. Ind. Microbiol. 1996; 16:68-72.

15. van Ginkel, C.G., Plugge, C.M., and Stroo, C.A. Reduction of chlorate with various energy substrates and inocula under anaerobic conditions. Chemosphere 1995; 31:4057-4066.

16. Attaway, H. and Smith, M. Reduction of perchlorate by an anaerobic enrichment culture. J. Ind. Microbiol. 1993; 12:408-412.

17. Logan, B.E., Zhang, H., Mulvaney, P., Milner, M.G., Head, I.M., and Unz, R.F. Kinetics of perchlorate- and chlorate-respiring bacteria. Appl. Environ. Microbiol. 2001; 67:2499-2506.

18. Achenbach, L.A. and Coates, J.D. Disparity between bacterial phylogeny and physiology. ASM News 2000; 66:714-716.

19. Achenbach, L.A., Michaelidou, U., Bruce, R.A., Fryman, J., and Coates, J.D. *Dechloromonas agitata* gen. nov., sp. nov. and *Dechlorosoma suillum* gen. nov., sp. nov., two novel environmentally dominant (per)chlorate-reducing bacteria and their phylogenetic position. Int. J. Syst. Evol. Microbiol. 2001; 51:527-533.

20. Coates, J.D., Michaelidou, U., O'Connor, S.M., Bruce, R.A., and Achenbach, L.A. "The diverse microbiology of (per)chlorate reduction." In *Perchlorate in the Environment*, E.T. Urbansky, ed. p. 257-270. New York, NY: Kluwer Academic/Plenum, 2000.

21. Bender, K.S., Rice, M.R., Fugate, W.H., Coates, J.D., and Achenbach, L.A. Metabolic primers for detection of (per)chlorate-reducing bacteria in the environment and phylogenetic analysis of *cld* gene sequences. Appl. Environ. Microbiol. 2004; 70:5651-5658.

22. Karr, E.A., Sattley, W.M., Jung, D.O., Madigan, M.T., and Achenbach, L.A. Remarkable diversity of phototrophic purple bacteria in a permanently frozen Antarctic lake. Appl. Environ. Microbiol. 2003; 69:4910-4914.

23. Smets, B.F., Siciliano, S.D., and Verstraete, W. Natural attenuation: extant microbial activity forever and ever? Environ. Microbiol. 2002; 4:15–317.

24. Bender, K.S., Shang, C., Chakraborty, R., Belchik, S.M., Coates, J.D., and Achenbach, L.A. Identification, characterization, and classification of genes encoding perchlorate reductase. J. Bacteriol. 2005; 187:5090-5096.

25. Shang, C., Bender, K.S., Achenbach, L.A., and Coates, J.D. Genetic system for the insertional mutagenesis of the (per)chlorate reducer *Dechloromonas aromatica*. Appl. Environ. Microbiol. 2005; (submitted):

26. Kengen, S.W., Rikken, G.B., Hagen, W.R., van Ginkel, C.G., and Stams, A.J.M. Purification and characterization of (per)chlorate reductase from the chlorate-respiring strain GR-1. J. Bacteriol. 1999; 181:6706-6711.

27. Berks, B.C., Sargent, F., and Palmer, T. The Tat protein export pathway. Mol. Microbiol. 2000; 35:260-274.

28. Coates, J.D. and Achenbach, L.A. Microbial perchlorate reduction: rocket-fuelled metabolism. Nature Microbiol. Rev. 2004; 2:569-580.

29. Danielsson-Thorell, H., Stenklo, K., Karlsson, J., and Nilsson, T. A gene cluster for chlorate metabolism in *Ideonella dechloratans*. Appl. Environ. Microbiol. 2003; 69:5585-5592.

30. McEwan, A.G., Ridge, J.P., McDevitt, C.A., and Hugenholtz, P. The DMSO reductase family of microbial molybdenum enzymes; molecular properties and role in the dissimilatory reduction of toxic metals. Geomicrobiol. J. 2002; 19:3-21.

Chapter 14

Field Demonstration of *In Situ* Perchlorate Bioremediation in Groundwater

P. B. Hatzinger[1], J. Diebold[2], C. A. Yates[3] and R. J. Cramer[3]

[1]*Shaw Environmental Inc., Lawrenceville, NJ 08648*
[2]*Shaw Environmental Inc., Pewaukee, WI 53072*
[3]*Indian Head Division, Naval Surface Warfare Center, Indian Head, MD 20640*

INTRODUCTION

A wide variety of microorganisms are capable of utilizing perchlorate (ClO_4^-) as a terminal electron acceptor while growing on one of several different organic or inorganic electron donors[1] (see also Chapters 12 and 13 in this volume). These bacteria reduce perchlorate initially to chlorate (ClO_3^-) and then to chlorite (ClO_2^-).[2,3] The chlorite is subsequently disproportionated into chloride (Cl^-) and oxygen (O_2).[4] Thus, the biodegradation process results in a detoxification of the perchlorate molecule.

Biological reactors have been successfully applied at both pilot-scale and full-scale to treat perchlorate-contaminated groundwater.[5-7] Full-scale fluidized bed bioreactors are presently operating at five locations, removing perchlorate from more than 30 million liters of groundwater per day. Moreover, two full-scale suspended growth bioreactors have been constructed to treat high concentrations of perchlorate in military and industrial wastewaters.[5] These *ex situ* systems have proven to be both reliable and economical. Moreover, biological treatment currently represents the only approach to both remove perchlorate from water and convert it to non-hazardous products (i.e., chloride and oxygen) in a single step.

The successful application of *ex situ* biological treatment systems for perchlorate, as well as the isolation and characterization of numerous pure cultures of perchlorate-degrading bacteria from natural environments, has prompted significant research concerning the potential for *in situ* perchlorate treatment through electron donor amendment to soils and groundwater. Current laboratory data suggest that perchlorate-reducing bacteria (PRB) are indigenous to many soils and groundwaters[8-10]. Studies also reveal that these microorganisms can be stimulated to metabolize perchlorate from concentrations as high as several hundred parts-per-million to below current analytical reporting limits (4 µg/L) by the addition of one of several different electron donors, such as acetate, lactate, citrate, ethanol, vegetable oil, or hydrogen gas.[7, 8-12] The wide distribution of PRB combined with their ability to metabolize perchlorate using common substrates suggests that *in situ* bioremediation may be a viable approach for treatment of perchlorate-contaminated groundwater. However, field data examining *in situ* perchlorate treatment in subsurface aquifers are currently very limited. This chapter describes the results from a pilot-scale field evaluation of *in situ* perchlorate bioremediation.

The field demonstration was performed at the Indian Head Division, Naval Surface Warfare Center (IHDIV) in Indian Head, Maryland. A groundwater recirculation design was tested to distribute electron donor (lactate) to indigenous PRB in a shallow aquifer at the site. Two field plots were installed, each consisting of two extraction wells, two injection wells, and nine groundwater monitoring wells. In one plot (test plot), extracted groundwater was amended with lactate and buffer and then re-injected into the aquifer. In the second plot (control plot), the groundwater was extracted and re-injected without substrate or buffer amendment. Perchlorate levels in the test plot declined by > 95 % in 8 of 9 monitoring wells during the demonstration from an initial average value of 171 mg/L. Five wells reached perchlorate levels < 1 mg/L and two wells were below the reporting level of 5 µg/L. Nitrate concentrations were also appreciably reduced throughout the test plot, with 7 of the 9 wells showing nondetectable levels within 7 weeks. Conversely, there was no significant reduction in either perchlorate or nitrate within the control plot.Perchlorate levels at the beginning of the demonstration averaged 127 mg/L, and these values were 118 mg/L after the 5-month study was complete. The data from this demonstration indicate that *in situ* perchlorate bioremediation can be a viable approach for treatment of perchlorate in aquifers containing localized, high concentrations of the contaminant.

SITE HISTORY AND CHARACTERIZATION

The Indian Head Division, Naval Surface Warfare Center is located on a 1,400-hectare peninsula bordered by the Potomac River and the Mattawoman Creek[13] (Figure 1) The facility, which is located approximately 50 km from Washington, DC in Charles County, MD, produces and tests energetic compounds for the US Navy. The demonstration was performed on the southeast side of IHDIV Building 1419, near the Mattawoman Creek (Figure 2). This building has been used historically to remove solid propellant from rockets, ejection seat motors, and other materials using a high pressure water washout - a process termed "hog out". Past discharge of aqueous wastewater resulting from this process has released perchlorate into shallow groundwater in the vicinity of the building.

Figure 1. Location of the Indian Head division, Naval Surface Warfare Center IHDIV).

Surficial Geology

The surficial geology in the general area of IHDIV is comprised of Pleistocene lowland deposits. These deposits contain gravel, medium to course grained sand, silt, and clay, with cobbles and boulders near the base of the formation.[14] The formation commonly includes reworked Eocene glauconite, variable-colored silts and clays, and brown to dark gray lignitic silty clay. Estuarine to marine fauna are present at select locations. The thickness of these deposits varies from 0 to 45 m. The portion of the peninsula nearest to the Potomac River is comprised of the Cretaceous Potomac Group. This formation consists of interbedded quartzose gravels, protoquartzitic to orthoquartzitic argillaceous sands, and white, dark gray and multicolored silts and clays. The thickness of the formation varies from 0 to approximately 240 m.

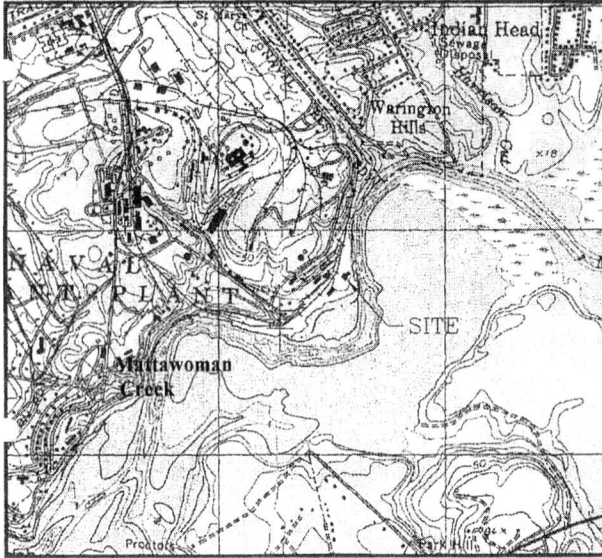

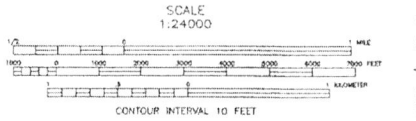

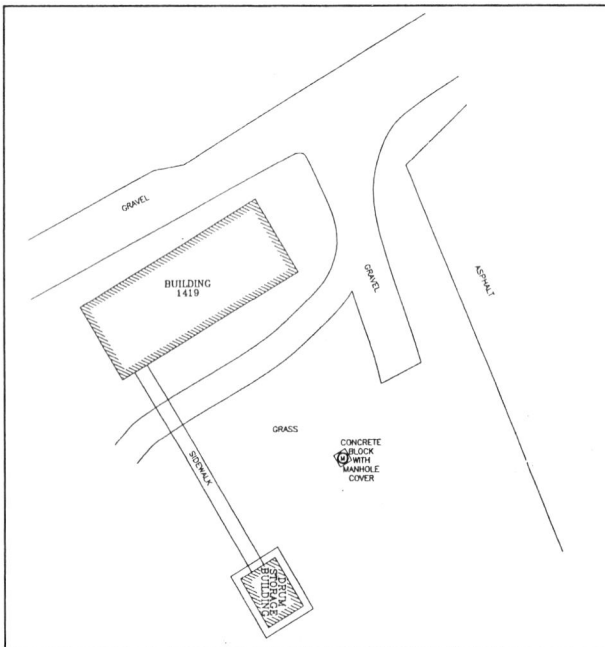

Figure 2. Location (top panel) and site layout (bottom panel) of the demonstration site. Scale is for the top panel.

Initial site characterization was conducted on the southeast side of Building 1419 using a direct-push (Geoprobe) rig. A site plan view showing the Geoprobe sampling locations is provided in Figure 3. Continuous sediment cores were obtained for geological analysis using this technique from surface to depths ranging from ~ 4.8 to 6.1 m (16 – 20 ft) below the ground surface (bgs). A total of 17 Geoprobe borings were initially taken. Geological cross sections based on these borings are provided in Figures 4 & 5. Vertical scales of the cross-sections in these figures are given in feet. The top 0.6 to 1.2 m (2 – 4 ft) consisted of fill material including organic solids, gravel, and silty sand. The underlying 3.4 to 4.0 m (11 – 13 ft) consisted of mottled light to olive brown clayey to sandy silts. The clay and sand fraction of the silts varied horizontally and vertically. Fine grained sand seams 2.5 to 5 cm in thickness were seen in many of the boring locations, but these seams were not continuous from boring to boring. At a depth of approximately 4.6 m bgs (15 ft bgs), a 0.3 to 0.5 m (1 – 1.5 ft) thick layer of sand and gravel was encountered. This layer was found to be continuous throughout the demonstration area. The sand and gravel layer is underlain by a gray clay layer, which extends to a depth of at least 6 m bgs (20 ft bgs), the deepest extent of the test borings. This clay layer is probably of the Potomac Group.

Perchlorate Concentrations and Site Geochemistry

Following the completion of each Geoprobe boring, a temporary 2.5-cm inner diameter (I.D.) polyvinyl chloride (PVC) well casing with a screened lower section (0.025-cm slot) was inserted into the open boring. Groundwater samples were then collected from each borehole for geochemical analysis using a peristaltic pump and plastic tubing. The samples were analyzed for perchlorate, nitrate, sulfate, pH, and dissolved oxygen (Table 1). The field investigation revealed a shallow, narrow plume of perchlorate contamination behind Building 1419 with levels ranging from below detection to approximately 430 mg/L (Figure 6). With a few exceptions, the pH of the site was below 5, and the dissolved oxygen levels were < 2 mg/L.

Based on the groundwater analysis from the 17 borings, six permanent monitoring wells were installed in the demonstration area (see Figure 2). These wells were designated MW-1 to MW-6. Borings were advanced using hollow-stem auger drilling. Split-spoon soil samples were collected at 0.8-m intervals from each boring ahead of the hollow-stem auger. The six borings were then completed as groundwater monitoring wells using 5-cm I.D. Schedule 40 PVC with 3-m well screens (0.025-cm slot). The bottom of the screened section for each well was set approximately 0.3 m into a gray clay layer encountered at approximately 5 m bgs throughout the demonstration area. The contaminated groundwater was perched above this layer.

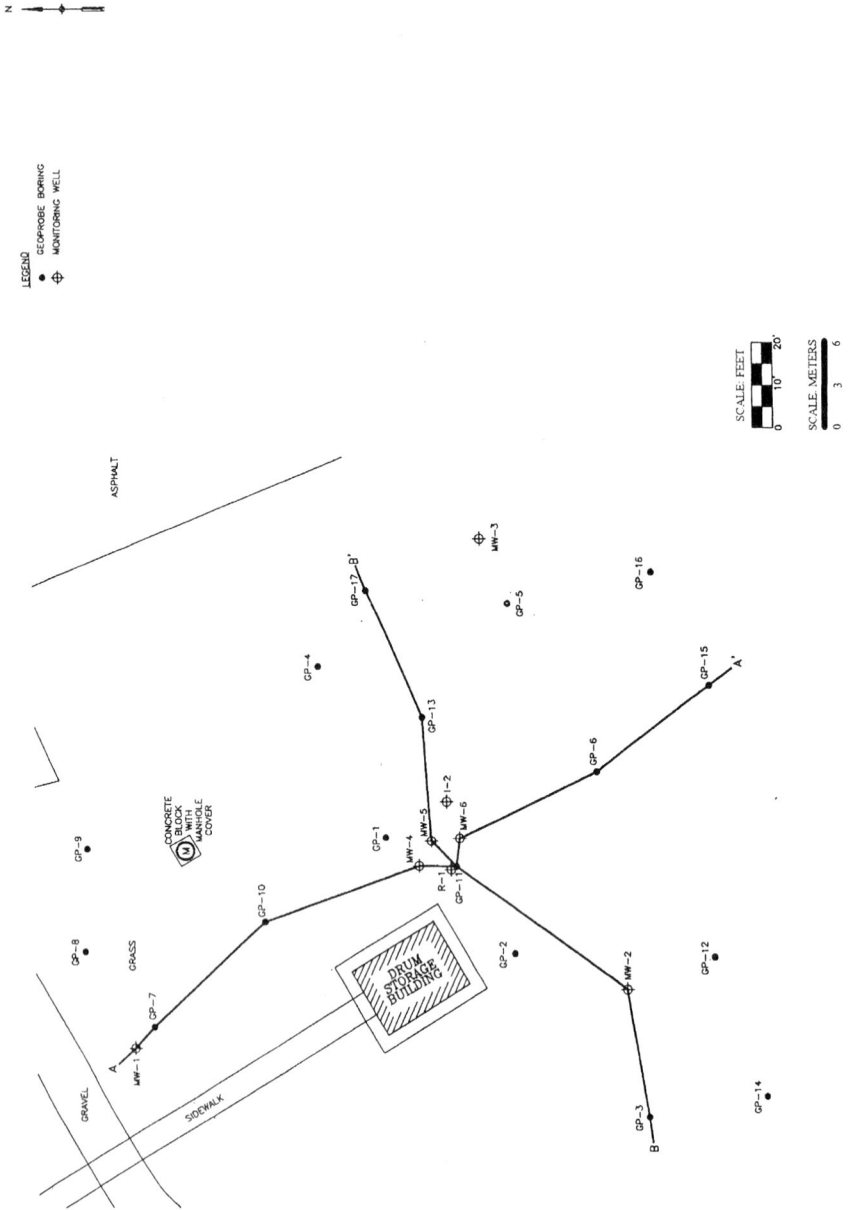

Figure 3. Geoprobe boring locations and cross-section plan view.

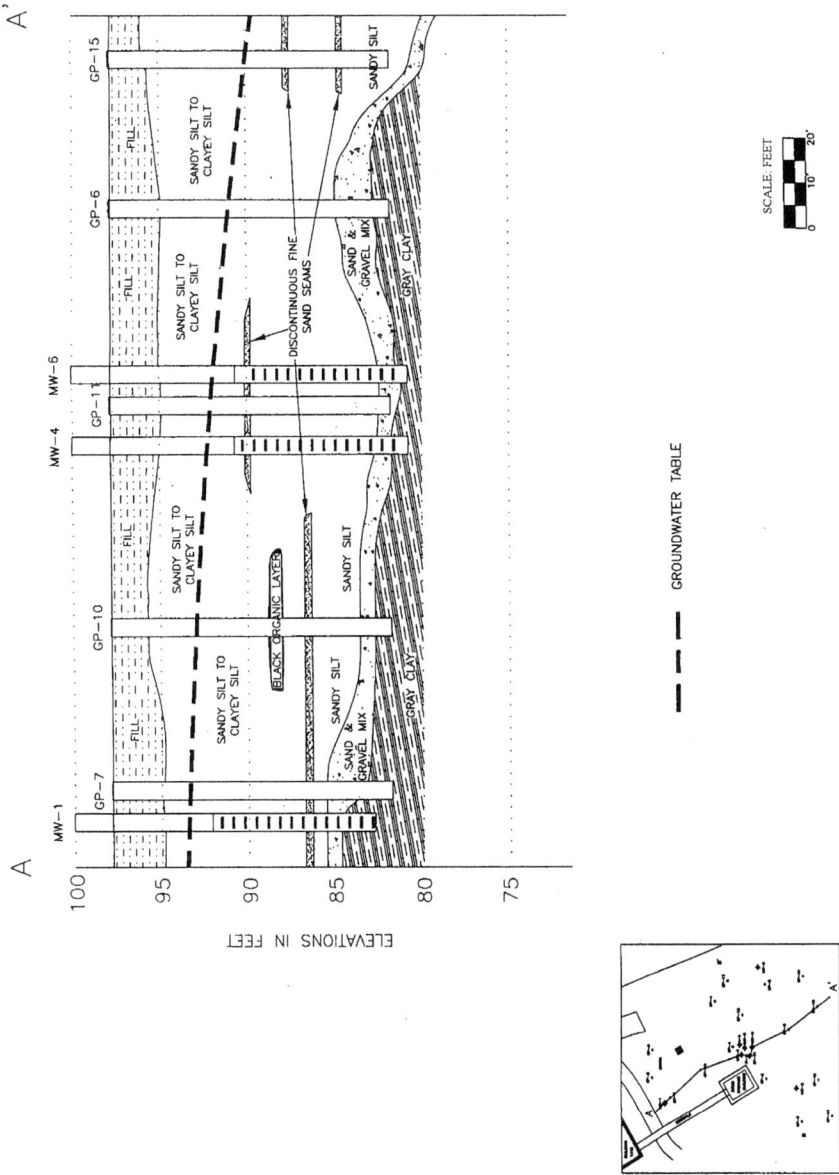

Figure 4. Geologic cross-section A – A'.

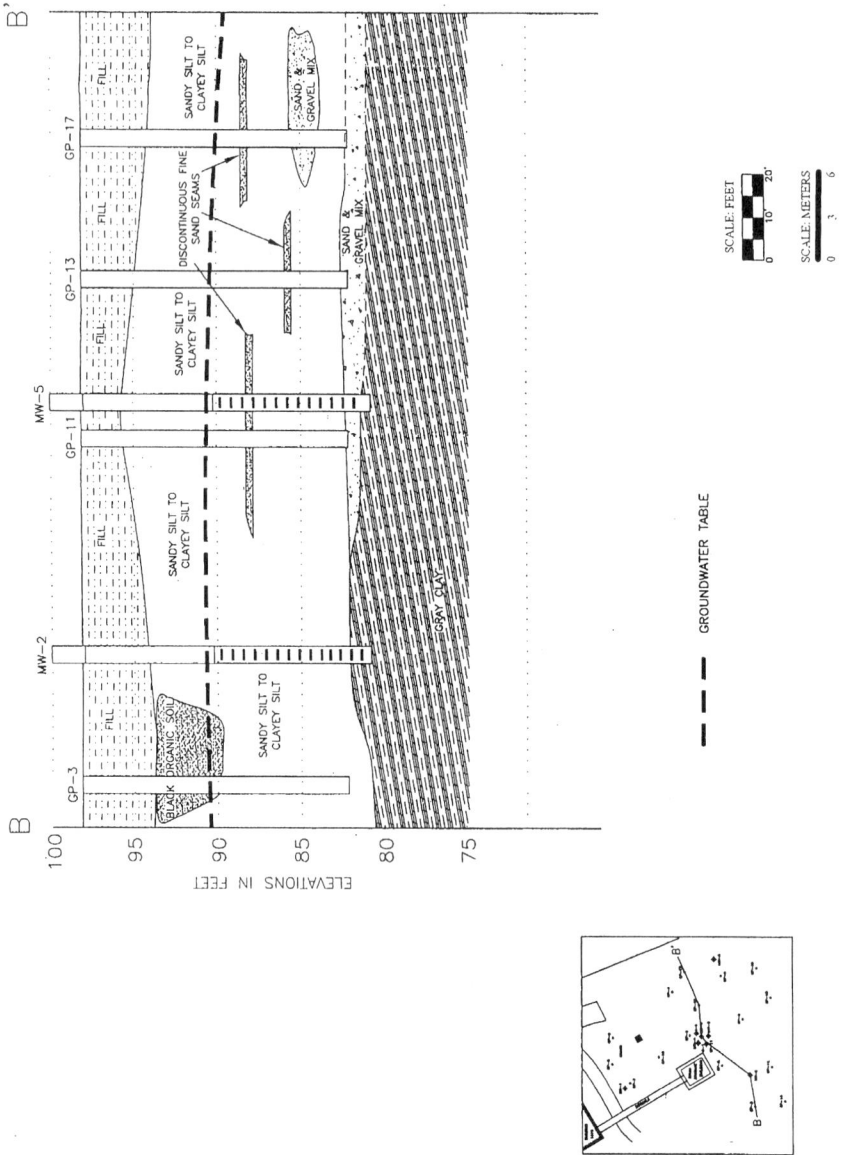

Figure 5. Geologic cross-section B – B'.

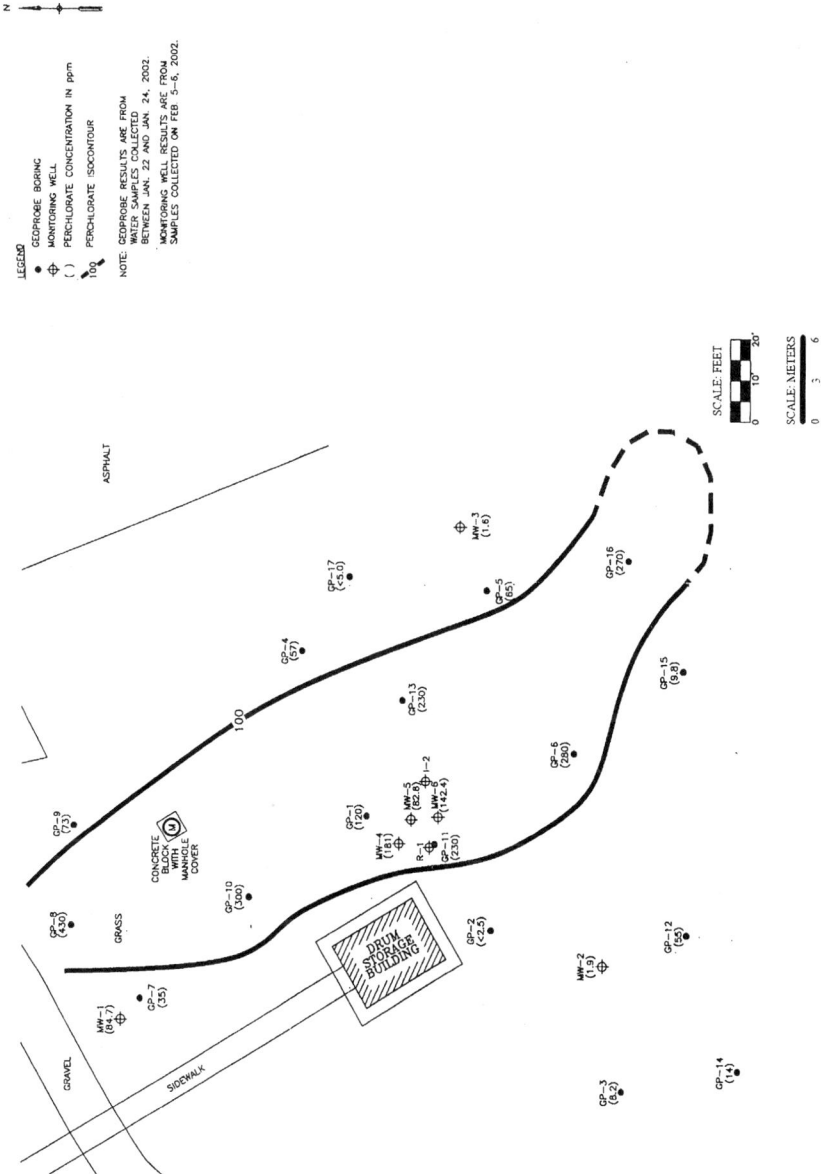

Figure 6. Perchlorate distribution (mg/L) in groundwater.

Table 1. Groundwater chemistry at 17 Geoprobe boring locations

Boring	Perchlorate (mg/L)	Nitrate as N (mg/L)	Sulfate (mg/L)	pH (SU)	DO (mg/L)[1]
GP-1	120	0.6	66	4.67	NA[2]
GP-2	<2.5	3.0	220	8.08	NA
GP-3	8.2	1.9	280	5.23	NA
GP-4	57	0.3	110	4.54	NA
GP-5	65	0.1	130	4.21	1
GP-6	280	11	69	5.62	1
GP-7	35	1.5	66	4.21	0.1
GP-8	430	14	62	4.57	NA
GP-9	73	0.4	56	4.44	0.8
GP-10	300	12	70	4.31	1
GP-11	230	14	72	4.71	0.8
GP-12	55	2.0	110	6.46	NA
GP-13	230	3.8	64	4.61	1.5
GP-14	14	1.5	250	4.97	NA
GP-15	9.8	<0.2	160	5.34	0.2
GP-16	270	2.8	74	4.16	1
GP-17	<5	<0.2	140	4.83	0.2

[1] Analysis performed by colorimetric field method (Chemettes). [2] NA: Not analyzed.

Local Hydrogeology

Groundwater elevations measured in the six monitoring wells indicated a groundwater flow direction to the southeast toward the Mattawoman Creek. The flow direction basically followed the surface topography. Depth to groundwater ranged from approximately 2.0 m to 3.1 m bgs. The average hydraulic gradient, as measured between wells MW-1 and MW-3, was 0.023 m/m.

Slug and Pump Testing

Slug tests were performed on monitoring wells MW-4, MW-5 and MW-6. These wells were selected due to their proximity to the planned test plot area. All slug test results were reduced using the Bouwer and Rice method for unconfined aquifers.[15] The slug test results indicated an average hydraulic conductivity (K) of approximately 0.006 cm per second (cm/sec) within the aquifer.

Using the slug test data, as well as results from a stepped pump test, a 12-hour continuous pump test was conducted. An initial flow rate of approximately 0.8 liters per minute was used at the start of the test. However, based on the observed rate of drawdown within the extraction well, the flow rate was adjusted down to approximately 0.6 liters per minute after approximately 4.5 hours of pumping. This reduction in flow rate stabilized the rate of decline in water level within the extraction well allowing for continuous pumping throughout the test. Drawdown levels were logged in both the extraction well and several nearby monitoring wells throughout the pump test to determine the influence on the aquifer of pumping in the vicinity of the extraction well. The drawdown data were reduced and analyzed using the Theis method for unconfined aquifers.[15] Based on the curve data, K value estimates ranged from 0.006 to 0.022 cm/sec.

Laboratory Microcosm studies

Aquifer solids and groundwater were obtained during the initial site assessment work. Sediments were collected from two locations (GP-1, GP-11) in 1.3-m length, 5-cm diameter acetate sleeves, which were sealed and stored at 4°C until use. The solids used in the study were from approximately 3.4 to 6.1 m bgs. Groundwater for microcosms was taken from three locations (GP-1, GP-11, GP-13) by placing a temporary well in each Geoprobe boring, then collecting groundwater using a peristaltic pump. The water was collected in clean 950-ml bottles, which were filled to the top, then stored at 4°C until use. Prior to beginning the laboratory studies, the groundwater from the three locations was combined, and the solids were removed from the acetate sleeves and homogenized.

Aquifer microcosms were prepared in sterile, 160-mL serum bottles by adding 25 g of wet sediment and 75 mL of groundwater to each bottle. One group of bottles was amended with 14 mg of carbonate to bring the slurry pH to approximately 7.3. The carbonate addition was based on previous titration tests (data not presented). The other set of bottles received no carbonate and remained at the site pH of ~ 4.5. Acetate, ethanol, lactate, or hydrogen gas were added to four bottles, two at each pH (i.e. duplicate bottles at site pH and duplicates adjusted to pH 7.3). The liquid electron donors (ethanol, acetate, and lactate) were added at a concentration of 250 mg/L, and hydrogen (a gaseous donor) was added to the bottle headspace in a 5-mL volume. In addition, two microcosm bottles at each pH received no electron donor and two adjusted to pH 7.3 received 1% formaldehyde to inhibit microbial activity (killed controls). The killed samples also received acetate as an electron donor. All samples were prepared with a nitrogen headspace. The bottles were incubated on a rotary shaker at 15°C. At various times of incubation, an 8-mL subsample of groundwater was removed from each bottle. The water was then passed through a 0.22-μm syringe filter to remove bacteria and sediment fines and placed at 4°C until analysis. The samples were analyzed for perchlorate by EPA Method 314.0 and for nitrate and sulfate by EPA 300.0 series methods.

The initial perchlorate levels in the microcosms averaged 116 mg/L, and the starting nitrate concentration was 6.3 mg/L as nitrate-N. The level of perchlorate in the microcosms increased from 116 mg/L at Day 0 (immediately after slurry preparation) to approximately 170 mg/L at Day 14. This increase was consistent among samples and appears not to be the result of analytical error.   Therefore, it is likely that this increase reflects perchlorate desorbing from the site sediments into solution.   Perchlorate losses were not observed in any of the microcosms that remained at pH 4.5, irrespective of the type of electron donor added (data not shown). Biodegradation of perchlorate was apparent in the pH-adjusted microcosms amended with lactate (Figure 7). Perchlorate concentrations declined from a high of 181 mg/L at Day 14 to less than 14 mg/L by Day 61 (the last sample collected) in the these bottles.   The pH-adjusted microcosms receiving acetate, ethanol, and hydrogen gas did not show appreciable losses of perchlorate during the course of the study. Perchlorate levels also did not decline in microcosms without added electron donor or in killed controls.

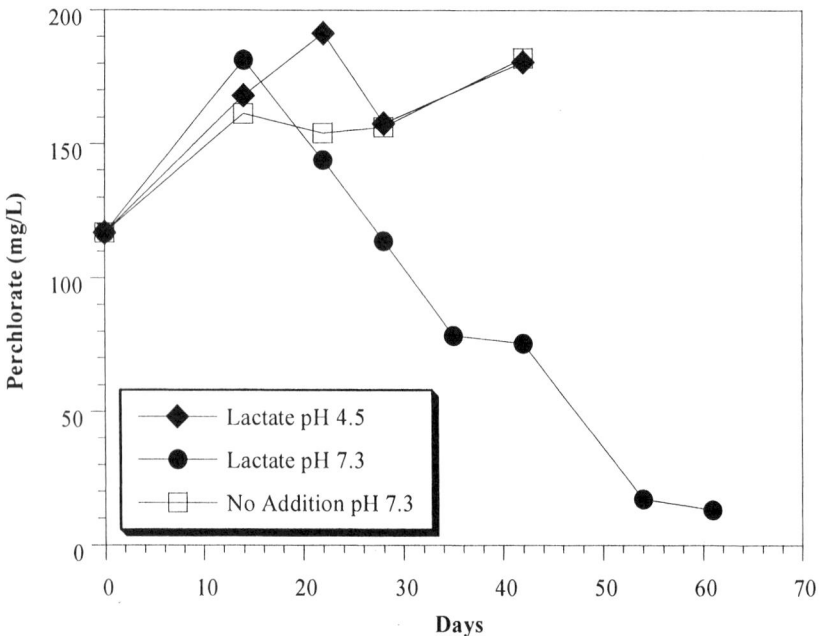

Figure 7. Influence of pH on biological perchlorate reduction in aquifer microcosms receiving lactate or no electron donor

The microcosm data suggested that indigenous PRB were present in the test plot location and that these bacteria could be stimulated to metabolize perchlorate using lactate as an electron donor, but only after the site samples were buffered to a more neutral pH. The results confirm previous studies at this location and at other sites which indicate that low pH (< 5.7) is inhibitory to biological perchlorate reduction[16].

Nitrate biodegradation was not apparent in any of the microcosms that remained at a pH of 4.5. In the microcosms adjusted to pH 7.3, however, nitrate was biodegraded to below detection (< 0.5 mg/L as nitrate-N) within 22 days in samples amended with each of the four electron donors (Figure 8). Nitrate biodegradation was not observed in the pH-adjusted samples that did not receive an electron donor or in killed control samples. These data suggest that the low pH is also inhibitory to biological nitrate reduction in these samples.

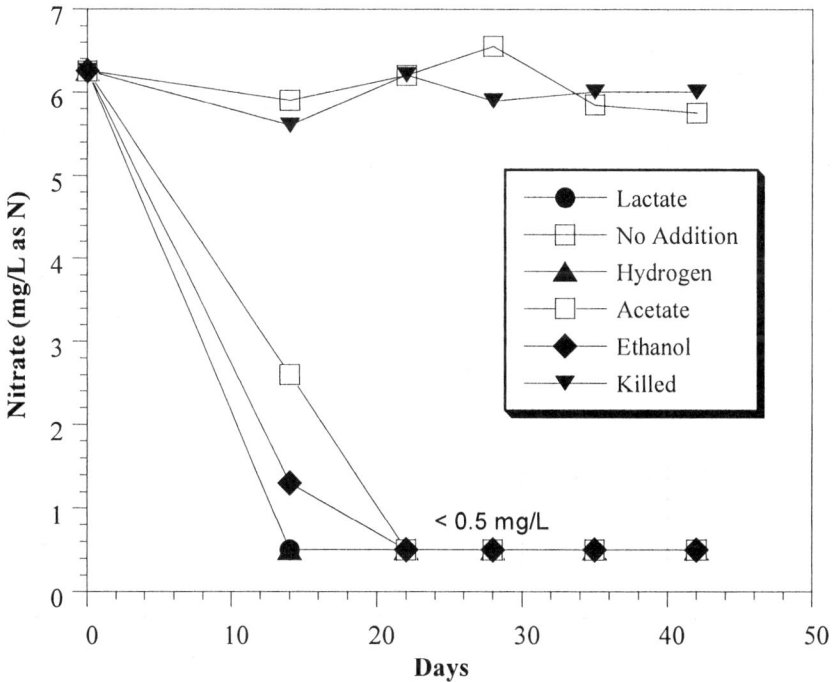

Figure 8. Nitrate levels in aquifer microcosms buffered to pH 7.3 and amended with different electron donors.

DEMONSTRATION SYSTEM DESIGN

Demonstration Objectives

The objectives of the field demonstration were as follows:

1. Show that electron donor (lactate) and buffer (carbonate/bicarbonate) can be effectively distributed throughout the contaminated aquifer using a groundwater extraction-injection design.

2. Demonstrate that perchlorate and nitrate can be biodegraded in the buffered aquifer using lactate as an electron donor.

3. Quantify the time required for perchlorate biodegradation and the levels of degradation achievable.

4. Identify key design and operational factors that influence full-scale application of *in situ* perchlorate bioremediation at this and other sites.

Recirculation Cell Design

A simple single-layer numeric model was developed to represent site conditions. The model was calibrated by simulating the pump test conditions and adjusting the K value for the aquifer until the drawdown levels observed in the model at distance were similar to those measured in the field during pump testing. This information was utilized to assess recirculation well layouts and anticipate operating conditions associated with the final field scale design.

Figure 9 shows the location, layout, and cross-sectional schematic of the test and control plots. A photograph of the demonstration plot is provided in Figure 10. Each of the plots was comprised of two injection wells, two recovery wells, and nine monitoring wells. There was 3.7-m between the paired injection and recovery wells in each plot. The installation details of all wells are provided in the next section. The treatment and control plots were located 6 m apart, with the test plot to the northeast of the control plot. In the test plot, the injection wells were on the west side of the cell (nearest the control plot) and the recovery wells on the east side of the cell (away from the control plot). This layout was reversed for the control plot. This configuration resulted in cross-gradient flow patterns within each cell (east-to-west in the control plot and west-to-east in the test plot). Moreover, the mounding created by the injection wells in the control plot prevented the amendments from the test plot from being introduced into the control plot cell.

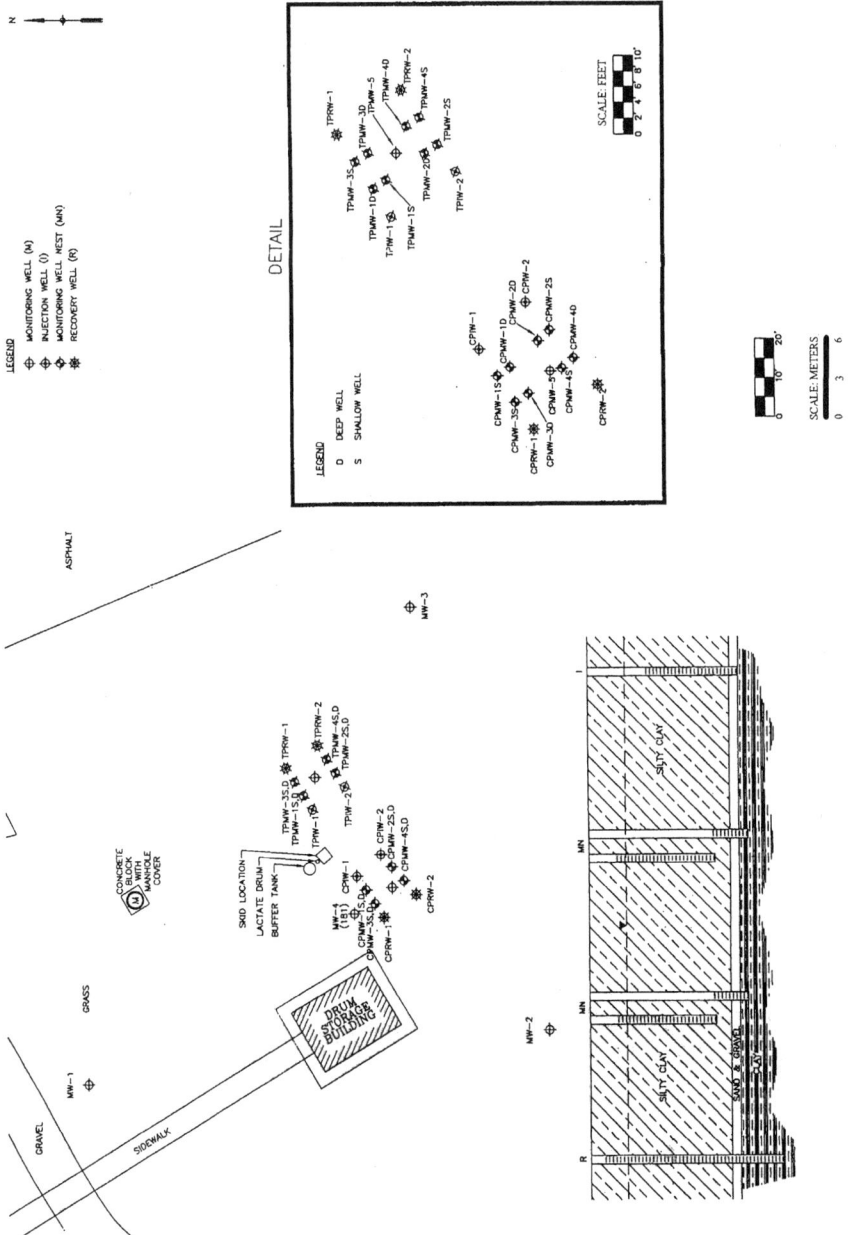

Figure 9. Test plot and control plot layout:Schematic and cross-section view.

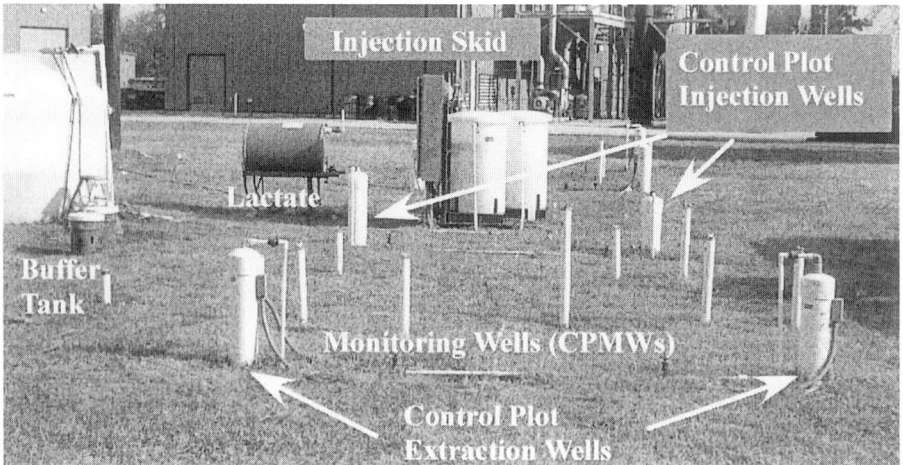

Figure 10. Photograph of the demonstration area looking northeast, with the well layout for the control plot provided in the foreground. The test plot is located directly behind the injection skid

The monitoring well layout in each plot included four pairs of nested wells and a single well. Each of the nests included one well screened within the saturated zone of the upper clayey silt layer (shallow monitoring wells), and one well with a screened interval intersecting the coarse sand and gravel layer located above the underlying clay soils found at the 4.0 to 4.9 m depth interval (deep monitoring wells). The two-screened sections overlapped by approximately 15 cm to ensure that no sand lenses were missed and that the system performance throughout the entire saturated zone was monitored. One well nest was located approximately 1.2 m from each injection well (in line with the paired recovery well), and the second was located about 2.4 m from the injection well. In addition to the nested wells, one fully-screened monitoring well was installed in the center of each plot.

An injection skid was designed to be integrated with the wells. The injection skid had separate transfer tanks, injection pumps, flow meters, and associated valves for the control and test plots. In addition, the test plot had two metering pumps installed to inject a pH solution and an electron donor-reagent to promote optimal aquifer conditions and stimulate biological activity. The injection skid was located between the control and test plots, as shown in Figure 10.

Demonstration System Installation

The injection and recovery wells were installed using a standard hollow stem auger drilling rig equipped with 26-cm outside diameter augers. Both the injection and recovery wells were constructed using 15-cm I.D. schedule 40 PVC. The injection wells were installed with approximately 2.4 m of screen (0.025-cm slot) set at the gravel/clay interface. The recovery wells were installed with approximately 4.6-m screens (0.025-cm slot) set 1.2-m into the clay layer. A sand pack was placed around each screened section and a bentonite plug (approximately 0.6 m thick for the recovery wells and 1.2 to 1.5-m thick for the injection wells) was placed above the sand pack.

The nested and fully penetrating monitoring wells were also installed using hollow stem auger drilling. The wells were constructed using 5-cm I.D. schedule 40 PVC well casing and screen materials. Screen lengths (0.025-cm slot) varied from approximately 2.1- to 2.4-m long for the shallow nested wells, 0.8- to 0.9-m long for the deep nested wells, and 3.0- to 3.4-m long for the fully penetrating wells. A sand filter pack was placed around the screened sections and a 0.6-m thick bentonite seal was placed around the upper portion of the well casing to prevent fluid infiltration or loss. The shallow and deep test plot monitoring wells were designated TPMW-1s to TPMW-4s and TPMW-1d to TPMW 4d, respectively.The fully penetrating well was designated TPMW-5. The control plot wells were given similar designations for each depth beginning with CPMW (i.e., CPMW-1s to CPMW-5).

System Operation

The four groundwater extraction pumps were placed in the wells and adjusted to pump at approximately 0.9 liters per minute each. Low-level floats were installed in each well to turn the pumps off if the extraction rate exceeded the recharge rate, causing significant drawdown. The water from the extraction wells in the test plot was pumped into a 200-L equalization tank. When this tank reached approximately 150 L, the water was amended with buffer and lactate and then re-injected into the formation. The electron donor was a 60 % solution (wt/wt) of food grade L-(+) lactic acid (sodium salt) supplied by Purac America Inc., Lincolnshire, IL. The concentrated buffer solution consisted of a 6.67% mixture containing either 80% bicarbonate (from $NaHCO_3$) and 20% carbonate (from Na_2CO_3) or 70 % bicarbonate and 30 % carbonate. The sodium carbonate and sodium bicarbonate were food-grade products purchased from Seidler Chemical Co., Newark, NJ. The same extraction/injection design was used for the control plot except that the water was re-injected without buffer or electron donor amendment. The injection pumps were set to operate at ~ 4 liters per minute per injection well.

DEMONSTRATION RESULTS AND DISCUSSION

Tracer Test

A conservative tracer test was performed to determine if each monitoring well in the test plot was hydraulically connected with the injection wells where buffer and electron donor were introduced into the formation. To perform this test, ~300 L of groundwater was pumped from the demonstration plot into a holding tank and then amended with sodium bromide to achieve a final bromide concentration of 250 mg/L. The bromide solution was then added as a slug to each of the two injection wells at a flow rate of approximately 7.5 liters per minute. Samples were collected from the bromide tank prior to injection, and then from each of the 9 monitoring wells in the test plot (TPMWs) after 1, 5, and 15 days. Samples were also analyzed for bromide during all subsequent groundwater monitoring events. Bromide was measured by ion chromatography (EPA method 300.0).

The bromide results are presented in Table 2. Bromide was detected (> 0.2 mg/L) in 4 of the 9 test plot monitoring wells (TPMWs) after 1 day and in 7 of the 9 wells after 5 days of system operation. The remaining two wells showed bromide concentrations above background levels by day 15 and 25 of operation for wells TPMW-4s and TPMW-2d, respectively. Thus, the results suggest that all wells in the test plot are hydraulically connected to the zone where buffer and electron donor are added to the aquifer.

Table 2. Bromide concentrations (mg/L) in test plot monitoring wells (TPMWs)

Day	TPMW-1s	TPMW-1d	TPMW-2s	TPMW-2d	TPMW-3s	TPMW-3d	TPMW-4s	TPMW-4d	TPMW-5
-7	< 0.2	< 0.2	< 0.2	< 0.2	< 0.2	< 0.2	< 0.2	< 0.2	< 0.2
1	1.6	3.3	< 0.2	< 0.2	1.5	< 0.2	< 0.2	<0.2	0.77
5	1.1	1.5	2.8	< 0.2	1.5	1.5	< 0.2	0.6	3.9
15	6.8	< 0.2	< 0.2	< 1.0	< 1.0	< 0.2	0.8	< 0.2	1.1
25	0.5	< 0.2	0.3	2.7	7.1	1.8	4.4	33	38
70	< 0.2	< 0.2	2	1.4	< 0.2	< 0.2	1.9	< 0.2	0.7
105	< 0.2	0.64	0.23	< 0.2	0.55	0.6	0.41	0.58	< 0.2
140	0.32	0.54	0.22	< 0.2	0.21	0.28	0.37	0.36	< 0.2

System Operation

A total volume of ~ 80,000 liters of groundwater was re-circulated through each plot during the course of the 140-day demonstration (Figure 11). The recirculation system was shut down after 111 days of operation, and one additional sampling event was performed on Day 140 to examine the residual effect of buffer and electron donor added to the aquifer. During the first month of the demonstration, the rate of water recirculation through the test plot was appreciably higher than through the control plot. During this period, approximately 25,000 L of water were pumped through the test plot compared to 8,000 L for the control plot. This difference was based on the yield of the aquifer formation in each of these zones.   After this time, however, the rate of pumping of the two plots was reasonably similar, as can be seen from the slope of the curves in Figure 10.   Increased rainfall in the late summer and early fall, including nearly 6 cm on Day 34, caused significant aquifer recharge and subsequently increased pumping rates during the demonstration. Over the course of the project, an average of 720 L of water per day was re-circulated through each cell.

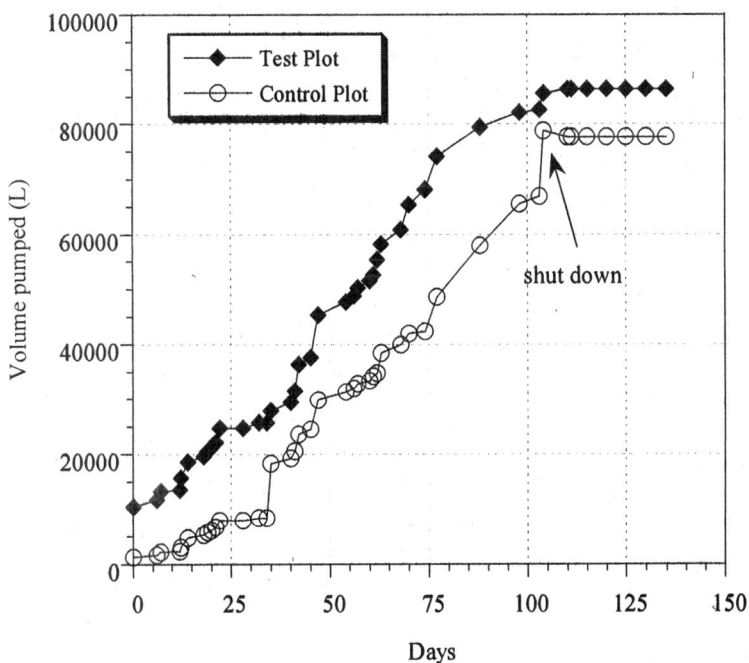

Figure 11. Groundwater quantities re-circulated through the test and control plots during the demonstration

The buffer pump was set to amend each 150 L of groundwater with ~2500 mg/L of the carbonate/bicarbonate mixture during re-injection. At two times during the early operation of the system (on Day 19 and Day 35) approximately 950 L of buffer was added to the aquifer. After each of these additions, the buffer pump was turned off and water was re-circulated for approximately one week through the test plot to disperse the buffer amendment throughout the formation. During the course of the demonstration, 4450 L of buffer was added to the aquifer. Approximately 3300 L of this buffer was a 6.67% solution containing 80% bicarbonate and 20 % carbonate. The remainder was a 6.67% solution containing a mixture of 70% bicarbonate and 30% carbonate. The latter solution, with a slightly higher ratio of carbonate, was added to the aquifer one month after the beginning of the demonstration to enhance the rate of pH increase. After the 1150-L addition was complete, the buffer was returned to an 80% bicarbonate and 20% carbonate mixture for the remainder of the field trial.

The lactate pump was set to supply electron donor at a flow rate of ~4.5 mL/min during re-injection of groundwater. Based on an injection time of 20 min per 150 L of groundwater, the concentration of lactate added to the injected water was expected to be ~ 380 mg/L. This concentration of lactate was calculated to provide a reasonable excess of electron donor in the formation based on the average concentrations of oxygen, nitrate, and perchlorate present throughout the test plot. An additional dose of electron donor (~ 11 L) was added to the aquifer during the early operation of the system on two occasions (on Day 19 and Day 35) in conjunction with the extra buffer addition. The lactate pump was turned off and the groundwater was re-circulated for one week to mix the electron donor after each of these additions. A total volume of 91 L of the 60 % lactate solution was added to the aquifer during the demonstration period (i.e., an average of 0.83 L/day). A total weight of 58 kg of lactate was added during the field study.

The pH and alkalinity of the water within the test plot were monitored throughout the demonstration to evaluate the effectiveness of the buffer addition to the aquifer. The concentrations of lactate and perchlorate, as well as nitrate and sulfate, were measured with time to assess the distribution and effectiveness of electron donor amendment to the aquifer for perchlorate remediation. The analytical results are summarized below

Groundwater Sampling

Baseline groundwater samples were collected from the test plot and the control plot 69 days (10 weeks) and 7 days (1 week) prior to the start-up of the injection system. During the demonstration, samples were taken from all

9 monitoring wells in the test plot on Days 14, 25, 49, 70, 105, and 140. The control plot wells were sampled on Days 14, 49, 105, and 140. Each well received dedicated sampling tubing at the start of the demonstration. The wells were sampled using a peristaltic pump, and each well was purged for 25 – 30 minutes prior to sampling. During most of the sampling events, a YSI 600 XL water quality meter with a flow cell was used to determine that key parameters (e.g., pH, conductivity) were stable prior to sample collection.

pH and Alkalinity

The pH of groundwater in the test plot and the control plot was measured using a field probe during sample collection and in the laboratory by EPA Method 150.1. Alkalinity was measured by titration according to EPA Method 310.1. The pH of the groundwater in each of the 9 TPMWs was observed to increase significantly during the course of the demonstration (Table 3). For example, the pH in TPMW-5 increased from 4.0, 7 days

Table 3. pH values in the test plot (TPMW) and control plot (CPMW) monitoring wells during active pumping

Day	TPMW-1s	TPMW-1d	TPMW-2s	TPMW-2d	TPMW-3s	TPMW-3d	TPMW-4s	TPMW-4d	TPMW-5
-7	5.99[1]	4.69	5.53	5.34	3.82	5.2	4.09	3.88	4.02
5	**5.81**	**5.2**	**5.74**	**5.15**	**4.09**	**4.8**	**4.17**	**4.43**	**4.56**
14	**6.45**	**5.73**	**6.33**	**5.05**	**3.91**	**5.2**	**4.39**	**3.98**	**3.97**
25	5.6	5.65	4.91	4.03	5.21	5.01	4.27	3.53	5.62
25	**6.32**	**6.14**	**5.3**	**4.46**	**5.54**	**5.29**	**4.55**	**3.85**	**5.97**
49	5.69	6.02	5.54	4.64	5.6	6.48	4.41	5.44	6.5
49	**5.78**	**6.34**	**5.82**	**4.62**	**5.9**	**6.42**	**4.79**	**5.76**	**6.46**
70	7.12	6.03	6.66	5.25	5.93	6.09	4.73	5.5	5.85
105	**6.51**	**6.44**	**6.9**	**6.24**	**5.74**	**6.49**	**5.93**	**6.27**	**6.28**

Day	CPMW-1s	CPMW-1d	CPMW-2s	CPMW-2d	CPMW-3s	CPMW-3d	CPMW-4s	CPMW-4d	CPMW-5
-7	5.33	5.58	5.18	4.19	4.37	4.63	5.5	5.6	4.24
14	**5.52**	**5.74**	**5.97**	**4.47**	**4.56**	**4.84**	**5.28**	**6**	**4.4**
49	6.16	5.03	5.19	3.74	4.6	4.41	5.9	6.17	4.75
49	**6.4**	**5.75**	**6.08**	**4.43**	**5.05**	**4.6**	**5.48**	**5.93**	**4.6**
105	**6.38**	**6.39**	**6.39**	**4.44**	**4.74**	**4.74**	**5.6**	**6.08**	**4.82**

[1]Values in bold are laboratory measurements (EPA 150.1) and those in plain text are by field probe.

before the start of the demonstration, to 6.5 at Day 49. Each well achieved a pH > 5.7 during the course of the demonstration. A significant pH increase was also observed in a in a few of the control plot monitoring wells during the demonstration, particularly CPMW-1s and CPMW-2s. This increase may reflect buffering from surface soils after significant rainfall events caused the water table to rise. However, the groundwater pH in most of the control plot wells remained near that recorded before pumping commenced.

The alkalinity in each of the test plot wells also showed a marked increase as buffer was added (Table 4); each well reaching in excess of 480 mg/L during the course of the study. For example, the alkalinity in TPMW-5, increased from < 2 mg/L (as $CaCO_3$) prior to the demonstration to 1600 mg/L on Day 105. Conversely, there was little change in the alkalinity of the CPMWs. The data show that the addition of the carbonate/bicarbonate buffer caused an appreciable increase in the alkalinity and the pH of the aquifer underlying the test plot.

Table 4. Alkalinity values (mg/L) in the test plot (TPMW) and the control plot (CPMW) with time

Day	TPMW-1s	TPMW-1d	TPMW-2s	TPMW-2d	TPMW-3s	TPMW-3d	TPMW-4s	TPMW-4d	TPMW-5
-7	92	5.4	60	15	<2.0	16	<2.0	<2.0	<2.0
25	508	200	91	3.9	130	95	14	< 4.0	640
49	160	530	220	69	240	600	49	470	162
70	3200	370	1670	270	710	690	64	320	410
105	680	390	390	740	250	720	480	1040	1600
140	1240	340	1420	150	590	490	340	510	600

Day	CPMW-1s	CPMW-1d	CPMW-2s	CPMW-2d	CPMW-3s	CPMW-3d	CPMW-4s	CPMW-4d	CPMW-5
49	150	59	84	20	20	25	34	120	20
105	120	110	89	2	5.9	5.9	26	29	3.9
140	110	110	110	7.9	20	7.9	28	31	7.9

Lactate

Lactate was measured in groundwater samples collected from the test plot using ion chromatography. The samples were analyzed on a Dionex DX-600 ion chromatograph equipped with a Dionex IonPac AS11-HC column. The sample method utilizes a gradient of sodium hydroxide increasing from 1 mM to 60 mM over a 40 min. run time. To ensure that lactate was not biodegraded prior to analysis, groundwater samples (20-mL volume) were passed through sterile 0.22-μm-pore-size cellulose acetate filters in the field. The water was collected in sterile 50-mL conical tubes and stored at 4°C until analysis.

Lactate was detected in groundwater from 7 of 9 TPMWs by Day 14, and all wells had measurable concentrations of lactate by Day 25 (Table 5). The lactate levels varied somewhat by well and with time; however, the electron donor was detected consistently above 10 mg/L in 8 of the 9 wells during the course of the demonstration, and each of the 8 wells had levels exceeding 100 mg/L at one or more sample points. At the end of the demonstration period on Day 140, 29 days after system shut-down on Day 111, lactate was below detection in 7 of 9 TPMWs. Among the test plot wells, TPMW-1d generally had the lowest concentration of lactate (< 7 mg/L on 5/6 samplings), and the groundwater collected from this well never exceeded 21 mg/L of the electron donor. This was also the one well in which perchlorate levels declined only marginally (43 %) during the demonstration (see next section) and in which nitrate never declined below 1 mg/L.

Table 5. Lactate concentrations (mg/L) in the test plot with time

Day	TPMW-1s	TPMW-1d	TPMW-2s	TPMW-2d	TPMW-3s	TPMW-3d	TPMW-4s	TPMW-4d	TPMW-5
14	139	6	34	37	249	249	0.5	0.5	376
25	15	21	96	35	85	463	652	562	390
49	38	3.8	68	248	97	159	44	297	114
70	410	2.2	170	21	15	130	12	40	11
105	83	0.18	56	16	2.9	35	21	7.1	15
140	110	< 0.5	230	< 0.5	< 0.5	< 0.5	< 0.5	< 0.5	< 0.5

Perchlorate

Perchlorate in groundwater was analyzed according to EPA Method 314.0. Perchlorate levels throughout the test plot showed a steady decline during the 5-month field demonstration (Figure 12). During the two baseline sampling events (69 days and 7 days before system start-up, respectively), perchlorate levels ranged from a low of 72 mg/L in well TPMW-3s to a high of 276 mg/L in TPMW-2d. The average perchlorate levels in the test plot during these sampling events were 171 mg/L (69 days before start-up) and 174 mg/L (7 days before start-up), respectively. By the end of the 20-week demonstration, perchlorate levels in two test wells (TPMW-1s and TPMW-2s) were below the PQL of 5 μg/L, one well was less than 20 μg/L (TPMW-5), and two additional wells were less than 1 mg/L. The reduction in aqueous perchlorate from the start of the demonstration was in excess of 99% for each of these wells. Of the remaining 4 wells in the test plot, two displayed perchlorate concentrations of less than 3.7 mg/L (TPMW-3s and TPMW-3d) at the end of the demonstration, and one (TPMW-4d) was less than 10 mg/L. However, perchlorate in groundwater from TPMW-4d had reached levels as low as 2 mg/L during system operation. The percent reduction in perchlorate in each of these wells exceeded 95% from the start to the end of the demonstration.

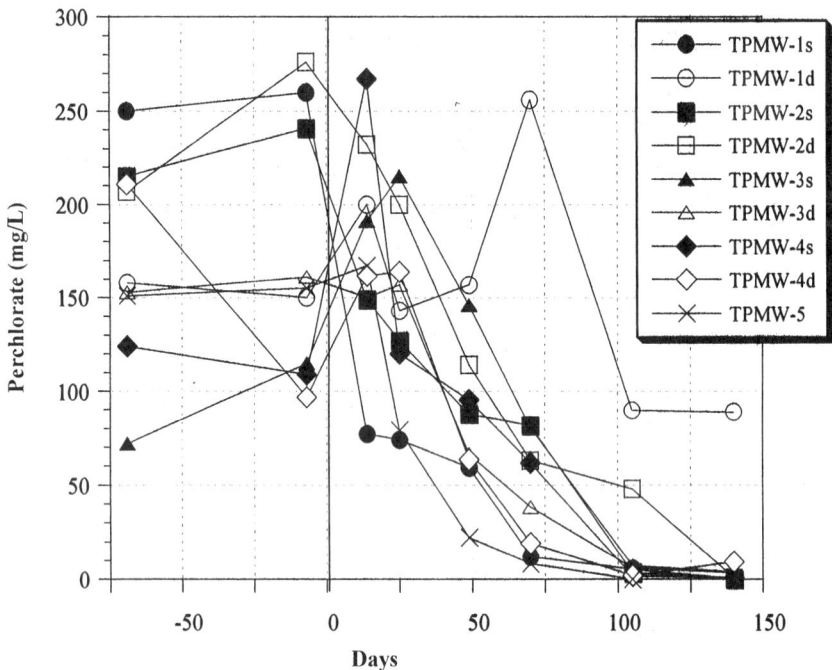

Figure 12.   Perchlorate levels in the test plot with time.

The only test plot well in which perchlorate levels did not decline precipitously during the demonstration was TPMW-1d. Perchlorate levels fell by only 43% in this well, ending at approximately 90 mg/L after 140 days. The bromide data suggest that the well was hydraulically connected to the injection wells, and the alkalinity and pH results show that this area received reasonable quantities of the buffer. However, the amounts of lactate measured in this well were consistently below those in the surrounding wells. TPMW-1d was on the upgradient side of the plot, and was in close proximity to one of the injection wells. It is possible that the injection well significantly impacted water flow in the region of TPMW-1d, preferentially bringing contaminated water from upgradient of the treatment area into the plot in the vicinity of TPMW-1d during injection cycles. In this case, the low lactate levels would reflect enhanced microbial consumption of the electron donor during degradation of the oxygen and nitrate introduced with the contaminated groundwater. These electron acceptors are known to inhibit biological perchlorate reduction, thus their continual introduction would also explain the persistence of perchlorate in this monitoring well.

Unlike the test plot, there was no consistent reduction in perchlorate levels in any of the wells in the control plot during the demonstration period (Figure 13). The average perchlorate concentration in the nine CPMWs 69 days prior to system start-up was 127 mg/L, and after 140 days of system operation, the

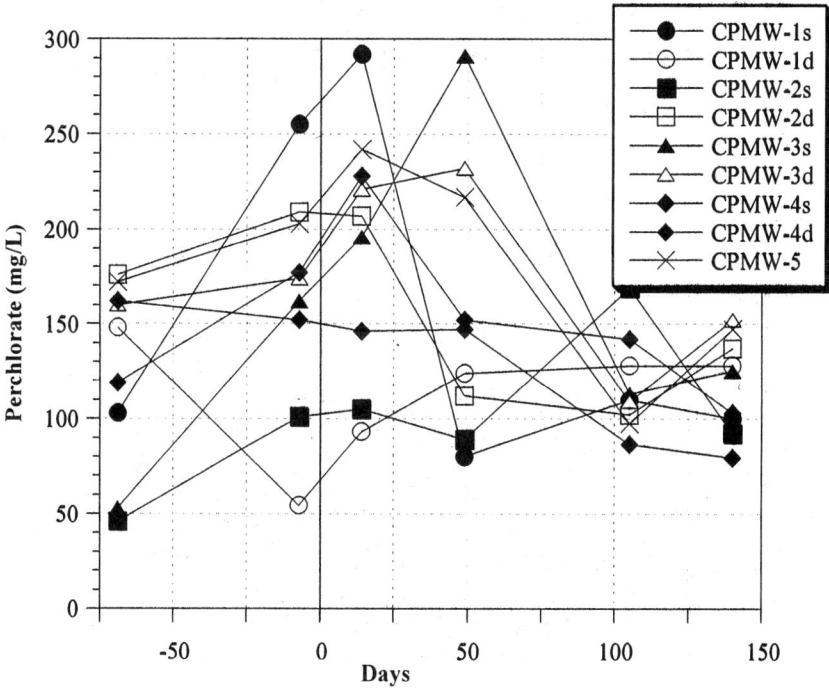

Figure 13. Perchlorate levels in the control plot with time.

concentration was 118 mg/L. A similar amount of water was re-circulated through both plots during the demonstration, but the water in the control plot received no amendments.

The data from the demonstration clearly show that the addition of buffer and electron donor to the test plot stimulated the microbial reduction of perchlorate in the aquifer. Losses of perchlorate to dilution or any other abiotic process would have been observed in both plots. The data also show that, even in an acidic aquifer with extremely high perchlorate levels, *in situ* biological reduction can effectively reduce perchlorate concentrations to less than 5 μg/L in a reasonably short period. Although a treatment level of 5 μg/L for perchlorate was not achieved in every well, a reduction in perchlorate levels exceeding 95 % was observed in 8 of the 9 TPMWs, including those screened in the shallow, less conductive zone in the aquifer. Based on the trends of perchlorate removal observed during the demonstration, it is likely that many of the other TPMWs would have reached non-detect levels of perchlorate with additional time of system operation.

Nitrate and sulfate

Although the focus of this demonstration was the biological reduction of perchlorate, levels of other common electron acceptors, including nitrate and sulfate, were monitored. Nitrate reduction (i.e., denitrification) occurs by a biological process similar to perchlorate reduction, and often occurs prior to perchlorate degradation. Nitrate is a regulated in the drinking water in the U.S. under the Safe Drinking Water Act. The compound, which commonly enters groundwater through agriculture activities, has a federal Maximum Contaminant Level (MCL) in drinking water of 45 mg/L as nitrate (10 mg/L as nitrate-N), a value that is much higher than that anticipated for perchlorate (i.e., 1 – 24 μg/L).[7,17] The biological reduction of sulfate occurs after perchlorate reduction, and produces hydrogen sulfide, which has a "rotten egg" odor that is undesirable in groundwater. Thus, one goal of *in situ* treatment systems for perchlorate and/or nitrate is to mix and distribute electron donor effectively so that sulfate reduction is minimized after reduction of the previous two electron acceptors is complete. This is readily accomplished in *ex situ* treatment systems (such as biological reactors), but more difficult for *in situ* applications.

Nitrate and sulfate were measured in groundwater samples by EPA method 300. The levels of nitrate in the test plot declined rapidly in several wells (Table 7). The levels of this contaminant average slightly above 2 mg/L as nitrate-N prior to the investigation in the test plot. Nitrate was below detection (< 0.2 mg/L nitrate-N) in 7 of 9 TPMWs by Day 49 of the study. As noted for perchlorate, TPMW-1d showed the slowest decline in nitrate concentrations. The starting levels of nitrate in the control plot wells were somewhat higher than in the test plot, averaging above 7 mg/L as nitrate-N at the commencement of the study. However, although there was some variability in nitrate levels from point to point in each well, there was no consistent reduction in nitrate levels across the control plot during the demonstration. After 140 days, the average concentration among the 9 wells remained above 7 mg/L as nitrate-N.

Table 7.   Levels of nitrate-N (mg/L) in the test plot (TPMW) and control plot (CPMW) with time.

Day	TPMW-1s	TPMW-1d	TPMW-2s	TPMW-2d	TPMW-3s	TPMW-3d	TPMW-4s	TPMW-4d	TPMW-5
-69	4.3	1.6	2.7	2.9	1.3	1.6	1.9	2.8	2.3
-7	3.6	2.1	2.8	4.3	1.6	0.88	1.1	1.3	2.1
14	< 0.2	5.4	< 0.2	3.2	2.2	0.52	3.4	1.6	1.0
25	< 0.2	0.7	< 0.2	1.4	0.8	0.6	1.1	1.7	0.4
49	< 0.2	1.0	< 0.2	< 0.2	2.3	< 0.2	< 0.2	< 0.2	< 0.2
70	< 0.2	3.7	< 0.2	< 0.2	0.4	1.7	< 0.2	< 0.2	< 0.2
105	< 0.2	0.31	< 0.2	0.55	0.84	< 0.2	< 0.2	< 0.2	< 0.2
140	< 0.2	0.64	< 0.2	0.9	0.21	< 0.2	< 0.2	< 0.2	< 0.2

Day	CPMW-1s	CPMW-1d	CPMW-2s	CPMW-2d	CPMW-3s	CPMW-3d	CPMW-4s	CPMW-4d	CPMW-5
-69	1.0	1.5	12	12	4.5	9.7	5.5	10	13
-7	0.96	0.9	2.2	13	16	13	6.9	5.2	12
14	1.2	< 0.2	3.1	8	13	11	11	3.1	12
49	1.2	0.61	2.6	5.9	10	5.8	4.9	0.7	4.9
105	6.4	2.8	7.5	8.5	6.5	5.4	6.3	0.42	6.9
140	4.6	5.6	6.1	14	9.7	14.5	0.58	3.7	7.9

There was a slight odor of hydrogen sulfide detected in some of the test plot wells during the demonstration, and the presence of a black precipitate was observed in a few wells on these occasions (presumably iron sulfide). During the short demonstration time, the goal was to supply adequate electron donor to achieve nitrate and perchlorate reduction, rather than to tightly control the process. If the demonstration were conducted for a longer period, the level of excess electron donor could have been minimized further. However, with the exception of one well (TPMW-2s), appreciable losses of sulfate were not observed during the course of the study in either plot (Table 8). The average concentration at the start of the demonstration in the 9 TPMWs was 174 mg/L and, at the end of the demonstration, the average was 240 mg/L. The only well that showed a significant decrease in sulfate concentration was TPMW-2s, where levels declined from 120 mg/ before the demonstration to 3.7 mg/L after 140 days. However, much of this decline occurred after the system was shut down on Day 105.

Table 8. Levels of sulfate (mg/L) in the test plot (TPMW) and control plot (CPMW) with time.

Day	TPMW-1s	TPMW-1d	TPMW-2s	TPMW-2d	TPMW-3s	TPMW-3d	TPMW-4s	TPMW-4d	TPMW-5
-7	85	97	120	79	230	93	320	250	290
49	170	106	140	710	260	46	400	290	270
70	120	63	150	150	370	63	370	225	200
105	46	71	91	130	330	21	540	290	480
140	89	89	3.7	72	450	110	640	360	350

Day	CPMW-1s	CPMW-1d	CPMW-2s	CPMW-2d	CPMW-3s	CPMW-3d	CPMW-4s	CPMW-4d	CPMW-5
49	67	89	150	99	60	68	105	77	110
105	99	120	110	99	120	95	130	82	110
140	120	110	150	86	109	74	79	150	120

DEMONSTRATION SUMMARY AND CONCLUSIONS

The general conclusions from this field demonstration are as follows:

1. The acidic aquifer in the vicinity of Building 1419 was effectively buffered using an aqueous mixture of carbonate and bicarbonate. The buffer increased local groundwater pH from values as low as 3.8 to values exceeding 5.9 for all test plot wells. The alkalinity in each of the wells reached in excess of 480 mg/L during the study.

2. The system design, which generated a recirculation cell within the aquifer, provided an effective distribution of buffer and electron donor throughout the saturated zone, even though the aquifer was characterized by regions with widely differing geology and conductivity.

3. *In situ* perchlorate biodegradation was rapidly observed using lactate as an electron donor. Perchlorate levels were reduced by > 95 % in 8 of the 9 monitoring wells within the test plot during the demonstration. In two wells, with starting perchlorate concentrations in excess of 210 mg/L, final perchlorate levels after 20 weeks of treatment were < the PQL of 5 µg/L. Conversely, there was no significant reduction in perchlorate levels in the control plot.

4. Nitrate-N levels in the test plot were reduced to below detection in 7 of the 9 monitoring wells within 7 weeks. The other two wells had nitrate-N concentrations less than 1 mg/L at the end of the 20-week study. There was no significant reduction in Nitrate-N in the control plot during the demonstration.

5. Sulfide was detected by odor in some of the test plot monitoring wells during the demonstration. However, analytical data revealed no appreciable reduction in sulfate levels throughout the test plot during the demonstration period. In future work at the site, tests should be performed to optimize electron donor delivery such that sulfate reduction is completely inhibited.

6. Data from the demonstration suggest that *in situ* bioremediation will be a viable option for perchlorate treatment in aquifers containing localized, high concentrations of the oxidant. These include source areas from hog out operations, demolition and open burn areas, and other regions where perchlorate or perchlorate-laden fuels were discharged.

ACKNOWLEDGEMENTS

The authors wish to acknowledge the Strategic Environmental Research and Development Program (SERDP) and the Indian Head Division, Naval Surface Warfare Center, Office of Naval Ordnance Safety and Security Activity (NOSSA) for providing financial support for this work. We also wish to thank Charles Condee, Darren Engbring, Bill Kosmer, Randi Rothmel, Matt Giovanelli, and Anthony Soto for their field and analytical support.

CREDITS

Some text and figures reprinted with permission of the Naval Ordnance Safety and Security Activity, Naval Sea Systems Command from material first published in Technical Report NOSSA-TR-2004-001, Field demonstration of in situ perchlorate bioremediation at building 1419, January 22, 2004.

REFERENCES

1.   Coates J.D., Achenbach L.A. Microbial perchlorate reduction: rocket-fuelled metabolism. Nature Rev Microbiol 2004; 2:569-580.

2.   Rikken G.B., Kroon A.G.M., van Ginkel C.G. Transformation of (per)chlorate into chloride by a newly isolated bacterium: reduction and dismutation.   Appl Microbiol Biotechnol 1996; 45:420-426.

3.   Kengen S.W.M., Rikken G.B., Hagen W.R., van Ginkel C.G., Stams A.J.M. Purification and characterization of (per)chlorate reductase from the chlorate-respiring strain GR-1.   J Bacteriol 1999; 181:6706-6711.

4.   van Ginkel C.G., Rikken G.B., Kroon A.G.M., Kengen S.W. M.   Purification and characterization of a chlorite dismutase: a novel oxygen-generating enzyme.   Arch Microbiol 1996; 166:321-326.

5.   Hatzinger P.B. Perchlorate biodegradation for water treatment. Environ Sci Technol 2005; *in press.*

6.   Xu J., Song Y., Min B., Steinberg L., Logan B.E. Microbial degradation of perchlorate: principles and applications. Environ Eng Sci 2003; 20:405-422.

7.   Hatzinger P.B., Whittier M.C., Arkins M.D., Bryan C.W., Guarini W.J.   In-situ and ex-situ bioremediation options for treating perchlorate in groundwater. Remediation 2002; 12:69-85.

8.   Wu J., Unz R.F., Zhang H.S., Logan B.E. Persistence of perchlorate and the relative numbers of perchlorate- and chlorate-respiring microorganisms in natural waters, soils, and wastewater. Biorem J 2001; 5:119-130.

9.   Waller A.S., Cox, E.E., Edwards, E.A. Perchlorate-reducing microorganisms isolated from contaminated sites.   Environ Microbiol 2004; 6:517-527.

10. Tipton D.K., Rolston D.E., Scow K.M. Transport and biodegradation of perchlorate in soils. J Environ Qual 2003: 32:40-46.

11. Cox E.E., McMaster M., Neville S.L. Perchlorate in groundwater: scope of the problem and emerging remedial solutions. Proceedings of the 36th Annual Engineering Geology and Geotechnical Engineering Symposium; 2001 Las Vegas, NV.

12. Hunter W.J.   Bioremediation of chlorate or perchlorate contaminated water using permeable barriers containing vegetable oil. Cur Microbiol 2002; 45:287-292.

13. NSWC Indian Head Division website (2005) Retrieved April 2005 from: http://www.ih.navy.mil/

14. Maryland Geological Survey. Geologic Maps of Maryland: Charles County (1968) Retrieved May 2005 from http://www.mgs.md.gov/esic/geo/cha.html

15. Weight, W.D., Sonderegger, J.L. *Manual of Applied Field Hydrogeology*. New York: McGraw-Hill, 2000.

16. Hatzinger, P. B. *In situ* bioremediation of perchlorate. Final report for SERDP Project CU-1163, Strategic Environmental Research and Development Program, Arlington, VA. Retrieved May 2005 from: http://clu-in.org/contaminant focus/default.focus/sec /perchlorate/cat/Treatment_Technologies.

17. Fan A.M., Steinberg V.E. Health implications of nitrate and nitrite in drinking water: an update on methemoglobinemia occurrence and reproductive and developmental toxicity. Reg Toxicol Pharmacol 1996; 23:35-43.

Chapter 15

Perchlorate Removal by Modified Activated Carbon

Robert Parette and Fred S. Cannon

The Pennsylvania State University, 212 Sackett Building, University Park, PA 16802

INTRODUCTION

Perchlorate's (ClO_4^-) prevalent use as the propellant in rocket fuel has led to the ClO_4^- contamination of considerable amounts of water throughout the United States.[1] Perchlorate contamination is especially a problem in the arid Southwestern U.S. where alternate perchlorate-free water sources are not always readily available. Perchlorate, while thermodynamically unstable, is kinetically non-reactive at low concentrations,[1] as typically found in ground and surface water conditions where contamination has occurred. In addition, perchlorate is extremely soluble in water and is a poor complexing agent,[1] which makes it a very mobile, persistent contaminant in the environment.

Perchlorate has been shown to inhibit iodide uptake by the thyroid gland; and in large doses, it has been linked to anemia and fetal brain damage[2]. While there is no nationally established drinking water standard for perchlorate, several states, including California, Massachusetts, and Texas have taken the initiative to establish statewide action levels for ClO_4^- as low as 1-6 parts per billion (ppb).[3,4]

Though ion exchange[5,6] and microbial methods[7,8] have been shown to be effective technologies to treat perchlorate contaminated water, the use of tailored granular activated carbon (GAC) to treat ppb perchlorate contaminated water has some advantages over these methods. Tailored GAC appears to be cost competitive compared to ion exchange (specific costs values can be obtained from tailored GAC and ion exchange suppliers) and the skeletal GAC can also be thermally reactivated once it is spent, allowing its reuse. If an ion exchange resin is regenerated for reuse, a brine solution is required which then subsequently must be handled. In contrast, ion exchange resins can not regenerated via a thermal process.

Microbial means may not be well suited for perchlorate concentrations in the low ppb range. Water utilities in the U.S. are also unaccustomed to using microbial methods to treat drinking water. Unlike microbial means, many water utilities are comfortable with the use of GAC. Powdered or granular activated carbon is used in over half of the water treatment facilities in the U.S.; and its prevalent use dates back more than 50 years. Annually, 300,000 tons of activated carbon is used by the water treatment industry worldwide[9].

In this chapter, the remediation of groundwaters contaminated with ppb levels of perchlorate by virgin and tailored granular activated carbon is discussed. Rapid small scale column tests (RSSCTs) were designed and operated with both virgin (as a control) and tailored GAC to simulate full-scale GAC contactors and appraise their feasibility as means to treat ppb level perchlorate contamination.

It was hypothesized that virgin GAC could be pre-loaded with cationic surfactants to appreciably increase the GAC's capacity for perchlorate. The cationic surfactants utilized in this research contained a quaternary ammonium functional group, which is comprised of a positively charged nitrogen atom bonded solely to carbon atoms. Quaternary ammonium-based compounds have a high pK_a, rendering the positive charge virtually independent of pH in natural waters.[10]

Once the cationic surfactant is pre-loaded, the positively charged quaternary ammonium head-group of the surfactant can then act as an ion exchange site where perchlorate can be adsorbed. The quaternary ammonium functional group is the active site in strong-base anion exchange resins[5,11], which have been shown to be effective in the removal of perchlorate from water.[5,6,12] The use of cationic surfactants has also been shown to significantly increase the removal of perchlorate, as well as nitrate and chromate, in ultrafiltration processes.[13,14,15] In addition, quaternary ammonium based cationic surfactants have been shown to form ion pairs with perchlorate in electrospray ionization mass spectrometry.[16]

MATERIALS AND METHODS

Activated Carbon

GAC obtained from two sources was used in this work: Superior Adsorbents, Inc. (SAI) of Emlenton, PA (a microporous bituminous GAC) and USFilter Westates of Sante Fe Springs, CA (UltraCarb, a mesoporous and microporous bituminous GAC). Unless otherwise identified, the GAC used was SAI GAC ground to 200 x 400 mesh size range (38 x 75 μm).

Groundwater

Groundwater was obtained from two sites in California and three sites in Massachusetts. The five waters were identified as C1, C2, M1, M2, and M3. Their source locations, perchlorate content, other anion content, and total conductivity are shown in Table 1. All groundwaters were received in high-density polyethylene 55-gallon drums. With the exception of water M3, the barrels were stored in ambient temperature laboratory conditions. In addition to perchlorate, M3 water contained organic contaminants of concern. The M3 barrels were stored at 4°C prior to use in RSSCT tests.

Table 1. Groundwaters used in RSSCT testing

Water Designation	Source Location	ClO_4^- (ppb)	Cl^- (mg/L)	NO_3^- (mg/L as NO_3^-)	SO_4^- (mg/L)	conductivity (μS)
C1	Redlands, CA	75	7.2	25.6	30	330
C2	Northern CA	12	not determined	not determined	not determined	1800
M1	Eastern MA	0.85	10.4	0.3	5.9	66
M2	Eastern MA	5.6	7.6	0.4	6.9	53
M3	Eastern MA	1	8.7	0.1	5.7	50

Cationic Surfactants

Cationic surfactants were obtained from several sources and used as received without further purification. Decyltrimethylammonium bromide (DTAB, $C_{10}H_{21}N(CH_3)_3Br$) and tributylheptylammonium bromide (THAB, $C_7H_{15}N(C_4H_9)_3Br$) were obtained from Fluka. Myristyltrimethylammonium bromide (MTAB, $C_{14}H_{19}N(CH_3)_3Br$) and cetyltrimethylammonium chloride (CTAC, $C_{16}H_{33}N(CH_3)_3Cl$) were obtained from Aldrich. Cetylpyridinium chloride (CPC, $C_{16}H_{33}NC_5H_5Cl$) was obtained from Acros. Tallowalkyltrimethylammonium chloride (T-50, $C_{16-18}H_{33-37}N(CH_3)_3Cl$) and dicocoalkyldimethylammonium chloride (2C-75, $(C_{12-16}H_{25-33})_2N(CH_3)_2Cl$) were obtained from Akzo Nobel's Arquad product line. While the other surfactants were homogenous, T-50 and 2C-75 were mixtures of surfactants that also contained water and isopropyl alcohol. The long chains in the surfactants that comprise T-50 were C_{14} (3%), C_{16} (32%), and C_{18} (65%). Two long chains were present in the surfactants that comprise 2C-75. The manufacturer determined these chains to be C_8 (1%), C_{10} (5%), C_{12} (54%), C_{14} (21%), C_{16} (11%), and C_{18} (8%).[17]

Analytical Methods

Concentrations of perchlorate were measured with the use of a Dionex DX-120 ion chromatograph (Sunnyvale, CA). The Dionex DX-120 was equipped with an AS40 autosampler, a 4-mm AS16 column, a 4-mm AG16 guard column, a 4-mm ASRS Ultra suppressor, and a DS4 detection stabilizer. The stabilizer used a current of 300 mA and a temperature of 35 °C. The elluent concentration used was 30 mM NaOH. A 1 mL sample loop was used when analyzing the two groundwaters from California. With the three groundwaters from Massachusetts, it was possible to use a 4 mL sample loop due to the low conductivity of the waters. With these conditions, perchlorate could be detected down to 1 ppb with the 1 mL sample loop and to 0.25 ppb with the 4 mL sample loop.

Chloride (Cl⁻), nitrate (NO_3^-), and sulfate (SO_4^{-2}) were also measured via ion chromatography utilizing a 25 μL sample loop and an elluent concentration of 7.5 mM NaOH. All other conditions were the same as those used to analyze samples for perchlorate.

Concentrations of cationic surfactants were measured utilizing the two-phase titration method of Tsubouchi et al.[18] This method was able to detect the surfactants down to concentrations of 0.1 mg/L. RDX and HMX concentrations were determined by HPLC with UV detection (EPA Method 8330) by Severn Trent Laboratory in Colchester, VT. The detection limit for both RDX and HMX was 0.25 ppb.

RSSCT Design

Na et al.[19] observed that perchlorate adsorbed onto non-tailored GAC in accordance with proportional diffusivity. Rapid small-scale column tests herein were deigned for proportional diffusivity, per the similitude equations of Crittenden et al.[20] With a GAC grain size range of 200 × 400 mesh (38 × 75 μm), RSSCT columns measured 13.5 cm in length. The column diameter should be at least fifty times greater than the diameter of the GAC grains to minimize wall effects. A column diameter of 0.5 cm was chosen, yielding a bed volume (BV) of 2.65 cm.[3] 1.37 – 1.45 g of activated carbon was used in each test. Flow rates used in the RSSCT tests simulated full-scale empty bed contact times (EBCTs) of 5 – 20 minutes when using 8 x 30 mesh GAC (600 μm x 2360 μm). Columns were machined internally within 1¼" acrylic rods. Waters HPLC pumps models 501 and 510 were adapted to provide flow through the columns and ¼" copper tubing connected the pumps to the columns.

GAC Tailoring

RSSCT columns were dry packed with virgin activated carbon. Glass wool was packed at the ends of the columns to hold the carbon in place. Thirty-three BV of a 0.4% surfactant solution (by weight of solid surfactant) was then recirculated through the GAC bed for a period of two days (unless otherwise identified) to pre-load the GAC with the respective cationic surfactant or polymer.

Temperature Control

For work performed with groundwaters from Massachusetts, the RSSCT columns were immersed in a cooling bath maintained at 8 − 9°C, the temperature required to hold the effluent from the columns between 11 − 12 °C, to simulate the temperatures of the actual groundwaters. The California groundwaters were naturally at temperatures between 18-20°C. Tests with these waters were operated at room temperature (20 ± 2°C).

PERCHLORATE ADSORPTION BY VIRGIN GAC

Na et al.[19] examined the adsorption of perchlorate from Redlands (C1) water onto virgin GAC utilizing RSSCTs containing 60 x 80 mesh (180 μm × 250 μm) GAC, so as to mimic 8 × 30 mesh (600 × 2360 μm) full-size grains. Based on their results, two vessels operating in series could treat water for 1 − 1 ½ months prior to initial detection of perchlorate in the effluent, when simulating a full-scale EBCT of 20 minutes through each vessel. The GAC would then have to be either replaced or thermally reactivated; and neither of these options would render the use of virgin GAC to be economically attractive.

In addition to repeating the experiment of Na et al. with 200 × 400 mesh GAC and a redesigned column, RSSCTs with virgin GAC were performed with four other waters to determine the length of time that the untailored GAC could adsorb perchlorate. The RSSCTs were operated to simulate a 20 minute EBCT in a full-scale GAC vessel that contained 8 × 30 mesh GAC. The RSSCT results for these 5 waters with virgin GAC are shown in Figure 1.[21, 22]

With C1 water, initial detection of perchlorate in the virgin GAC RSSCT effluent was observed at 1000 BV. These results agree closely with the results observed by Na et al.[19]. C2 water, from a site in Northern California, contained only 12 ppb ClO_4^- (roughly 20% of the perchlorate in C1) yet virgin GAC was only able to treat 700 BV prior to the detection of perchlorate in the column effluent. This drop in performance, relative to C1 water, is attributable to higher concentration of salts in the C2 water. While competing ions were not measured directly for C2 water, this can be inferred by the much higher conductivity for C2 water (1800 µS vs. 330 µS for C1 water).

Although virgin GAC removed perchlorate to below detectable levels for only a limited duration (700-1100 BV) when processing either of the California groundwaters, the removal duration was considerably longer for the three Massachusetts groundwaters tested. In addition to having low perchlorate concentrations (0.85 – 5.6 ppb), these waters had much lower concentrations of competing ions. The RSSCT tests for the Massachusetts waters were performed with UltraCarb.

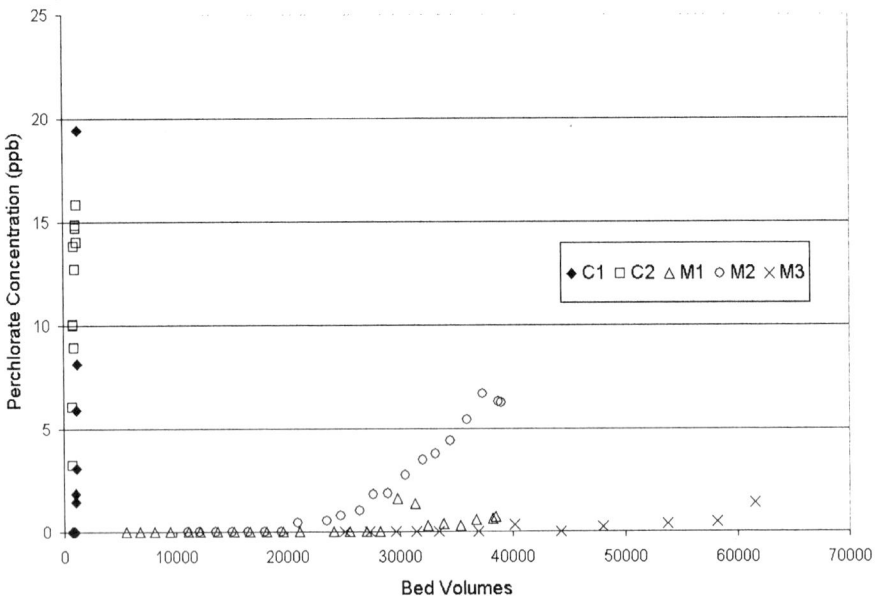

Figure 1. Adsorption of perchlorate from various waters onto SAI GAC

While both M1 and M3 water contained approximately the same concentration of perchlorate and have similar conductivities, UltraCarb adsorbed perchlorate more effectively from M3 water. This was likely caused by the higher nitrate and sulfate concentrations in M1 water. With M1 water (0.85 ppb), virgin UltraCarb treated 30,000 BV prior to both initial and full perchlorate breakthrough. With M3 water, perchlorate was detected at 40,000 BV and remained at 0.25 ppb through 54,000 BV. Full breakthrough (1 ppb) occurred at 60,000 BV (a 100% increase relative to M1). With M2 water (5.6 ppb), perchlorate was removed to below detectable levels for 20,000 BV. Perchlorate in the effluent reached 1 ppb at 26,000 BV; and full breakthrough occurred at 37,000 BV.[22]

It should be noted that UltraCarb had approximately the same adsorptive capacity as SAI GAC when treating C1 water, as shown in Figure 2. Therefore, the authors herein surmise that the large increase in BVs of Massachusetts groundwater treated (in comparison to the California groundwaters) was not related to the type of virgin GAC used, but rather to the water conductivities and concentrations of competing ions.

The adsorption of perchlorate onto both SAI GAC and UltraCarb compared favorably with full-scale data. This is also shown in Figure 2. Research was conducted in Redlands, CA at the Texas St. Water Treatment Plant (C1 water). One 5000 gallon vessel at the plant (10' diameter and 8' height) was filled with 20,000 lbs of bituminous GAC (AquaCarb from USFilter Westates). During normal operation two vessels would be used in series. However, for this study, a single vessel was used individually making the scale effectively half-scale. The flow rate through the vessel was 225 gallons

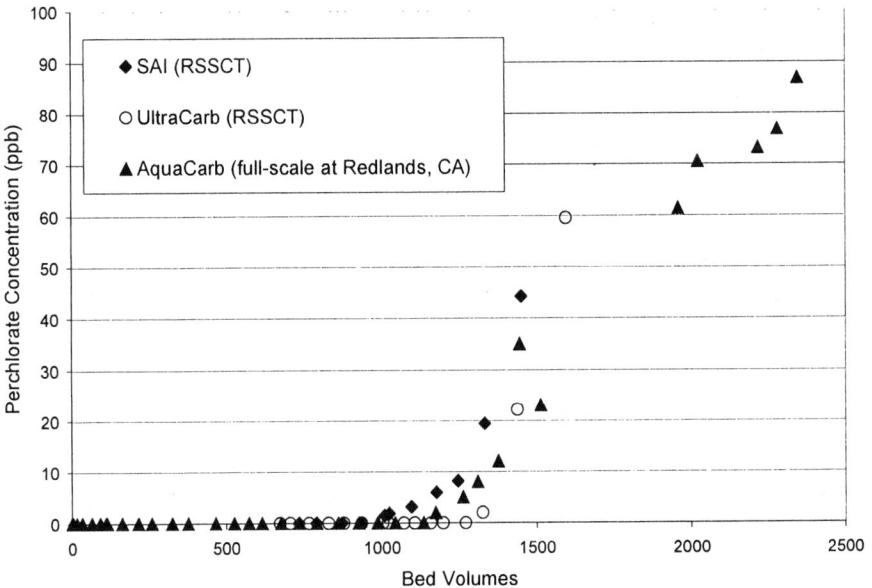

Figure 2. Adsorption of perchlorate from C1 water onto SAI and UltraCarb GAC.

per minute, corresponding to an EBCT of 22 minutes, and the study employed 8 × 30 mesh (600 μm × 2360 μm) GAC. Perchlorate breakthrough was detected at about the same time in the RSSCTs column (1000 BV for SAI GAC and 1250 BV for UltraCarb) as observed at full-scale in Redlands (1150 BV).

Overall, the use of virgin GAC as an effective adsorbent for perchlorate contaminated groundwater appears conditional at best. Virgin GAC may be a viable treatment option for ppb level perchlorate contamination if the concentration of other competing anions in the water is low. At full-scale, with the use of a 20 minute EBCT and 8 x 30 mesh GAC, the BVs treated to initial detection of perchlorate would correspond to 14 months for M1 water, 9.3 months for M2 water, and 18.5 months for M3 water.

Influence of Temperature and Water pH on Perchlorate Removal

As mentioned earlier, the temperature in RSSCT tests was controlled to simulate the temperatures of the actual groundwaters. Surface charge will vary with temperature as most surfaces are more negative at higher temperatures and more positive at lower temperatures.[23] The point of zero charge (pH_{pzc}) for UltraCarb at various temperatures was determined by the mass titration method described by Menendez and coworkers.[24] The point of zero charge is defined as the pH at which the net charge on the surface is zero.[25] The pH_{pzc} was measured at 9.8 throughout the 4-12°C range, 9.63 at 20°C, and 9.4-9.5 between 25-37°C.[22]

In RSSCT testing, the adsorption of the negatively charged perchlorate ion was significantly influenced by changes in temperature. In RSSCTs employing a 20 minute EBCT, perchlorate from C1 water was adsorbed onto UltraCarb GAC at temperatures of 4, 20 and 30°C. At 4°C, the GAC had the greatest capacity for perchlorate with breakthrough observed at 1500 BV. Perchlorate breakthrough at 20°C was noted at 1250 BV. Increasing the temperature from 20°C to 30°C resulted in a large decrease in perchlorate capacity with breakthrough observed after only 650 BV.[22]

The pH is another factor which can have a significant effect on the adsorption of perchlorate onto virgin GAC. Activated carbon has a greater positive charge as the pH is lowered[26]. This enhanced positive charge brought on by a drop in pH increased the adsorption of perchlorate onto GAC. Likewise, should the pH increase (become more basic), the GAC surface became more negatively charged and subsequently less perchlorate adsorbed. To test this, perchlorate from C1 water was loaded onto GAC at pH values of 8, 6, and 4. HCl was used to adjust the water to the desired pH value. The results from this experiment are shown in Figure 3.

As Figure 3 illustrates, the adsorption of perchlorate can be greatly affected by the influent pH. Breakthrough at pH 8 for virgin GAC occurred at 1200 BV. Lowering the influent pH to 6 nearly doubled the BV treated to breakthrough, extending it from 1200 to 2000 BV. At pH 4, the increased adsorption of perchlorate was even more dramatic. At pH 4, perchlorate breakthrough occurred at approximately 3100 BV. This accounts for an 2.6 times increase in BV treated to breakthrough when compared to pH 8.

While decreasing the pH increased the capacity for perchlorate, it is not practical for use at full-scale for several reasons. Due to buffering in the water and the large quantity of water, this process would be extremely expensive. In addition to the amount of acid required, approximately an equal amount of base would then be necessary to raise the pH back to a range in which the water could be purveyed to the drinking water system. This would again add considerable cost to the process. Also, while the pH of groundwater is typically stable, a slug of water with a higher pH contacting a GAC would be a concern. This scenario could cause previously adsorbed perchlorate to desorb.

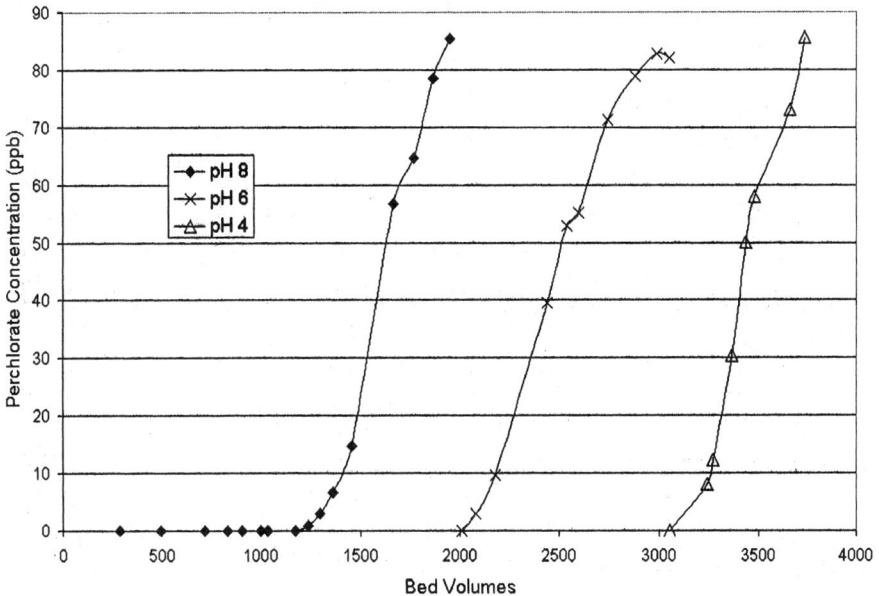

Figure 3. Effect of change in pH on the adsorption of perchlorate from C1 water onto GAC (20-22°C)

IRON PRE-LOADED AND AMMONIA TAILORED GAC

Na et al.[19] and Chen et al.[27] were somewhat successful in modifying GAC so that it became a more effective adsorbent for perchlorate. These tests were also conducted with groundwater from Redlands, CA (water C1). Na and coworkers tailored the GAC with a combination of iron (ferric chloride) and oxalic acid. The most favorable condition found for prolonging the service life of the GAC was when the preloading solution contained 500 mg/L of Fe (as Fe) and 580 mg/L of oxalic acid. This improved the perchlorate adsorptive capacity of the GAC by approximately 50%.

Chen and coworkers[28] utilized gaseous ammonia (NH_3) at temperatures between 500 – 800°C to tailor the GAC. The GAC was heated to the desired temperature in a nitrogen atmosphere prior to treatment with ammonia. The most effective treatment conditions were found to be treatment with ammonia at 700°C for 60 minutes. GAC treated in this manner was able to remove perchlorate for four times longer than could be treated by virgin GAC prior to the initial detection of perchlorate in the column effluent.[27] Specifically, when employing C1 water, this ammonia tailored GAC removed perchlorate to below detection levels for 4000 BV. Ammonia thermal tailored GAC could also be effectively regenerated with either carbon dioxide (CO_2) or ammonia.[29] After reactivation, 90 – 100% of the capacity for perchlorate could be restored to a previously exhausted GAC. This capacity was then able to be restored through at least three cycles of thermal reactivation.

CATIONIC POLYMER PRE-LOADED GAC

Polydiallyldimethylammonium chloride (polyDADMAC) is widely used at water treatment facilities as a coagulant aid; and many polyDADMAC products are approved by the National Sanitation Foundation (NSF) for use in drinking water plants. For these reasons, polyDADMAC was chosen as a potential tailoring agent for GAC to attempt to improve its capacity for perchlorate. The structure of polyDADMAC is shown in Figure 4.

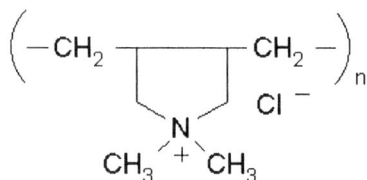

Figure 4. Structure of polyDADMAC.

The quaternary ammonium group carries a positive charge and is located in the ring structure. Each repeating unit has 161.7 atomic mass units (amu), therefore there is one positive charge per every 161.7 amu. The number of subunits (n) can vary greatly. On the low end, polyDADMAC can have a MW of less than 100,000. On the high end, the MW can be greater than 500,000.

PolyDADMAC in 4 molecular weight ranges was obtained from Calgon Chemical and pre-loaded onto SAI GAC. PolyDADMAC with a MW less than 100,000 was considered very low. Between 100,000 and 200,000 MW was classified as low. A molecular weight range between 200,000 and 350,000 was classified as medium and between 400,000 and 500,000 was considered high. Following tailoring (1-day recirculation time), C1 water was passed through the columns of 60×80 mesh ($180~\mu m \times 250~\mu m$) GAC to evaluate the effect of the polyDADMAC loading on perchlorate adsorption. These tests simulated a 20 minute EBCT for 8×30 mesh GAC, and the results are shown in Figure 5.

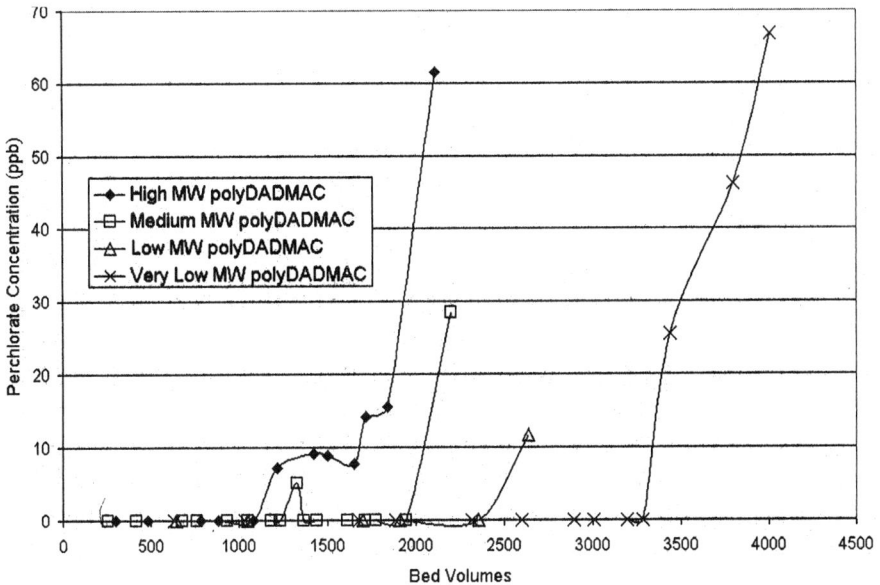

Figure 5. Effect of polyDADMAC pre-loading on the adsorption of perchlorate from C1 water.

As the molecular weight of the polyDADMAC was decreased, the GAC was able to treat more BV before perchlorate breakthrough was observed. For the high MW polyDADMAC, initial perchlorate breakthrough was noted at 1200 BV, approximately the same location observed for virgin GAC. However treatment with the high MW polyDADMAC appeared to increase the overall capacity for perchlorate versus that of the virgin GAC. Medium molecular weight polyDADMAC extended the BV to initial breakthrough to 2000 BV. GAC treated with low MW polyDADMAC showed an initial perchlorate breakthrough point of 2400 BV, double that of the virgin GAC. The very low MW polyDADMAC produced the most favorable results extending breakthrough to 3400 BV. Extending the loading time to 2 or 5 days did not alter the results significantly. With a 2-day pre-loading time, GAC tailored with very low MW polyDADMAC was able to treat 3800 BV to initial breakthrough. Extending the pre-loading period to 5 days resulted in 4000 BV treated prior to initial perchlorate breakthrough.

A limited study into the desorption of polyDADMAC from the GAC was performed. Effluent TOC levels were monitored from the column containing very low MW polyDADMAC. TOC levels in the column effluent dropped to influent levels by 500 BV. The other MW ranges of polyDADMAC were not investigated.

CATIONIC SURFACTANT PRE-LOADED GAC

Cationic Surfactant Loading

It was noted that tailoring with low molecular weight blends of polyDADMAC led to greater perchlorate capacity in GAC than with their high molecular weight counterparts. With this in mind, the authors investigated tailoring with quaternary ammonium monomer species. These alkyl monomers contain the same functional group as polyDADMAC, however due to their small size (molecular weight of a few hundred vs. in the hundreds of thousands for polyDADMAC), the monomers should have access to a large proportion of the GAC pore structure that was not available to the polymer.

The initial mass loading and corresponding milliequivalents (meq) per gram GAC for each surfactant are shown in Table 2. All 7 surfactants were pre-loaded onto the GAC in similar quantities on both a mass basis (0.21 – 0.26 g of surfactant per gram GAC) and on the basis of meq positive charge (0.56 – 0.82 meq/g carbon).[21]

Table 2. Extent of GAC preloading by the various cationic surfactants.

surfactant	MW	chemical formula	Initial loading (g surfactant/g GAC)	Initial meq N / g GAC
DTAB	280.3	$C_{10}H_{21}N(CH_3)_3Br$	0.21	0.75
MTAB	336.4	$C_{14}H_{19}N(CH_3)_3Br$	0.24	0.71
THAB	364.5	$C_7H_{15}N(C_4H_9)_3Br$	0.21	0.58
CTAC	320	$C_{16}H_{33}N(CH_3)_3Cl$	0.26	0.82
T-50	350	$C_{16-18}H_{33-37}N(CH_3)_3Cl$	0.26	0.76
2C-75	450	$(C_{12-16}H_{25-33})_2N(CH_3)_2Cl$	0.25	0.56
CPC	358	$C_{16}H_{33}NC_5H_5Cl$	0.25	0.69

It should be noted that the surfactant loadings shown in Table 2 are initial values only and can not necessarily be used to predict the ability of the tailored GAC to adsorb perchlorate without considering other factors. It was thought that some surfactant would desorb from the tailored activated carbon during the processing of water, resulting in the loss of potential adsorption sites for perchlorate. Desorption is discussed in a later section.

In addition to surfactant leaching from the activated carbon, biodegradability of the surfactants is an important aspect which must be considered when selecting a tailoring agent. Though biodegradability may not influence results greatly in the relatively short RSSCTs, it is important to consider for full-scale applications over a longer period of time. Others showed that in a 10-day biochemical oxygen demand (BOD) test, 91% of DTAB, 59% of MTAB, 35% of CTAC, and 3% of CPC were degradaded.[30] Octadecyltrimethylammonium chloride (which comprises 65% of T-50) and the dialkyldimethylammonium surfactants that comprise 2C-75 showed no degradation in the 10-day test[30]. Tailoring agents with longer alkyl chains (less desorption anticipated) and low biodegradability were expected to perform well as tailoring agents.

Adsorption of Perchlorate onto Surfactant Pre-loaded GAC

To examine the extent to which a given surfactant could increase the GAC's capacity for perchlorate, SAI GAC was pre-tailored with each surfactant, and C1 water was then processed through the column. The increased capacities for perchlorate, resulting from the preloading of cationic surfactants, are shown in Figures 6 and 7. RSSCT tests with DTAB, THAB, MTAB, and

CTAC tailored GAC utilized nearly the same EBCT (simulated 20 – 22 min for 8 × 30 mesh GAC grains) as used previously in the virgin GAC experiment. These results are shown in Figure 6. CTAC-tailored GAC was able remove perchlorate to below detection for 34,000 BV, 30 times longer than for virgin GAC. MTAB tailored GAC was able to remove perchlorate to below detectable limits for 29,000 BV. THAB, though having only seven carbon atoms in its longest chain, performed nearly as well as MTAB, with perchlorate breakthrough observed at 27,500 BV. DTAB was the least effective of the surfactants with initial perchlorate breakthrough observed at 12,000 BV.[21]

In addition to tailoring unused virgin GAC, previously exhausted GAC (relative to perchlorate) was also "post-tailored" to prolong the potential service life of this GAC. Conventional GAC from the Redlands, CA Texas St. Plant that had been saturated at full-scale with perchlorate was crushed and sieved, then post-tailored with MTAB. Following this treatment, the GAC was then able to remove perchlorate for a duration of 15,000 BV of C1 water, as illustrated in Figure 6. In comparison, when MTAB was pre-loaded on fresh GAC, initial perchlorate breakthrough occurred at 28,000 BV. The post-tailored GAC offered about 60% the capacity of GAC that was tailored prior to use. The 15,000 BV to initial breakthrough would represent more than one year of service life at Redlands, indicating that material that was deemed to be of little value still could be quite useful.

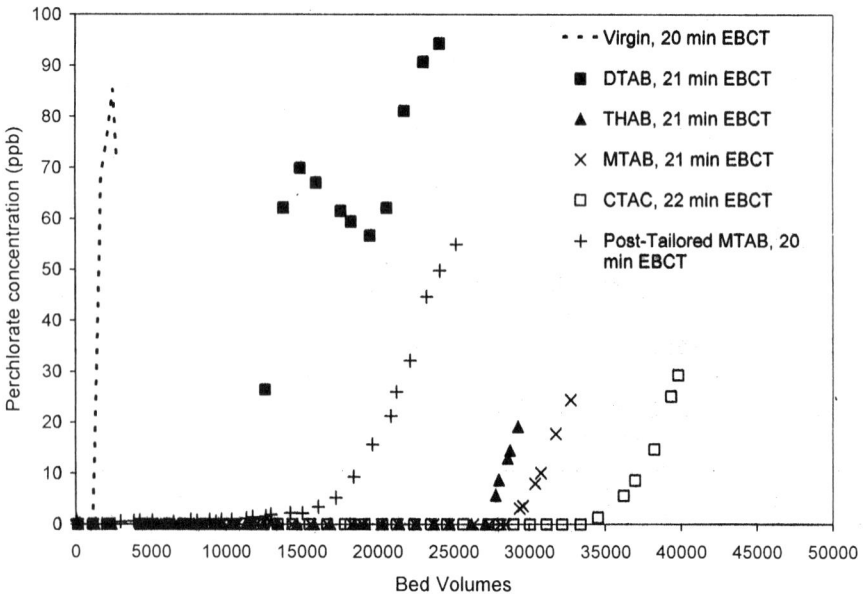

Figure 6. Adsorption of perchlorate from C1 water onto cationic surfactant pre-loaded GAC, 20-22 minute EBCT.

GACs pre-loaded with the cationic surfactants T-50, 2C-75, and CPC were used in RSSCTs with a simulated EBCT of 7-8 minutes (for 8×30 mesh GAC), as shown in Figure 7. This would be equivalent to a 14.3 minute EBCT for 12×40 mesh ($425\mu m \times 1700$ μm), a 10.1 minute EBCT for 20×40 mesh (425 μm $\times 850$ μm) GAC, or an 8.4 minute EBCT for 20×50 mesh (300 μm $\times 850$ μm). Smaller particle size corresponds to greater headloss; and ion exchange media can typically be in the 20×50 mesh range. Lowering the EBCT shortened the laboratory duration of these runs from 4 – 5 weeks to 1 – 2 weeks and if a lower EBCT was effective, processes utilizing cationic surfactant tailored GAC could potentially be designed with smaller vessels, reducing capital costs.

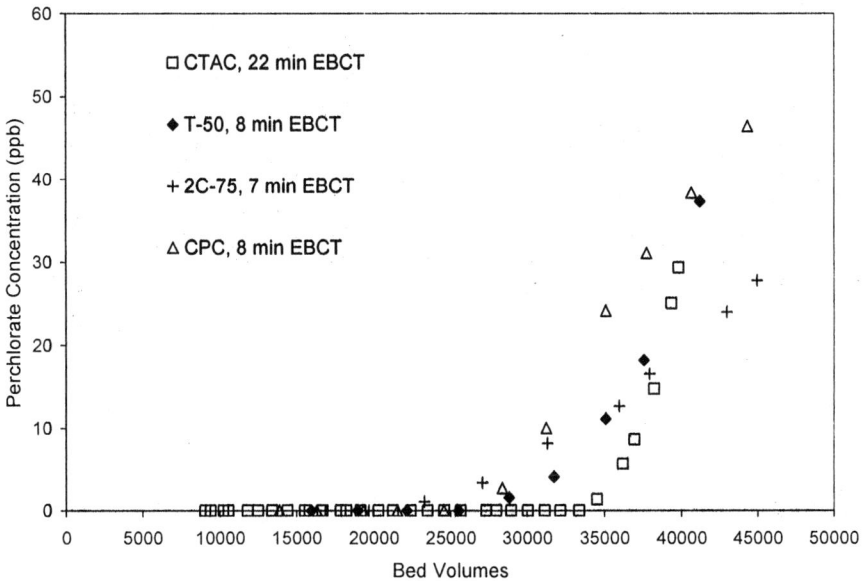

Figure 7. Adsorption of perchlorate from C1 water onto cationic surfactant pre-loaded GAC, 7-8 minute EBCT.

GACs tailored with T-50 and CPC were able to treat 27,000 BV of C1 water prior to the detection of perchlorate in the RSSCT effluent, though the slope of the breakthrough profile for CPC was steeper than that of T-50-tailored GAC. Though GAC tailored with 2C-75 processed only 23,300 BV before initial detection of perchlorate, the breakthrough profile for 2C-75 was less steep (higher overall capacity) than either that of the T-50 or CPC-tailored GAC. The slopes of the breakthrough curves for the RSSCTs that employed a 7-8 minute EBCT are considerably more gradual than for the RSSCTs that employed a 20-22 minute EBCT. CTAC tailored GAC with a 22 minute EBCT is shown for comparison.[21]

The ability of CTAC pre-loaded GAC to remove perchlorate was also tested with C2, M2, and M3 water. The results from these RSSCT tests are shown in Figures 8 and 9. In Figure 8, GAC pre-loaded with CTAC was able to treat 10,000 BV of C2 water prior to perchlorate breakthrough with full breakthrough of 12 ppb observed at 30,000 BV when operated with a simulated 5 minute EBCT (for 8 x 30 mesh GAC). This once again illustrates that greater concentrations of competing anions will lead to a reduced capacity to for perchlorate adsorption.

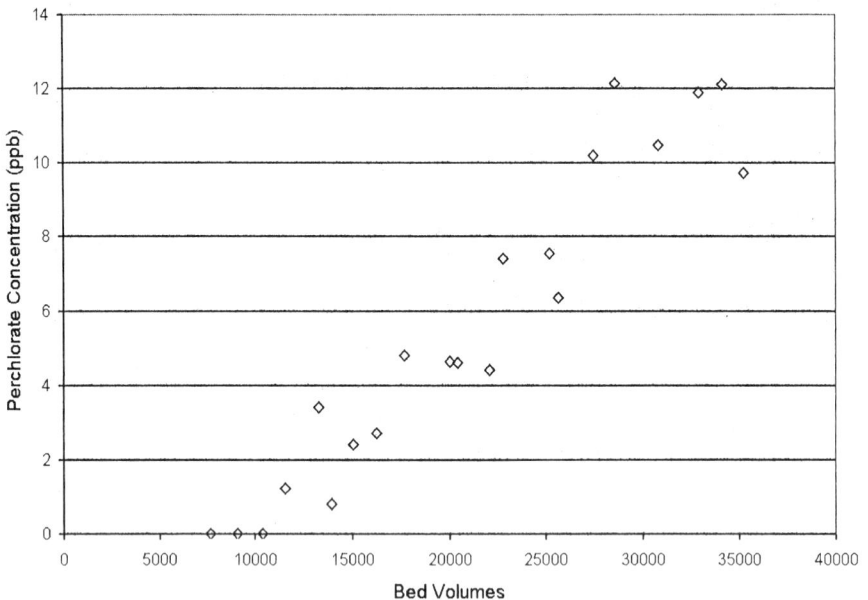

Figure 8. Adsorption of perchlorate from C2 water onto CTAC pre-loaded GAC (background conductivity 1800 μS)

As was the case for the adsorption onto virgin GAC, CTAC-tailored GAC was able to treat much more M2 and M3 water prior to the detection of perchlorate than either of the California groundwaters. As shown in Figure 9, UltraCarb pre-loaded with a CTAC processed about 170,000 BV of M2 water (5.6 ppb ClO_4^-) with a simulated 5-minute EBCT (for 8 × 30 mesh GAC) before perchlorate was detected in the effluent. Perchlorate in the RSSCT effluent was measured at 1 ppb after 210,000 BV and full breakthrough (5.6 ppb) corresponded to 270,000 BV. With M3 water (1.0 ppb ClO_4^-), CTAC-tailored UltraCarb was able to treat 270,000 BV with a simulated 10-minute EBCT before perchlorate was detected in the RSSCT effluent. Between 270,000 BV and the 394,000 BV, the concentration of perchlorate was measured between non-detect and 0.74 ppb.[22]

With the Massachusetts waters, tailored GAC would likely be more attractive economically and operationally compared to virgin GAC as the tailored GAC could be in service for multiple years. For example, with a 5-minute EBCT and 8 x 30 mesh GAC, a single bed processing M2 water with a 5 minute EBCT could operate for two years until 1 ppb perchlorate breakthrough occurred. With a 20-minute EBCT it could operate for 8 years. With 12 x 40 mesh GAC and a 3.5 minute EBCT, it could operate for 1.4 years; and with 20 x 50 mesh GAC and a 2 minute EBCT, it could operate for 0.8 years. Combining two tailored beds in series would further prolong these durations. The service-life estimates do not include biological processes, particulates, and other factors that may impact the GAC bed life at full-scale.

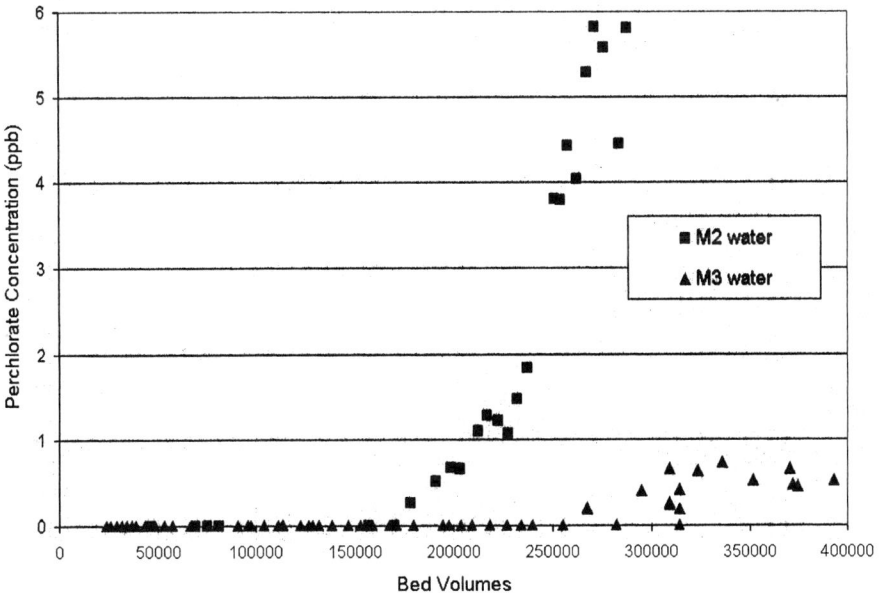

Figure 9. Adsorption of perchlorate from M2 and M3 water onto CTAC-tailored UltraCarb.

DESORPTION OF SURFACTANTS FROM GAC

The perchlorate capacity of GAC could be greatly increased by first pre-loading the carbon with cationic surfactants. However, it was important to determine whether these surfactants could desorb from the activated carbon during subsequent water treatment service. Effluent samples from the RSSCTs that contained GAC tailored with CTAC, T-50, 2C-75, and CPC (treatment of C1 water from Figures 7) were collected and analyzed for the quaternary ammonium group via the two-phase titration method of Tsubouchi et al.[18] This allowed the cationic surfactants investigated to be detected down to approximately 0.1 mg/L (100 ppb). These four surfactants were chosen because they were the most effective from both a performance and cost standpoint. No washing of the GAC was performed between the conclusion of the tailoring process and the start of the RSSCTs in order to appraise the maximum desorption that could possibly occur. This means that some of the surfactant remained lodged between GAC grains; as compared to merely sorbed within GAC grains as would be achieved with full-scale fluidized-bed surfactant loading.

The effluent concentrations of the desorbed cationic surfactants are shown in Figure 10. Of the four surfactants tested, CTAC-tailored GAC exhibited the greatest desorption and dislodging. By comparing the desorption and dislodging in Figure 10 to the initial mass loading shown in Table 2, it was

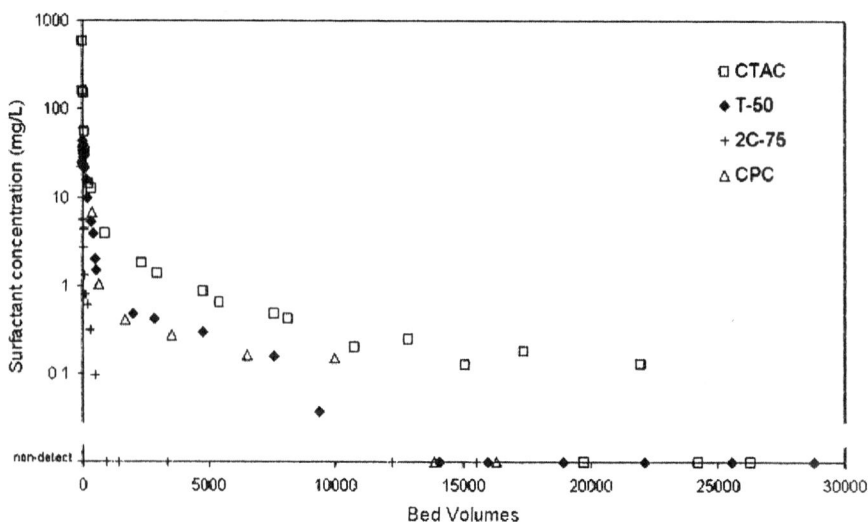

Figure 10. Desorption of cationic surfactants from tailored GAC with C1 water[21] (Reprinted from Water Research with permission from Elsevier).

discerned that 21.2% of the initial CTAC that was either adsorbed or enmeshed within the GAC bed, subsequently leached from the GAC. CTAC was detected in the effluent through 22,000 BV. Less T-50 (6.5%) and CPC (8.1%) leached out of these beds than did CTAC. Both T-50 and CPC were below detection after 14,000 BV. GAC tailored with 2C-75 exhibited the least desorption or dislodging, with only an estimated 0.6% of all 2C-75 leached from the surface. Concentrations of 2C-75 fell to below detectable values at 900 BV. This indicates that the two long alkyl chains of 12-14 carbon atoms each helped to anchor the surfactant more effectively than did surfactants that contained one (longer) alkyl chain.[21]

Unlike the adsorption of perchlorate onto both virgin and tailored GAC, the desorption of cationic surfactants appears to be independent of the groundwater used. Desorption of CTAC from GAC was monitored while processing C1 water (SAI carbon), C2 water (SAI carbon), and M3 water (UltraCarb). The ionic content of these three waters varied greatly. M3 water had a very low conductivity of only 50 μS. The conductivity of C1 water measured 330 μS. The conductivity of C2 water was much higher, measuring 1800 μS. As demonstrated by Figure 11, the desorption of CTAC from GAC follows the same trend regardless of which type of groundwater was used.

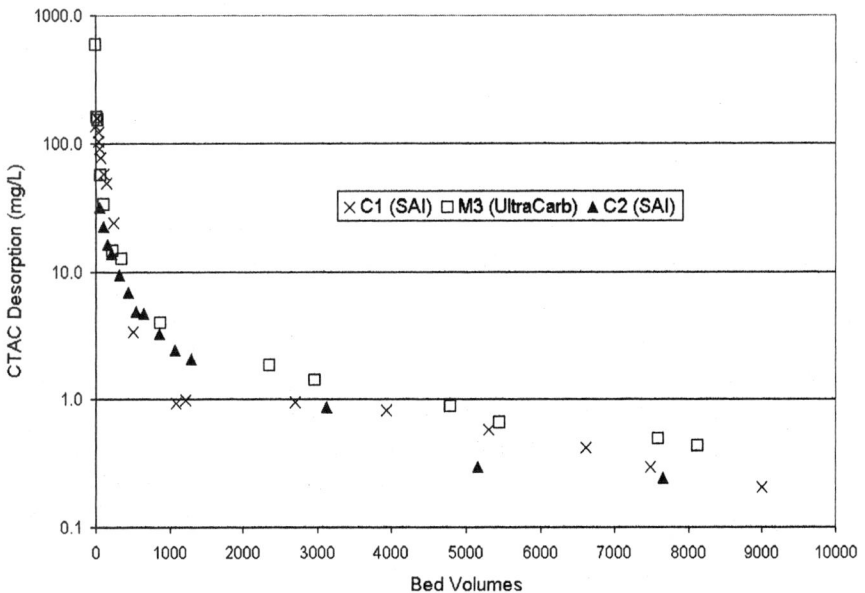

Figure 11. Desorption of CTAC from tailored GAC.

Although the desorption of surfactants from GAC did not appear to differ when natural waters were used, this did not hold true when deionized/distilled water was passed through a column that contained cationic surfactant tailored GAC. The desorption of the surfactant 2C-75 from GAC with deionized/distilled water was compared to the amount that desorbed when C1 water was passed through a RSSCT containing 2C-75 pre-loaded GAC. This is illustrated in Figure 12. With C1 water, it was estimated that only 0.6% of the surfactant that had been loaded onto the GAC was subsequently desorbed or was dislodged. However, an estimated 25.3% of the surfactant desorbed or was dislodged when deionized/distilled water was used. This large increase in desorption may be attributable to the higher solubility of cationic surfactants in distilled water as the addition of salts lowers the solubility of surfactants.[31]

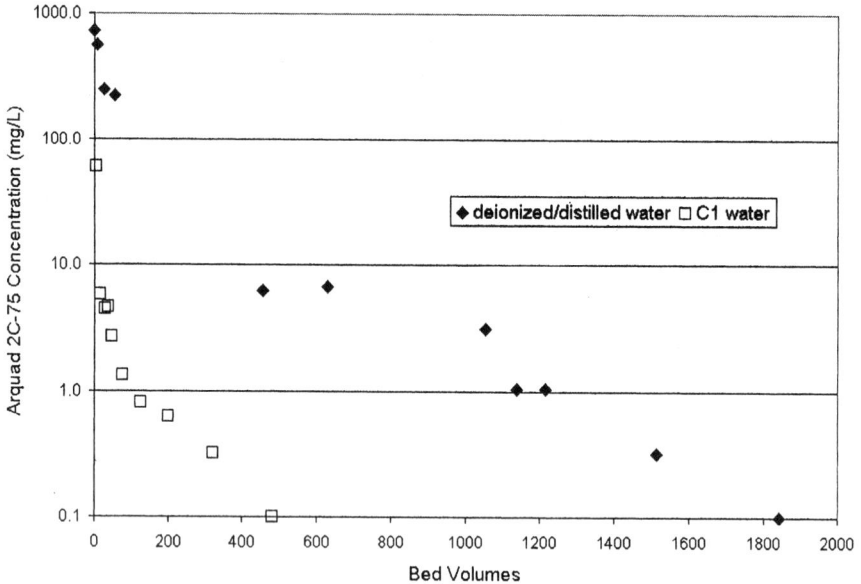

Figure 12. Desorption of Arquad 2C-75 from tailored GAC with deionized/distilled water vs. C1 water.

To test whether quaternary ammonium compounds in solution could interfere with perchlorate analysis via ion chromatography, a 5 ppb ClO_4^- solution was spiked to 1 mg/L CTAC (which amounted to a CTAC to ClO_4^- molar ratio of 62.5 to 1) and placed on a shaker table for one day at 20°C to mix. When this sample was analyzed for perchlorate, the value monitored was within 20-25% of a control which contained the same amount of perchlorate and chloride (but no CTAC). Thus we have determined that for our analytical conditions, the presence of CTAC did not influence whether perchlorate breakthrough could be discerned.

Adsorption of Leached Surfactants onto a Virgin GAC Polishing Bed

To mitigate the leaching of surfactants from the tailored GAC, a virgin GAC polishing bed was added in series to the tailored GAC to adsorb any leached surfactant. The polishing bed had the same dimensions as the tailored bed. CTAC was chosen for this investigation as it exhibited the greatest desorption of the four surfactants studied. As shown in Figure 13, the polishing bed successfully removed the leached surfactant to below detectable concentrations. For comparison, the desorption of CTAC from the tailored bed alone is shown again in Figure 13.[21]

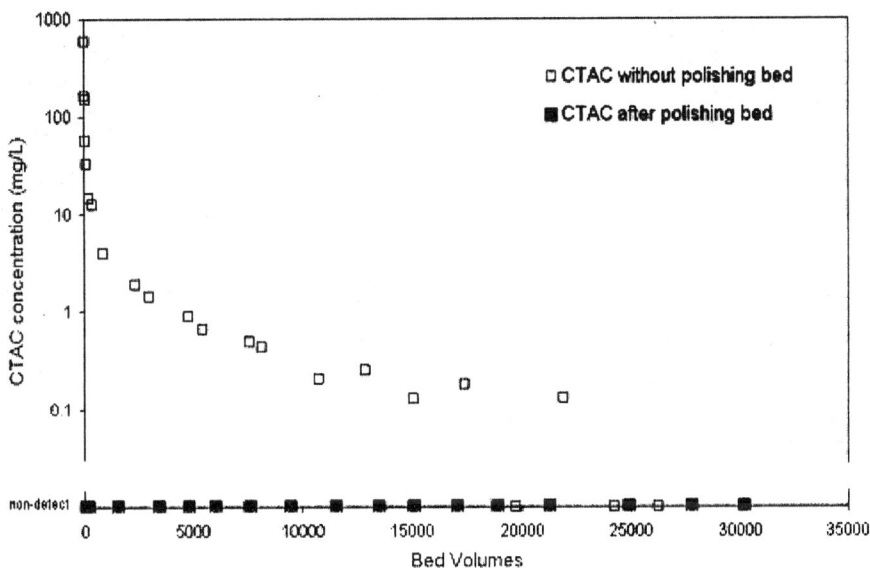

Figure 13. Effectiveness of polishing bed in controlling leaching from CTAC tailored GAC[21] (Reprinted from Water Research with permission from Elsevier).

COADSORPTION OF ORGANICS

GAC pre-loaded with cationic surfactants significantly increased the capacity for perchlorate. To test how this pre-loading affected the coadsorption of organics, the authors compared virgin GAC to CTAC pre-loaded GAC when processing M3 water that contained 1 ppb ClO_4^-, 5.5 or 6.6 ppb RDX, and 0.6 ppb HMX.

As shown in Figure 14, virgin UltraCarb had a good capacity for adsorbing these nitro-organics. RDX was detected at 0.36 ppb after 308,000 BV of M3 water had been processed through the virgin GAC. The effluent RDX increased to 0.84 ppb at 339,000 BV, 1.2 ppb at 371,000 BV, and 1.9 ppb at 404,000 BV. HMX was not detected through 404,000 BV.

While the virgin UltraCarb had good capacity for RDX and HMX, the CTAC tailoring process diminished the capacity for these organics. These results are also shown in Figure 14. RDX (0.4 ppb) was detected in the effluent from the CTAC-tailored UltraCarb at approximately 7,800 BV. The concentration of RDX in the effluent rose to 6 ppb at 90,000 BV. HMX (0.3 ppb) was detected at 116,000 BV and remained at this level until the RSSCT was ceased at 140,000 BV. Although the CTAC-tailored UltraCarb has some capacity for RDX and was able to remove HMX completely for over 100,000 BV, Figure 14 clearly illustrates that CTAC-tailored GAC can not be used as a stand-alone technology to equivalently remove both perchlorate and these nitro-organics from M3 water when they contaminate water at these relative proportions.[22]

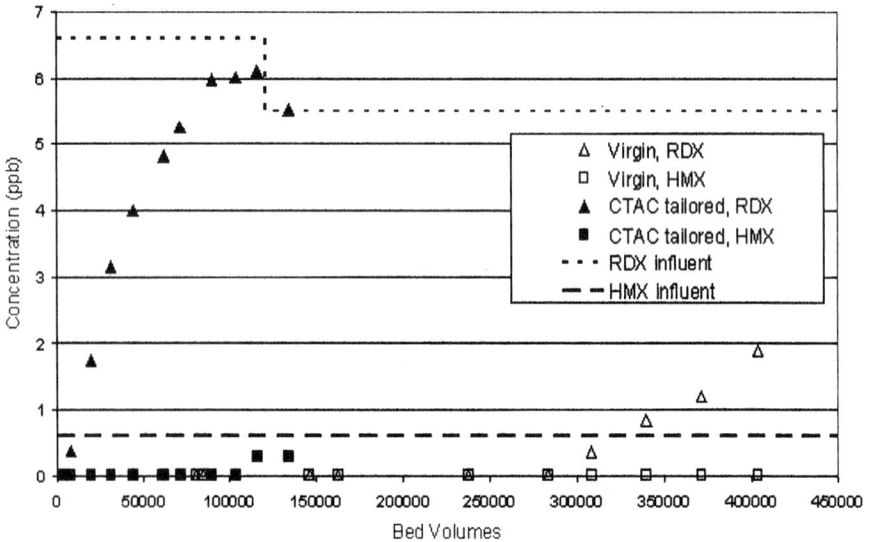

Figure 14. Adsorption of RDX and HMX from M3 water onto virgin and CTAC-tailored UltraCarb[22] (Reprinted from Water Research with permission from Elsevier).

To remove both perchlorate and the nitro-organics from M3 water, the authors surmised that a CTAC-tailored UltraCarb bed could be followed in series with a polishing virgin GAC bed. For the lead bed, the authors post-tailored virgin UltraCarb that had become exhausted with perchlorate with CTAC (this was the spent GAC that remained after the Figure 1, M3 water test). In follow sequence with this CTAC post-tailored GAC, the authors employed a virgin GAC bed. The dimensions of this polishing bed were the same as the lead bed. M3 water was then processed through the two columns.

The effluent concentrations of RDX, HMX and ClO_4^- from the second of the two beds are shown in Figure 15. RDX was detected at 256,000 BV (where BV represents the flow through the first of the two columns), at a concentration of 0.6 ppb. The concentration of RDX exceeded 2 ppb at approximately 315,000 BV. No HMX was observed in any effluent sample through 320,000 BV. Perchlorate was first detected in the effluent from the polishing bed at 195,000 BV. From 195,000 BV to the end of the RSSCT run at 322,000 BV, the concentration of perchlorate fluctuated between non-detect and 0.65 ppb.[22]

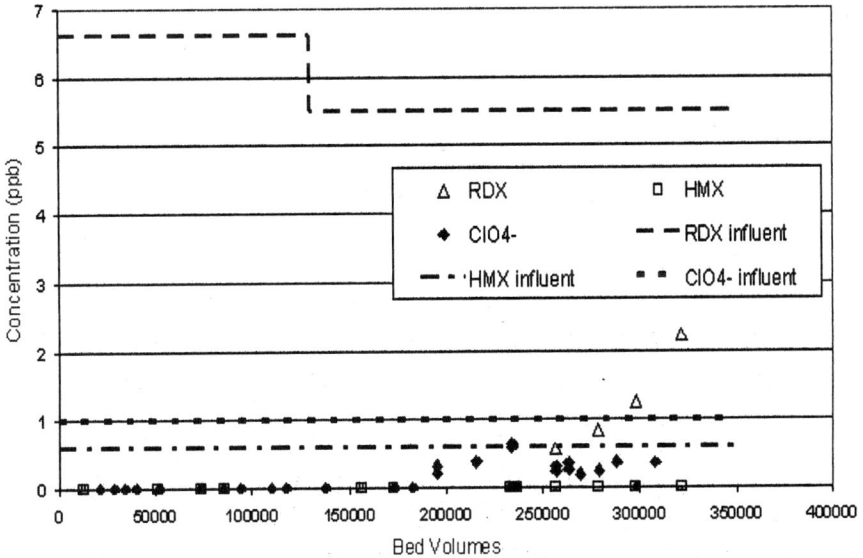

Figure 15. RDX, HMX, and ClO_4^- breakthrough from CTAC post-tailored UltraCarb followed by virgin GAC UltraCarb bed[22] (Reprinted from Water Research with permission from Elsevier).

If it is assumed that the post-tailored UltraCarb contributed very little to RDX adsorption, then the polishing bed was able to remove RDX for 80% as long as the virgin UltraCarb bed had. This loss in capacity is likely attributable to the desorption of CTAC from the post-tailored bed and subsequent adsorption onto the polishing bed. Initially, the GAC was pre-loaded with 0.24 grams of CTAC per gram GAC. However, it was estimated that 22% of CTAC that was initially pre-loaded desorbed or dislodged from the post-tailored GAC and was subsequently adsorbed by the polishing bed. All effluent samples taken from the polishing bed exhibited below-detectable concentrations of CTAC.[22]

In the case of M3 water, it appears that the correct ratio of a CTAC tailored bed to a polishing bed is approximately 1:1 when the goal is to adsorb both perchlorate and RDX/HMX for the same duration under these relative concentrations. CTAC-tailored UltraCarb (Figure 9) was able to remove perchlorate from the M3 water to below detection for 270,000 BV. When coupled with a polishing bed, this should add at least 40,000 BV of perchlorate removal capacity (based on adsorption of perchlorate from M3 water onto virgin GAC as shown in Figure 1), though the increased capacity for perchlorate by the polishing bed will be higher as desorbed CTAC from the first bed adsorbs onto the top of the polishing bed, thus rendering that section of the bed enhanced in its ability to remove perchlorate. Likewise, virgin UltraCarb was shown to remove RDX for 308,000 BV (Figure 14). While the capacity of the polishing bed for nitro-organics will be reduced as CTAC is adsorbed, the results (Figure 15) indicated that the polishing bed's capacity for RDX was not reduced by greater than 20%. Therefore, the polishing bed alone should be able to remove RDX for about 250,000 BV (as shown in Figure 15). As CTAC tailored UltraCarb provided some capacity for RDX through 90,000 BV (as shown in Figure 14), this should push the RDX capacity for the two column system to the 300,000 BV range, consistent with the time frame of perchlorate removal.

THERMAL REACTIVATION OF EXHAUSTED TAILORED GAC

Thermal reactivation is the most commonly used regeneration method for activated carbon. This process provides a means to restore the capacity of the carbon at a cost less than would be incurred by replacing the exhausted GAC. Thermal reactivation is normally done off-site in a rotary kiln, fluidized bed, or multiple hearth furnace. Three main steps are involved in conventional thermal reactivation. The GAC is first dried at 105°C. Following drying, the GAC is heated in an inert gas environment at 650 –

850°C to pyrolyze the adsorbed material. Finally, the charred material is removed via steam, CO_2, or a combination of the two at $700 - 900°C$ to reopen the pores[32].

GAC saturated with perchlorate can be effectively regenerated via a thermal method. Perchlorate salts will decompose between $400 - 500°C$;[33] therefore the high temperatures during thermal reactivation are ideal for eliminating perchlorate. Research by Chen et al.[29] has shown that thermal reactivation with either NH_3 or CO_2 could restore $90 - 100\%$ of the perchlorate capacity in ammonia tailored GAC.

With respect to cationic surfactant pre-loaded GAC that has become saturated with perchlorate, while the conditions of thermal reactivation are ideal for destroying perchlorate, the surfactants that were pre-loaded onto the GAC are also lost in the process as determined by preliminary experiments. Therefore, the GAC must be reloaded with cationic surfactants following reactivation if the GAC is to be used again as an adsorbent for perchlorate.

In lab testing, GAC pre-loaded with 2C-75 (0.25 grams 2C-75 per gram GAC) was used to treat University Park, PA tap water that had been spiked with perchlorate to 800 ppb. This RSSCT column simulated a 20 minute EBCT with full-scale 8 x 30 mesh GAC. In the first adsorption cycle, perchlorate was removed to below detection for 10,000 BV. The GAC was then regenerated in a Thermogravimetric Analyzer (Cahn Instruments, Cerritos, CA) by heating to 800 °C in N_2 and then treating with CO_2 at 800°C for 20 minutes. The GAC was cooled in a nitrogen atmosphere.

Following thermal reactivation, the GAC was reloaded with 2C-75 (0.15 grams per gram GAC) and the GAC was once again used to treat tap water spiked to 800 ppb ClO_4^-. In the second adsorption cycle which simulated a 9 minute EBCT with full-scale 8 x 30 mesh GAC, the 2C-75 pre-loaded GAC was able to treat 7,500 BV prior to perchlorate detection in the effluent. The first and second adsorption cycles are shown in Figure 16. This drop-off in performance is likely attributable to not reopening all the pores during thermal reactivation. In the first adsorption cycle, the GAC was pre-loaded with 0.25 grams 2C-75 per gram GAC and this resulted in 0.3 mmol ClO_4^- removed per mmol of N that was pre-loaded. Following thermal reactivation the GAC could only be pre-loaded with 0.15 grams 2C-75 per gram GAC when using the same loading protocol; reloading resulted in 0.4 mmol ClO_4^- removed per mmol N reloaded. It is projected that the thermal reactivation and reloading process could be further optimized so that more of the GAC's surfactant pre-loading capacity can be restored.

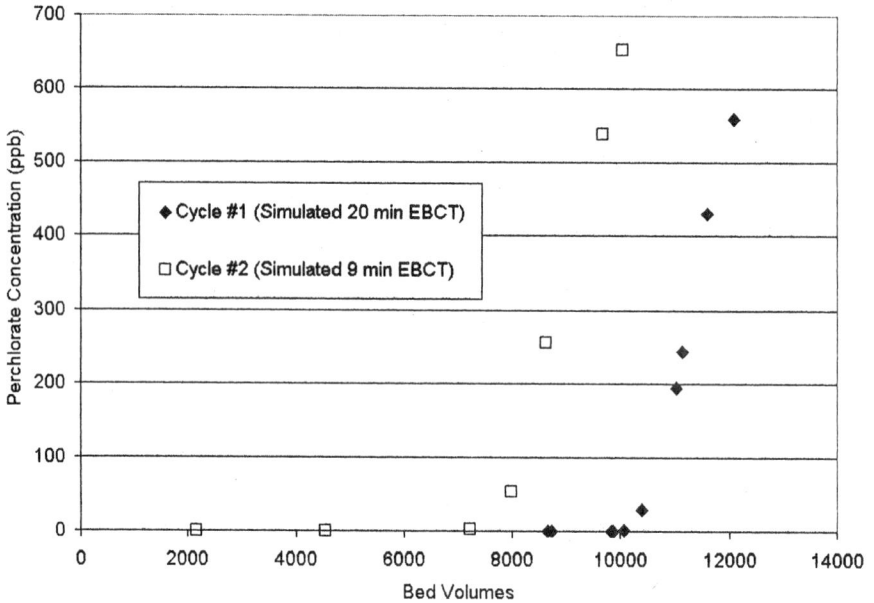

Figure 16. Adsorption of perchlorate from University Park tap water spiked to 800 ppb ClO_4^- onto 2C-75 Surfactant Tailored GAC through Two Cycles. GAC was re-tailored following thermal reactivation of GAC used in Cycle #1. Pre-loading with 0.25 g 2C-75/g GAC in first cycle and 0.15 g 2C-75/g GAC before the second.

CONCLUSIONS

- Virgin GAC viably removed perchlorate from the three Massachusetts waters (M1, M2, and M3) tested. 20,000 – 40,000 BV of these groundwaters could be treated prior to detection of perchlorate in the RSSCT effluent. These waters contained 0.85 – 5.6 ppb ClO_4^- and had conductivities of 50 – 66 μS. At full-scale (8 × 30 mesh GAC) with a 20 minute EBCT, it is projected that this carbon could remove perchlorate to below detectable limits (below 0.25 ppb) for 9-19 months. Virgin GAC was less effective with the California groundwaters, which contained much higher concentrations of competing ions. Perchlorate could be removed to below detectable levels for 1,000 BV with C1 water and for 700 BV with C2 water.

- RSSCT columns operated with virgin GAC and C1 water, designed in accordance with proportional diffusivity, effectively simulated full-scale data (with C1 water) obtained at the Redlands, CA, Texas. St. Plant.

- In addition to the effect of competing ions, the adsorption of perchlorate onto virgin GAC can be greatly influenced by both pH and temperature. pH and temperature both affect the surface charge of the GAC. GAC becomes more positively charged as both the pH and temperature are lowered.

- By first pre-loading the GAC with very low MW polyDADMAC, the BVs treated prior to initial breakthrough could be increased nearly four times when C1 water was processed through the RSSCT column.

- Pre-loading the GAC with quaternary ammonium cationic surfactants significantly increased the GAC's ability to remove perchlorate. GAC pre-loaded with cetyltrimethylammonium chloride (CTAC) was able to treat 34,000 BV of C1 water prior to perchlorate breakthrough when operated to simulate a 20 minute empty bed contact time (EBCT) with 8 × 30 mesh GAC, increasing the perchlorate capacity by 30 times compared to virgin GAC.

- With C1 water, GAC pre-loaded with tallowtrimethylammonium chloride (T-50) and cetylpyridinium chloride (CPC) showed initial perchlorate breakthrough around 27,000 BV when operated to simulate an 8 minute EBCT with 8 × 30 mesh GAC. Dicocodimethylammonium chloride (2C-75) tailored GAC exhibited initial breakthrough at 23,300 BV when operated at a simulated 7 minute empty bed contact time. These GAC's had nearly the same ratio of meq adsorbed perchlorate per meq of quaternary ammonium sites available (7.3 – 10.1%). Thus, it is expected that the BVs to perchlorate breakthrough would have been similar to CTAC had these three RSSCTs (T-50, 2C-75, and CPC) been operated with a simulated EBCT of 20 minutes.

- The use of a virgin GAC polishing bed effectively reduced the desorption of surfactants to below detectable concentrations of 0.1 mg/L

- CTAC pre-loaded GAC was able to treat 170,000 BV of M2 water and 270,000 BV M3 water before perchlorate was detected in the RSSCT effluent. This represents an increase in the time to perchlorate breakthrough by 8.5 times over virgin activated carbon for M2 water and 6.75 times for M3 when operated to simulated 5 and 10 minute EBCTs with full-scale 8 x 30 mesh GAC respectively.

- Cationic surfactant pre-loading greatly decreased the capacity of the GAC for removing the nitro-organic chemicals RDX and HMX from M3 water. Virgin GAC was able to treat more than 300,000 BV before 0.3 ppb RDX was observed in the effluent (with 6 ppb RDX in the influent). However, CTAC pre-loading decreased the time to 0.3 ppb breakthrough to 7,800 BV for RDX. By combining a CTAC pre-loaded "lead" bed with a virgin UltraCarb polishing (follow) bed, perchlorate and RDX could by simultaneously removed for approximately the same duration.

- GAC pre-loaded with 2C-75 (0.25 g/g GAC) was able to treat 10,000 BV of University Park, PA tap water spiked with perchlorate to a concentration of 800 mg/L prior to detection of ClO_4^- in the RSSCT effluent. The exhausted GAC was then thermally reactivated with CO_2 at 800°C and reloaded with 2C-75 (0.15 g/g GAC). In the second adsorption cycle, perchlorate was detected in the RSSCT effluent at 7,500 BV.

ACKNOWLEDGEMENTS

The authors would like to thank the American Water Works Association Research Foundation (AWWARF), the Consortium for Premium Carbon Products from Coal (CPCPC), the Department of Defense, and the City of Redlands, CA for funding that made this research possible. In-kind support was provided by USFilter Westates. We would also like to thank Weifang Chen and Adam Redding for assistance with the laboratory work, Jim Graham of USFilter Westates for assistance with the projections to full-scale systems layouts, and Katherine Weeks and Beth DuPlessie of AMEC Earth & Environmental with assistance in laying out the M1, M2, and M3 studies.

REFERENCES

1. Gullick RW, LeChevallier MW, Barhorst TS. Occurrence of perchlorate in drinking water sources. Journal AWWA 2001; 93(1): 66-77.

2. Li FX, Squartsoff L, Lamm SH. Prevalence of thyroid diseases in Nevada counties with respect to perchlorate in drinking water. Journal of Occupational and Environmental Medicine 2001; 43(7), 630-34.

3. Renner R. Texas lowers perchlorate guidelines. Environmental Science & Technology Online. Retrieved 2001 from: http://pubs.acs.org/subscribe/journals/estag-w/2001/oct /policy/rr_texas.html

4. Massachusetts Department of Environmental Protection. Retrieved 2004 from: http://www.mass.gov/dep/ors/files/perchlor.htm

5. Clifford DA, Tripp AR, Roberts DJ. Nitrate and perchlorate chemistry and treatment. AWWA Inorganic Contaminants Workshop. Reno, NV, 2004.

6. Gu B, Ku YK, Brown GM. Treatment of perchlorate-contaminated groundwater using highly selective, regenerable ion-exchange technology: a pilot-scale demonstration. Remediation 2002; 12(2): 51-68.

7. Min B, Evans PJ, Chu AK, Logan BE. Perchlorate removal in sand and plastic media bioreactors. Water Research 2004; 38(1): 47-60.

8. Xu J, Song Y, Min B, Steinberg L, Logan BE. Microbial degradation of perchlorate: principles and applications. Environmental Engineering Science 2003; 20(5): 405-21.

9. Patrick, JW. *Porosity in carbons*. New York: John Wiley & Sons, 1995.

10. Laughlin, R. "Aqueous phase science of cationic surfactant salts." In *Cationic Surfactants: Physical Chemistry*. Surfactant Science Series 34, D Rubingh and P Holland ed., New York: Marcel Dekker, 1991.

11. Wachinski, AM and JE Etzel. *Environmental Ion Exchange*. New York: Lewis Publishers, 1997.

12. Viera, AR. *The removal of perchlorate from waters using ion-exchange resins*. MS Thesis. University of Nevada, Las Vegas, 2000.

13. Gzara L, Dhabbi M. Removal of chromate anions by micellar-enhanced ultrafiltration using cationic surfactants. Desalination 2001; 137(1-3): 241-50.

14. Morel G, Graciaa A, Lachaise J. Enhanced nitrate ultrafiltration by cationic surfactant. Journal of Membrane Science 1991; 56(1): 1-12.

15. Yoon J, Yoon Y, Amy G, Cho J, Foss D, Kim T. Use of surfactant modified ultrafiltration for perchlorate (ClO_4^-) removal. Water Research 2003; 37(9): 2001-12.

16. Magnuson ML, Urbansky ET, Kelty CA. Determination of perchlorate at trace levels in drinking water by ion-pair extraction with electrospray ionization mass spectrometry. Analytical Chemistry 2000; 72(1): 25-9.

17. Akzo Nobel. Retrieved 2004 from: http://www.surface.akzonobelusa.com/cfm/Properties_Search.cfm

18. Tsubouchi M, Mitsushio H, Yamasaki N. Determination of cationic surfactants by two-phase titration. Analytical Chemistry 1981; 53(12): 1957-9.

19. Na C, Cannon FS, Hagerup B. Perchlorate removal via iron-preloaded GAC and borohydride regeneration. Journal AWWA 2002; 94(11): 90-102.

20. Crittenden JC, Reddy PS, Hand DW, Arora H. *Prediction of GAC performance using rapid small-scale column tests*. American Water Works Association Research Foundation and American Water Works Association. 1989.

21. Parette R, Cannon FS. The removal of perchlorate from groundwater by activated carbon tailored with cationic surfactants. Water Research. Submitted for publication August, 2004.

22. Parette R, Cannon FS, Weeks K. Removing low ppb level perchlorate, RDX, and HMX from groundwater with CTAC-pre-loaded activated carbon. Water Research. Submitted for publication March, 2005.

23. Kosmulski M. *Chemical Properties of Material Surfaces.* Surfactant science series 102. New York: Marcel Dekker, 2001.

24. Menendez JA, Phillips J, Xia B, Radovic LR. On the modification and characterization of chemical surface properties of activated carbon: in search of carbons with stable basic properties. Langmuir 1996; 12(18): 4404-10.

25. Noh JS, Schwarz JA. Estimation of the point of zero charge of simple oxides by mass titration. Journal of Colloid and Interface Science 1989; 130(1): 157-63.

26. Moore, B. *Enhancing Cincinnati, Ohio reactivated granular activated carbon for protecting against organic spills and removing disinfection by product precursors.* Ph.D. Thesis, The Pennsylvania State University, 2000.

27. Chen WF, Cannon FS, Rangel-Mendez JR. Ammonia-tailoring of GAC to enhance perchlorate removal- Part II: Perchlorate adsorption. Carbon 2005; 43: 581-90.

28. Chen WF, Cannon FS, Rangel-Mendez JR. Ammonia-tailoring of GAC to enhance perchlorate removal- Part I: characterization of NH3 thermally tailored GAC. Carbon 2005; 43: 573-80.

29. Chen WF and Cannon FS (2005). Thermal regeneration of ammonia-tailored granular activated carbon exhausted with perchlorate. Carbon 2005; submitted for publication.

30. Masuda F, Machida S, Kanno M. Studies on the biodegradability of some cationic surfactants, Proceedings of VII International Congress on Surface-active Substances, 129-38, 1976.

31. Jonsson, B. *Surfactants and polymers in aqueous solution.* New York: John Wiley and Sons, 1998.

32. Suzuki, M.; D. M. Misic, O. Koyama and K. Kawazoe. Study of thermal regeneration of spent activated carbons: thermogravimetric measurement of various single component organics loaded on activated carbons. Chemical Engineering Science 1978; 33: 271-279.

33. Nikitina, Z. K. and V. Y. Rosolovskii. Catalytic decomposition of sodium chlorate and perchlorate by oxygen compounds of sodium. Russian Journal of Inorganic Chemistry 1995; 40(3): 379-385.

Chapter 16

Titanium Catalyzed Perchlorate Reduction and Applications

Baohua Gu, Peter V. Bonnesen, Frederick V. Sloop, and Gilbert M. Brown

Environmental Sciences and Chemical Sciences Divisions
Oak Ridge National Laboratory, Oak Ridge, TN 37831

INTRODUCTION

Catalyzed chemical and electrochemical techniques to reduce perchlorate (ClO_4^-) to chloride are possible, although their direct use in drinking water treatment has been very limited. This limited use is partly due to the introduction of chemicals; chemical oxidation byproducts; and the slow reduction kinetics of perchlorate as a result of the high activation energy required, as discussed in Chapter 1. Common reductants such as zero-valent iron metals, dithionite, sulfite, and thiosulfate do not react with ClO_4^- ions at any observable rate at ambient temperature and pressure. However, perchlorate has been reported to have been reduced by several transition metal ions such as titanium(III), vanadium(II)/(III), molybdenum(III), and ruthenium(II) [1, 2] and iron(II) at an elevated temperature [3] in aqueous solution. Although electrochemical technologies are well established for other industries (e.g., electroplating of metals and electrolysis of brines), they have not yet found direct application in drinking water treatment, [2] perhaps because of the relatively high capital and energy costs for degrading contaminants at low concentrations.

However, the catalyzed electrochemical technique may well be suited for degrading ClO_4^- in small waste streams such as concentrated brines used for regenerating spent anion-exchange resins. [4] In particular, a novel application of a catalyzed electrochemical technique has recently been reported to directly reduce ClO_4^- sorbed on a spent ion-exchange resin bed, thereby regenerating the resin as illustrated in Figure 1. [5] In this case, a typical fixed resin bed is regenerated by catalyzed electrochemical reduction of ClO_4^- sorbed on the resin in a recirculation electrochemical cell. Reduced Ti(III) serves as a reducing agent and, once it is oxidized as Ti(IV), it is re-reduced at the cathode. This technique essentially eliminates the generation of secondary wastes because a recycle mode is used. It is particularly suited for destroying perchlorate and regenerating those highly selective resins, thereby significantly decreasing both the operational and capital costs by ion exchange. This chapter will cover both the theoretical and practical aspects of titanium-catalyzed electrochemical reduction of perchlorate and its potential applications for water treatment.

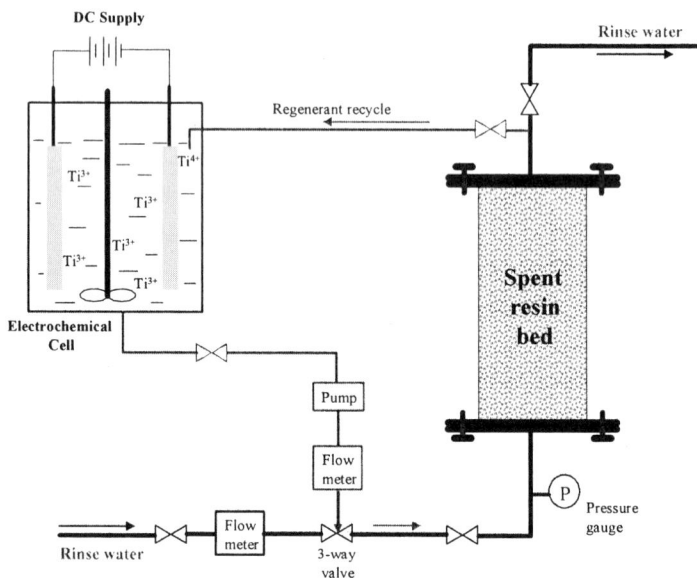

Figure 1. Schematic representation of the catalyzed electrochemical reduction of sorbed perchlorate and the regeneration of the spent resin bed.

PERCHLORATE REDUCTION KINETICS BY TI(III)

Literature reviews indicate that ClO_4^- reduction can occur only when it is accompanied by an oxygen atom transfer reaction[1, 2] because the perchlorate anion has no low-lying electronic energy levels available to accept an electron in a single electron transfer reaction. This explains why only certain transition metal ions such as Ti(III), V(III), V(II), Mo(III), Re(V), and Ru(II) are capable of reducing ClO_4^-, because they can accept an oxygen atom from perchlorate, or the redox reaction is formally an oxygen atom transfer reaction. The reaction may be written as either a one- or a two-electron transfer process depending on the characteristics of the reducing agent.

$$ClO_4^- + e^- \rightarrow ClO_3 + O^{2-} \qquad (1)$$

$$ClO_4^- + 2e^- \rightarrow ClO_3^- + O^{2-} \tag{2}$$

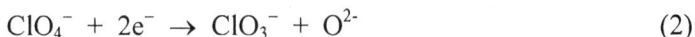

Perchlorate reduction by Ti(III) was first reported by Rothmund in 1909 (cf. [2]). He showed that Ti(III) reduces ClO_4^- to chloride in an acidic aqueous solution at ambient temperature. Duke and Quinney [6] have made a kinetic study of the reduction of perchlorate with Ti(III) in chloride media. They postulated a complicated multi-path reaction mechanism dependent on the formation of an activated complex of perchlorate with a variety of Ti(III) species followed by disproportionation of the complex into an oxotitanium(IV) species and the ClO_3 radical, as shown in Eq. (3).

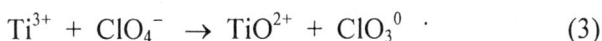

$$Ti^{3+} + ClO_4^- \rightarrow TiO^{2+} + ClO_3^0 \cdot \tag{3}$$

The overall reaction is given as [6]

$$8Ti^{3+} + ClO_4^- + 8H^+ \rightarrow 8Ti^{4+} + Cl^- + 4H_2O \cdot \tag{4}$$

The kinetics of the reduction of ClO_4^- by the aquo-Ti(III) ion was also studied in great detail by Cope et al.,[7] and the reaction rate is expressed as

$$-d[Ti^{3+}]/dt = [k + k'(H^+)][Ti^{3+}][ClO_4^-] \tag{5}$$

where $k = 1.9 \times 10^{-4}$ M^{-1} s^{-1}, and $k' = 1.25 \times 10^{-4}$ M^{-2} s^{-1}. Obviously, the reduction rate depends on both the initial concentrations of Ti(III) and ClO_4^- and the H^+ concentration. Similarly, Liu et al. [8] investigated the reduction kinetics of ClO_4^- by the complex of Ti(III) with the ligand N-(hydroxyethyl)ethylenediamine-N,N',N''-triacetic acid (HEDTA). This Ti(III)-ligand complex is reasonably stable in air and will react with ClO_4^- over a matter of about one hour. Recent studies also indicate that the rate of perchlorate reduction depends not only on the acidity but also on the concentration of ethanol.[9] An ethanolic medium (>92%) increases the rate of the titanous-perchlorate reaction by several orders of magnitude, relative to fully aqueous media. The effectiveness of ethanol in speeding up such reactions is attributed to its fostering the formation of perchlorato complexes in a slightly less polar environment than water. Note that, in all these reactions, the perchlorate is reduced to chloride and Ti(III) is oxidized to the corresponding Ti(IV) species or crystalline TiO_2 precipitates because of its low solubility in water.

The reduction of perchlorate by Ti(III)-oxalate complexes was studied in detail in our laboratory in an attempt to elucidate the reaction kinetics under different experimental conditions. Excess amounts of oxalic acid are used as complexing agents for both reduced Ti(III) and oxidized Ti(IV) species so as to prevent it from precipitation as TiO_2, which is undesirable when the process is used to destroy ClO_4^- and to regenerate spent resins sorbed with ClO_4^- (illustrated in Figure 1). This is because TiO_2 precipitates could potentially lead to pore clogging and therefore blockage of the surface exchange sites of the resin. The choice of oxalic acid as a complexing agent is also advantageous because reactions between Ti(III) and oxalic acid form a monovalent anionic complex, $Ti(C_2O_4)_2^-$, which could be sorbed by the anion-exchange sites and thus displace and reduce sorbed ClO_4^- ions on the resin surfaces.

Laboratory experiments were first performed to determine the reactions and stability of Ti(III)-oxalate complexes. As can be expected, the Ti(III) species are unstable in air and can rapidly oxidize to Ti(IV) or TiO_2 species even in the presence of 0.2 M sulfuric acid (Figure 2a) or both 0.2 M oxalic acid and sulfuric acid (Figure 2b). The UV-visible absorbance at ~500 nm (in H_2SO_4) and at 400 nm (in oxalic acid) decreased consistently over time with a concomitant increase in UV absorption at ~290 nm due to the formation of Ti(IV) species. The complexation between Ti(III) and oxalic acid is evidenced by the spectral shift from ~500 nm in H_2SO_4 to 400 nm with the addition of oxalic acid (Figure 2b). The oxidation of Ti(III)-oxalate complexes appeared to follow the zero-order rate kinetics as evidenced by the time-dependent decrease of the UV-visible absorbance at 400 nm (Figure 3).

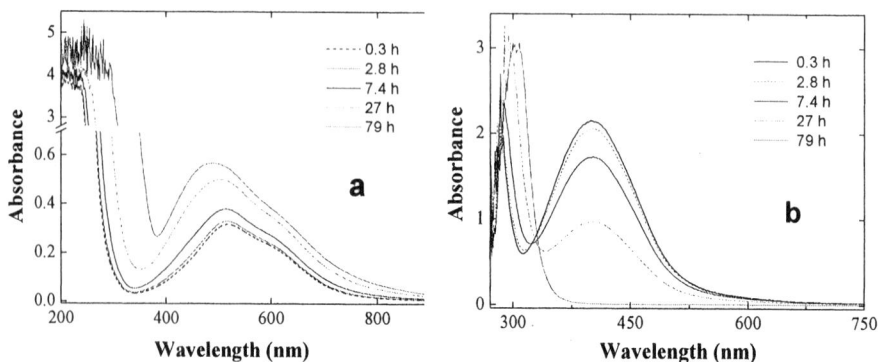

Figure 2. UV-visible spectral changes with exposure time to atmospheric oxygen: (a) 50 mM Ti(III) in 0.2 M sulfuric acid and (b) 5 mM Ti(III) in 0.2 M oxalic acid and 0.2 M sulfuric acid.

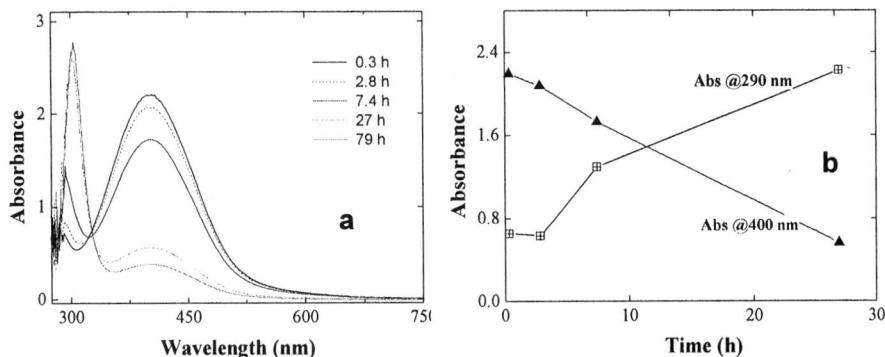

Figure 3. Oxidation kinetics of Ti(III) (5 mM) in 0.2 M oxalic acid. The absorbance peak of the Ti(III)-oxalate complex at 400 nm decreased consistently with the exposure time in air.

Accordingly, reactions between ClO_4^- and Ti(III)-oxalate complexes were then performed under continuous N_2 purging, in which the Ti(III)-oxalate complexes are stable. The reaction rate with Ti(III) (0.01 M) was monitored by following the changes in absorbance at 400 nm. Because a large excess amount of ClO_4^- (0.1 to 0.5 M) was added, and its concentration remained relatively unchanged compared with that of Ti(III), results indicate that the reduction of ClO_4^- in homogeneous solution by Ti(III)-oxalate complexes followed the pseudo–first-order rate kinetics, i.e.,

$$Rate = k[Ti(III)] \qquad (6)$$

A plot of $\ln(C/C_0)$ of Ti(III) and the reaction time yielded a straight line at different ClO_4^- concentrations (Figure 4). The calculated rate constants were 2.3×10^{-5} and 9.6×10^{-5} sec^{-1} in 0.1 and 0.5 M ClO_4^- solutions, and the corresponding half-lives were 8.5 and 2 h, respectively. If second order rate constants are calculated, the average of these values (1.9×10^{-4} and 2.3×10^{-4} $M^{-1}s^{-1}$) is in close agreement with the acid independent term obtained by Cope, et al.[7] for aquo Ti(III). Therefore, the reaction between Ti(III) and ClO_4^- is relatively slow and too sluggish for practical use directly for drinking water treatment, as stated earlier. Additionally, the groundwater or drinking water usually contains dissolved oxygen and nitrate, which also react with Ti(III) and therefore a further decrease in the rate of reduction of ClO_4^- is expected. On the other hand, the addition of ethanol greatly increases the reduction rate of ClO_4^- and, as shown in Figure 4, the reaction half-life decreases to ~50 min in the presence of 50% ethanol. These results are consistent with previous studies by Bishop and Evans[10] and Earley et al.[9]

The reaction between Ti(III)-oxalate complexes and ClO_4^- could be written as

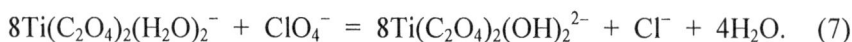

$$8Ti(C_2O_4)_2(H_2O)_2^- + ClO_4^- = 8Ti(C_2O_4)_2(OH)_2^{2-} + Cl^- + 4H_2O. \quad (7)$$

The reaction byproducts are Cl^- anions and soluble Ti(IV)-oxalate complexes, and no TiO_2 precipitates were observed.

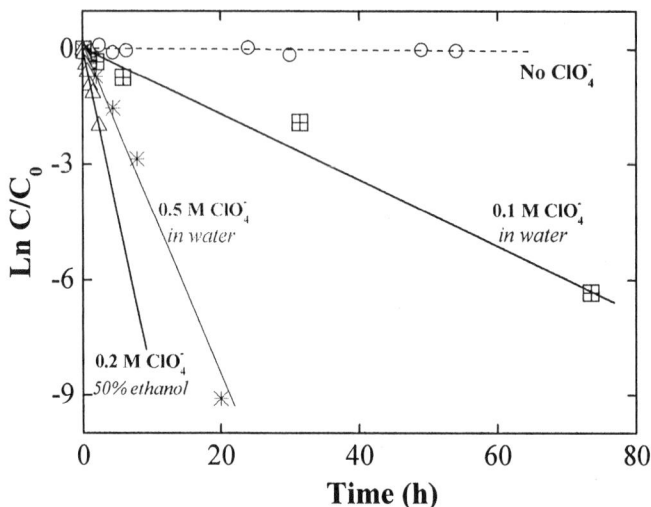

Figure 4. Reaction kinetics between Ti(III) (0.01 M) and ClO_4^- at varying concentrations in the presence of 0.2 M oxalic acid under nitrogen atmosphere.

ELECTROCHEMICAL REDUCTION OF PERCHLORATE AND RESIN REGENERATION

Although it is impractical to use Ti(III) directly for drinking water treatment, the chemistry of perchlorate reduction by Ti(III) offers an attractive process for resin regeneration because perchlorate is strongly and selectively sorbed to many Type-I strong-base anion exchange resins such as the bifunctional anion exchange resins and the monofunctional Purolite A-520E resin.[11,12] As a result, regeneration of these resins loaded with ClO_4^- is a challenging task, and a new paradigm of resin regeneration is needed for large-scale field applications. Tetrachloroferrate ($FeCl_4^-$) ions formed in a mixture of ferric chloride ($FeCl_3$) and hydrochloric acid (HCl) can effectively displace sorbed ClO_4^- and thus regenerate spent resins loaded with ClO_4^-.[13] However, this process does not reduce or destroy ClO_4^-, and an additional destruction step is necessary to completely degrade ClO_4^-, as described in detail in Chapter 10.[3] From a thermodynamic standpoint, perchlorate is a highly oxidized species, and a regeneration method based on the reduction of ClO_4^- to a lower oxidation state of chlorine is a logical approach. The batch kinetic experiments (Figure 4) also suggest that this process is technically feasible.

Because the reduction of each mole of ClO_4^- requires 8 moles of Ti(III) (Eq. 7), a relatively large quantity of Ti(III) solution is needed for the regeneration process. To minimize the amount of Ti(III) needed for regeneration and the secondary waste production, an electrochemical cell is designed to reduce Ti(IV) to Ti(III), which can then be reused for the reduction of ClO_4^-, as illustrated in Figure 1. A platinum wire is used as a cathode for the reduction of Ti(IV) to Ti(III), and a graphite rod is used as the anode. Again, oxalic acid is used to bind Ti(III) in the presence of a large excess of oxalate, creating the net anionic complex $[Ti(C_2O_4)_2(H_2O)_2]^-$. This anionic Ti(III)-oxalate complex is electrostatically attracted to the positively charged exchange sites of the resin, thereby facilitating the reduction of sorbed ClO_4^- ions. The oxidized Ti(IV) is also complexed by oxalate, preventing the premature precipitation of this species in the pores of the resin. Note that oxalic acid has a long history of use in the removal of scale and metal ion precipitates, and its high solubility in water makes any excess easy to remove from the resin bed.

Laboratory experiments were subsequently conducted for the regeneration of a spent resin column that had previously been loaded with ClO_4^-. The column (10×22 mm) was packed with the experimental bifunctional anion-exchange resin (Purolite D-3696). A solution consisting of 0.02 M Ti(IV), 0.5 M oxalic acid, and 0.6 M sulfuric acid was first reduced electrochemically at the platinum electrode (at -0.40 V vs the reference electrode), and a nearly complete reduction of Ti(IV) to Ti(III) took place within ~2 h. The reduced Ti(III)-oxalate solution was then circulated through the resin column for the reduction of ClO_4^-, and oxidized Ti(IV) or Ti(IV)-oxalate species were continuously reduced back to Ti(III) in the electrochemical cell (Figure 1). The Ti(III)-oxalate solution was circulated through the resin column for ~65 h to ensure a nearly complete reduction of ClO_4^- sorbed on the resin bed because of a relatively slow reaction between ClO_4^- and Ti(III) on the basis of the batch kinetic studies (with a half-life on the order of 2–8 h) (Figure 4). The reaction could be even slower because Ti(IV)-oxalate complexes must be able to diffuse effectively within the resin beads (or the internal porous structure) to react with adsorbed ClO_4^- anions.

At the completion of the reduction phase, the resin column was washed with 1 M HCl, followed by washing with purified water to remove excess amounts of oxalic acid.

To evaluate the performance of the regenerated resin, the column was subjected to ClO_4^- breakthrough tests at a constant flow rate of 30 mL/min, and the breakthrough curves of ClO_4^- before and after regeneration were compared, as described previously.[11,13] The initial influent ClO_4^- concentration was 10 mg/L, prepared in a background solution consisting of 3 mM $NaHCO_3$, 1 mM $CaCl_2$, 0.5 mM $MgCl_2$, 0.5 mM Na_2SO_4, and 0.5 mM KNO_3. Results (Figure 5) indicate that, although the breakthrough of ClO_{4-} did not exactly match the initial breakthrough curve (before regeneration), the regenerated resin column performed reasonably well in removing ClO_4^-. Approximately 70% of the anion-exchange sites on the resin bed were recovered on the basis of the amounts of ClO_4^- sorbed before and after the electrochemical regeneration of the column. When this experiment was duplicated using a separately packed column, similar results were obtained (Figure 5).

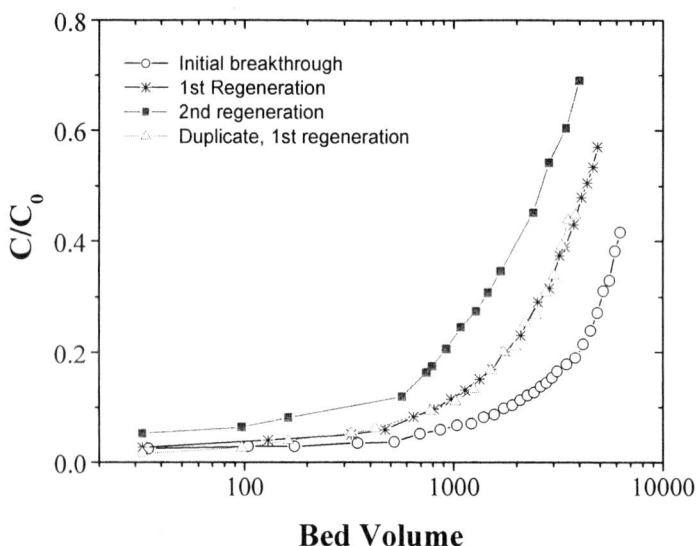

Figure 5. Comparisons of the breakthrough curves of ClO_4^- in a freshly packed resin column (initial breakthrough) with those spent resin columns after electrochemical regeneration. The initial influent ClO_4^- concentration was 10 mg/L.

However, when the column was subjected to the second regeneration cycle, the reactivity of the regenerated resin appeared to deteriorate rapidly to ~50% of the original ClO_4^- removal capacity (Figure 5), suggesting that either sorbed ClO_4^- ions were not completely reduced, or the exchange sites on resin surfaces were blocked. Indeed, grayish mineral precipitates were observed within the resin column, and this observation could be attributed to the formation of crystalline TiO_2 or hydrous $Ti(OH)_4$ phases. A contributing factor to the difficulty in regenerating the resin may also be the sorption of negatively charged Ti(IV) and Ti(III)-oxalate complexes and oxalate anion itself on the resin, decreasing its sorption capacity for ClO_4^-.

Additional electrochemical regeneration studies were performed using the same resin, although the breakthrough experiments were performed at a higher initial ClO_4^- concentration (50 mg/L) to accelerate the run time. Additionally, a solution of 0.6 M oxalic acid and 0.5 M H_2SO_4 was used first to condition the resin column before or after an input of Ti-oxalate complexing solution because of a concern that direct contact between the Ti-containing solution and oxygenated water containing no complexing agent in the column may cause the precipitation of TiO_2. Results indicate, however, that this procedure made little difference with respect to increasing the sorption capacity of the regenerated resin (Figure 6). The breakthrough of

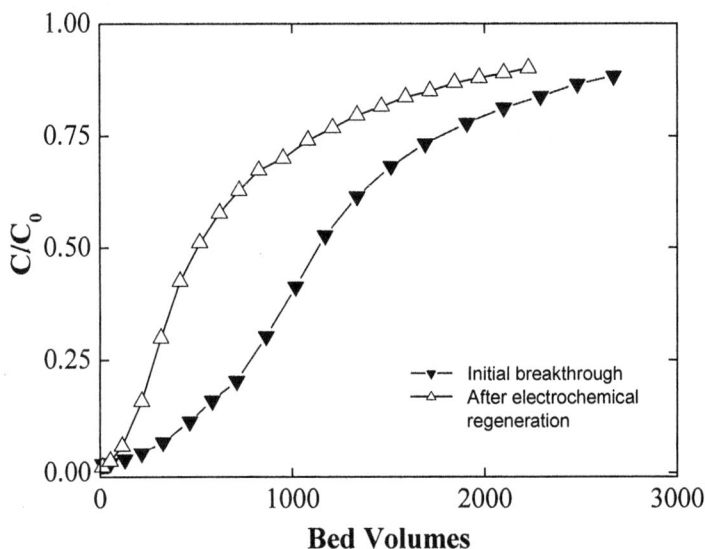

Figure 6. Comparisons of the breakthrough curves of ClO_4^- in a freshly packed resin column (initial breakthrough) with the spent resin column after electrochemical regeneration. The initial influent ClO_4^- concentration was 50 mg/L.

ClO_4^- occurred much earlier in the electrochemically-regenerated column than in the freshly prepared resin column (the initial breakthrough); only ~60% of the original exchange sites on the resin were recovered after regeneration. Apparently, about 40% of adsorbed ClO_4^- or the titanium-oxalate complexes were not completely washed off the resin beads, explaining an overall reduced adsorption capacity of the resin after regeneration.

Nevertheless, these experiments (Figures 5 and 6) provided the proof-of-principle that catalyzed electrochemical reduction of ClO_4^- and thus the regeneration of the spent resin column are technically feasible. Further studies are warranted in order to find the optimal conditions and chemical reagents that could make catalyzed electrochemical regeneration more efficient and complete for large-scale applications. For example, ascorbic acid, citric acid, and hydroxyethylethylenediaminetriacetic acid (HEDTA) could also be used as compexing agents for Ti(III) or Ti(IV) species. The complex of Ti(III)-ascorbate is expected to be positively charged so that it should have less affinity for being adsorbed by the anion-exchange resin and should be readily desorbed. A drawback is, however, that the positively charged Ti(III)- or Ti(IV)-ascorbate complexes may be unable to diffuse into the micropores of the resin beads to react with sorbed ClO_4^- on the resin. The use of Ti(III) complexed by HEDTA may be an attractive alternative since the complex is neutral. It will be neither attracted nor repelled by the positive charges of the interior of the micropores of the resin.

Additionally, methylrhenium dioxide (CH_3ReO_2) recently has been shown to react with ClO_4^- and other halite or perhalite ions to abstract an oxygen atom.[14-16] The kinetics reported in the literature indicate this species reacts with ClO_4^- about four orders of magnitude faster than the Ti(III)-complexes such as Ti(III)-oxalate. The net reaction is a reduction of ClO_4^- to chloride where the first step involving reduction of ClO_4^- to chlorate is rate limiting.

$$4CH_3ReO_2 + ClO_4^- \rightarrow 4CH_3ReO_3 + Cl^- \cdot \qquad (8)$$

Similarly, the conversion of the oxidized CH_3ReO_3 to a reduced CH_3ReO_2 species could be accomplished by an electrochemical process that would allow the amount of rhenium to be held to a minimum. Both CH_3ReO_3 and CH_3ReO_2 are neutral species and thus will be neither attracted nor repelled by positively charged exchange sites on the resin, although the effectiveness of CH_3ReO_2 for reducing sorbed ClO_4^- ions on the resin is yet to be investigated.

STUDIES WITH ^{36}Cl RADIO-LABELED PERCHLORATE

To provide insight into the incomplete regeneration of resin by Ti(III) reduction, we designed a series of batch experiments to measure the release of ClO$_4^-$ by Ti(III) reduction of perchlorate sorbed on the resin. These experiments utilized radioactive ^{36}Cl perchlorate so that Cl$^-$ anions produced by the reduction could be measured independently of the total Cl$^-$ concentration in solution. This radioactive isotope of chlorine emits an energetic beta as it decays, and it was used for determining the fate of perchlorate reduced by Ti(III) in the resin. The ^{36}Cl-labeled perchlorate was synthesized in our laboratory by oxidation of Na^{36}Cl (Isotope Products Laboratories, Burbank, CA) by an electrochemical technique using a platinum gauze electrode, based upon previously published procedures.[17-19] The oxidation to perchlorate was incomplete, as is usually the case, and resulted in a roughly 90:10 ratio of ^{36}ClO$_4^-$ to ^{36}ClO$_3^-$ (based on precipitating the perchlorate using tetraphenylarsonium chloride and counting the residual ^{36}Cl activity due to chlorate.) The absence of chloride was verified by the absence of any silver chloride precipitate being formed upon silver nitrate treatment. This mixed perchorate/chlorate tracer solution was used without further attempt to enrich it in perchlorate.

Perchlorate sorption on the Purolite D-3696 resin was first studied to obtain an initial K$_d$ value. The experiment was performed as follows: 200 mg D-3696 resin (dry weight equivalent) was weighed into 30-mL Teflon FEP centrifuge tubes in duplicate. To each tube was added 5 mL of a solution containing cold (carrier) NaClO$_4$ at a concentration of 5 mM, and sodium chlorate at a concentration of 0.6 mM (to maintain a roughly 90:10 ^{36}ClO$_4^-$ to ^{36}ClO$_3^-$ ratio as was in the tracer). This solution was spiked with the tracer solution described earlier so that about 1.4% of the total Cl content was ^{36}Cl, and the total ^{36}Cl activity in the solution was 0.094 µCi/mL. The tubes were agitated for 24 h on an orbital shaker at laboratory ambient temperature (22±1°C). After equilibration, the solution was completely withdrawn from the resin using a fine-tipped pipette, filtered through a 0.2 micron PTFE filter, and an aliquot (1 mL) was added to 10 mL of Packard Ultima Gold XR cocktail for liquid scintillation analysis of ^{36}Cl activity. Calculation of the sorption and distribution coefficients (K$_d$) of the ^{36}Cl from the solution before and after treatment with the resin gave an average K$_d$ of ~5600 mL/g, which appeared to be much lower than those reported for perchlorate sorption (Chapter 10).[11,20] The results indicate a >99.5% uptake of the ^{36}Cl-labeled chlorate tracer from the solution, or a loading of ~4.7% of the available anion-exchange sites on the resin (assuming an anion-exchange capacity of 2.5 mmol/g dry weight of resin).

Following the sorption experiment, the resin loaded with ClO_4^- and ClO_3^- (to ~4.7% capacity) was subjected to stripping (or regeneration) by the Ti(III) reduction technique. The samples were flushed with argon for at least 10 min using an argon needle inlet/needle outlet and septum on the tube. Five mL of amber-colored Ti(III)-oxalate regenerant solution (0.5 M oxalate, 0.02 M Ti, 0.6 M sulfuric acid) was added into the tube under argon, using a gas-tight syringe. The tubes were agitated again for 24 h, and the solutions remained amber colored during the contact. The tubes were then opened, and the entire solution was removed from the resin using a fine-tipped pipette and filtered into a polypropylene tube using a 0.2-μm PTFE Acrodisc syringe filter. Duplicate 0.5-mL aliquots were added to 10 mL of Ultima Gold XR cocktail for liquid scintillation counting of ^{36}Cl radioactivity. Note that negligible quench was observed from mixing 0.5 mL of sample solution and 10 mL of cocktail for ^{36}Cl analysis.

These stripping procedures were repeated twice after removal of the first Ti(III) stripping solution. A new septum was placed back onto each tube, which was flushed with argon for at least 10 min, and a fresh 5-mL aliquot of Ti(III)-oxalate solution was again added under argon purging to reduce sorbed ClO_4^-.

Results from the ^{36}Cl radioassay indicate that only ~21.8% of the ClO_4^- on average was stripped off the resin on the first treatment cycle with Ti(III). Less was removed from the subsequent two treatments (4.7%, and 2.4%, respectively, for the second and third stripping experiments). For a total of a 72-h period, about 29% of the ClO_4^- was removed on average from the resin. Agreement between replicate resin samples was very good (with ±10%). These observations suggest that either the sorbed ClO_4^- anions were not completely reduced by Ti(III) reduction, or the $^{36}Cl^-$ anions (formed by the reduction of ClO_4^-, Eq. 7) remained on the resin to maintain a charge balance. In other words, an anion always needs to be at the cationic exchange site. As the chloro-oxyanions are reduced to chloride, they are going to stay in the resin unless they are replaced by some other anions such as Ti(III)-oxalate complex, oxalate, and sulfate anions.

To further evaluate whether reduced $^{36}Cl^-$ anions remain sorbed on the resin, the resins were agitated with 5 mL of 1 M NaCl for 1 h so that many of the $^{36}Cl^-$ anions would be displaced by mass reaction. However, the amount of ^{36}Cl radioactivity observed in solution was very low, and only ~0.67% ^{36}Cl radioactivity was recovered. The recovery was much less than would have been expected from isotopic dilution of $^{36}Cl^-$ with non-radioactive Cl^- anions, suggesting that most of the reduced $^{36}Cl^-$ should have been replaced by other anions. The two samples were further treated with 5 mL of 1-M HCl solution after removing the NaCl solution. The resins were shaken for an additional 17 days, and the HCl solution was then removed from the resin, filtered, and analyzed for ^{36}Cl radioactivity. However, only an additional 0.6% ^{36}Cl radioactivity was recovered. Therefore, the total recovery of ^{36}Cl was only about 30.2% by a sequential reduction with Ti(III)-oxalate and stripping with NaCl and HCl. These results thus suggest that the sorbed ClO_4^- anions on the resin were not completely reduced by the Ti(III)-oxalate reduction, which may be partially attributed to their slow reaction kinetics and/or inaccessibility of Ti(III)-oxalate to all the exchange sites to react with the sorbed perchlorate.

These resin samples (after HCl stripping) were further evaluated for sorption of unlabeled ClO_4^- and ^{36}Cl-radio-labeled $^{36}ClO_4^-$, since less than 4% of the anion exchange sites were sorbed with perchlorate on the resin (after sequential reduction and stripping with Ti(III), NaCl and HCl). Solutions of 5 mL of 5-mM unlabeled ClO_4^- were first equilibrated with the resins in the same manner as described earlier. The solutions were removed, filtered, and counted as before. Very little ^{36}Cl radioactivity (0.06%) was recovered by displacement with unlabeled ClO_4^- in the solution because of the large excess of available anion exchange sites. Approximately 8–9% of the anion exchange sites were sorbed with perchlorate at this point.

Additional perchlorate uptake experiments (with the same samples used earlier) were performed using the the 90:10 $^{36}ClO_4^-$ to $^{36}ClO_3^-$ tracer solution in an attempt to reevaluate the K_d for perchlorate sorption. However, one of the duplicate samples was equilibrated with the same radio-labeled perchlorate solution (5 mL of 5-mM perchlorate/0.6-mM chlorate; 0.094 µCi per mL ^{36}Cl), and the other sample was equilibrated with a solution containing five times more concentrated perchlorate (i.e., 5 mL of 25-mM perchlorate/3-mM chlorate; 0.47 µCi per mL of ^{36}Cl). The solutions were again agitated on the orbital shaker for 24 hours, filtered, and analyzed as above. Results indicate that perchlorate uptake (by ^{36}Cl assay) was close to, but less than, the initial uptake, with a K_d value of 3400 mL/g in the first duplicate sample. The uptake of $^{36}ClO_4^-$ was 99.3% of the ^{36}Cl (or an additional perchlorate loading of 4.65% on the resin). The total ClO_4^- loading increased to ~13% of the anion exchange sites on the resin. The estimated K_d value for the second duplicate sample (with 25-mM perchlorate) was much lower, about 170 mL/g, because of a high concentration of ClO_4^- in solution and a high loading of perchlorate on the resin (nearly 30% of total anion exchange capacity).

Results of the batch sorption and stripping experiment using ^{36}Cl-labeled ClO_4^- are consistent with the column electrochemical regeneration studies, in which we were unable to achieve 100% regeneration of the resin by the Ti(III)-oxalate reduction technique. Evidently, there was an incomplete recovery of the ^{36}Cl, even after a dilution exchange with first excess Cl$^-$ anions and then perchlorate. It is still unclear if the incomplete regeneration can be partially attributed to the slow reaction kinetics and/or inaccessibility of Ti(III)-oxalate to the exchange sites to react with sorbed ClO_4^-, or a subsequent clogging of micro-pores within the resin beads (due to the precipitation of TiO_2).

SUMMARY AND IMPLICATIONS

This work provides a proof-of-principle demonstration that Ti(III)-catalyzed electrochemical techniques could potentially be used for reduction of ClO_4^- in small waste streams, such as the regeneration of selective anion-exchange resins that are loaded with ClO_4^-. The technique may not be directly applied for the treatment of large volumes of ClO_4^--contaminated water at relatively low concentrations because of its slow reaction kinetics and the use of chemical reagents. Further studies are needed to optimize the reaction conditions in order to achieve a complete reduction of ClO_4^- and the regeneration of spent resin beds. Alternative complexing and reducing agents may be used to enhance the reaction completeness of sorbed ClO_4^- in the resin and to overcome potential clogging of micropores within the resin beads resulting from the precipitation of TiO_2.

ACKNOWLEDGMENTS

This project was partially supported by Corporate Environmental Safety and Health of Lockheed Martin Corporation and U.S. Department of Energy (DOE). Oak Ridge National Laboratory is managed by UT-Battelle, LLC, under contract DE-AC05-00OR22725 with DOE.

REFERENCES

1. Taube, H. In *Mechanistic aspects of inorganic reactions*. Rorabacher, D. B., R. Endicott, J. F., Eds., ACS Symposium Series No. 198, 1982, pp 151.

2. Urbansky, E. T. Perchlorate chemistry: Implications for analysis and remediation. *Bioremed. J.*, 1998; 2:81–95.

3. Gu, B., Dong, W., Brown, G. M., Cole, D. R. Complete degradation of perchlorate in ferric chloride and hydrochloric acid under controlled temperature and pressure. *Environ. Sci. Technol.*, 2003; 37:2291–2295.

4. Calgon Carbon Corporation. Removal of perchlorate and other contaminants from groundwater at JPL, Report to Jet Propulsion Labortory, Pasadena, CA. NAS7.000218. 1999.

5. Gu, B., Brown, G. Regeneration of anion exchange resins by catalyzed electrochemical reduction. 2002:US Patent 6,358,396.

6. Duke, F. R., Quinney, P. R. The kinetics of reduction of perchlorate ion by Ti(III) in dilute solution. *J. Am. Chem. Soc.*, 1954; *76*:3800–3803.

7. Cope, V. W., Miller, R. G., Fraser, R. T. M. Titanium III ion as a reductant in electron-transfer reactions. *J. Chem. Soc. A.*, 1967; *2*:301–306.

8. Liu, B. Y., Wagner, P. A., Earley, J. E. S. Reduction of perchlorate ion by (N-(hydroxyethyl)ethylene-diaminetriacetato)aquotitanium(III). *Inorg. Chem.*, 1984; *23*:3418–3420.

9. Earley, J. E. S., Tofan, D. C., Amadei, G. A. In *Perchlorate in the Environment*; Urbansky, E. T., Ed.; Kluwer/Plenum: New York, 2000, pp 89-98.

10. Bishop, E., Evans, N. The analytical kinetics of the titanium(III)-perchlorate reaction - homogeneous reaction kinetics. *Talanta*, 1970; *17*:1125-1130.

11. Gu, B., Brown, G. M., Alexandratos, S. D., Ober, R., Dale, J. A., Plant, S. In *Perchlorate in the Environment*; Urbansky, E. T., Ed.; Kluwer/Plenum: New York, 2000, pp 165–176.

12. Gu, B., Ku, Y., Brown, G. Sorption and desorption of perchlorate and U(VI) by strong-base anion-exchange resins. *Environ. Sci. Technol.*, 2005; *39*:901–907.

13. Gu, B., Brown, G. M., Maya, L., Lance, M. J., Moyer, B. A. Regeneration of perchlorate (ClO_4^-)-loaded anion exchange resins by novel tetrachloroferrate ($FeCl_4^-$) displacement technique. *Environ. Sci. Technol.*, 2001; *35*:3363-3368.

14. Abu-Omar, M. M., Appelman, E. H., Espenson, J. H. Oxygen-transfer reactions of methylrhenium oxides. *Inorg. Chem.*, 1996; *35*:7751–7757.

15. Abu-Omar, M. M., Espenson, J. H. Facile Abstraction of Successive Oxygen Atoms from Perchlorate Ions by Methylrhenium Dioxide. *Inorg. Chem.*, 1995; *34*:6239–6240.

16. Abu-Omar, M. M., McPherson, L. D., Arias, J., Bereau, V. M. Clean and efficient catalytic reduction of perchlorate. *Angew. Chem. Int. Ed. Engl.*, 2000; *39*:4310–4313.

17. Deane-Drummond, C. E. Rapid method for the preparation of $^{36}ClO_3^-$ from $^{36}Cl^-$ by electrolysis. *Int. J. Appl. Radiat. Isot.*, 1981; *32*:758–759.

18. Ruiz-Cristin, J., Chodera, A. J., Briskin, D. P. A modified method for the production of $^{36}ClO_3^-$ for use in plant nitrate transport studies. *Anal. Biochem.*, 1989; *182*:146–150.

19. Tromballa, H. W. Preparation and determination of ^{36}Cl-labelled chloride, chlorate, and perchlorate. *Radiochem. Radioanal. Let.*, 1970; *5*:285–292.

20. Gu, B., Ku, Y., Brown, G. M. Treatment of perchlorate-contaminated water using highly-selective, regenerable ion-exchange technology: a pilot-scale demonstration. *Remediation*, 2002; *12*:51–68.

Chapter 17

Membrane and Other Treatment Technologies – Pros and Cons

Ping Zhou, Gilbert M. Brown, and Baohua Gu

Environmental and Chemical Sciences Divisions, Oak Ridge National Laboratory, Oak Ridge, TN 37831

INTRODUCTION

Although ion exchange and biological reduction are currently among the most widely used treatment technologies for perchlorate contamination,[1,2] other treatment technologies, such as membrane filtration and electrodialysis, may be useful under certain environmental conditions. This chapter provides a brief overview of membrane-based technologies and their underlying principles, mechanisms by which they remove perchlorate from contaminated water, factors that affect their performance, and technical limitations. Comparisons between membrane and other treatment technologies are reviewed and tabulated as general guidance for selecting the best water treatment technology under site-specific conditions.

Membrane-based technologies may be divided into three categories: i) reverse osmosis (RO), ii) nanofiltration (NF) and ultrafiltration (UF), and iii) electrodialysis (ED). These technologies have been utilized for decades in water deionization and desalinization.[3] In membrane-based processes, water is forced through a semi-permeable membrane while dissolved salts are unable to pass through the membrane (Figure 1). Membrane permeability to different anions and cations may be adjusted by engineering the membrane pore size and selectivity. In general, two streams are produced in the membrane process: the filtrate or permeate, which is nearly deionized water, and the brine concentrate or rejectate, which contains all rejected salts or dissolved materials including perchlorate. The membrane is the key component of the technology and is usually made of organic polymers. As such, membrane fouling and scaling by alkaline earth and transition metals or metal oxides, dissolved organic matter, silicates and suspended solids sometimes present a problem, depending on their concentrations in the feed water. While the use of membrane technology for treatment of perchlorate-contaminated water is still under investigation, it is generally anticipated that this process could be effective and ideal for point-of-use treatment, such as home water purification and some small-scale water treatment systems.[4, 5]

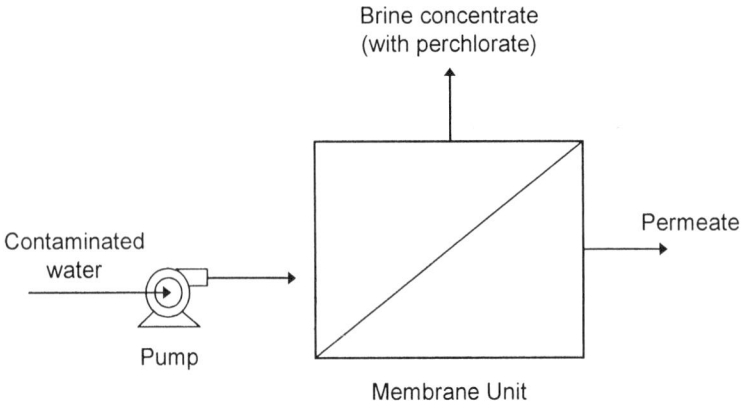

Figure 1. Schematic representation of a membrane filtration system for treatment of perchlorate-contaminated water.

REVERSE OSMOSIS

Reverse osmosis (RO) is a physical separation technique based on the principle of osmosis. In this technology, high pressure (the driving force) is applied on one side of the semi-permeable membrane to reverse the osmosis process, forcing water molecules to pass through the membrane out of the contaminated water. The semi-permeable nature of the membrane allows water to pass more readily than dissolved ions such as perchlorate and most other salts, leaving them behind. Since RO membranes can have pore diameters as small as 3 Å, they can remove constituents down to less than 0.1 nm in size[6] and are capable of rejecting most salts and organics that have a molecular weight cut-off (MWCO) greater than 150 to 250 daltons. The removal of ions, such as ClO_4^-, by RO may be enhanced because charged species are more likely to be rejected by membranes than uncharged species.

RO can concentrate all dissolved and suspended solids, and the permeate usually contains a very low concentration of dissolved solids. Two streams of water are generated in the RO system: treated water and concentrated or salty water containing perchlorate and other salts or minerals that usually requires further treatment prior to disposal. Most RO technologies use crossflow filtration, which provides a level of self-cleaning and has been widely used for the desalination of seawater.

The use of RO systems for treating perchlorate-contaminated water has been limited to bench-scale studies.[2,7] In 1999, a treatability study was conducted by U.S. Filter using a groundwater sample from the Jet Propulsion Laboratory (JPL) site in Pasadena, California.[7] The treatability study used two stages of RO, with the concentrated perchlorate "rejectate" stream from the first stage serving as the influent to the second stage. RO was also investigated as a polishing step for the first stage permeate (the dilute effluent). The first stage received 13 gal/ft^2/day (gfd) of raw groundwater with 800 μg/L ClO_4^- at a pressure of 190 to 200 psi. Two different types of RO membranes were tested in the first stage (only): cellulose acetate (CA) and thin film composite (TFC). With an 80% recovery rate (i.e., recovering 80% of the influent as permeate), the TFC membrane reduced the ClO_4^- concentration to 12 to 16 μg/L, considerably better than the CA membrane which reduced the ClO_4^- concentration to 650 μg/L. Concentrated rejectate (3600 μg/L ClO_4^-) from the first stage TFC system was fed to the second stage RO, where it passed through a TFC seawater membrane at 230 psi and a membrane flux rate of 10 gfd. (Rejectate from the CA system was also passed through a second stage, but analytical problems produced anomalous results that will not be discussed further here.) The second stage permeate contained only 17 to 18 μg/L ClO_4^- at a recovery rate of 50%. To ensure that RO could reduce the permeate ClO_4^- concentration to below the detection limit of 4 μg/L, permeate with 27 to 38 μg/L ClO_4^- (higher concentrations than actually observed in first stage permeate) was fed to a TFC RO unit at a membrane flux of 13 gfd and at 200 psi. An 80% recovery and the nondetect goal were achieved at the feed ClO_4^- concentration of 27 μg/L. Higher influent perchlorate concentrations and higher recovery rates produced effluents (permeates from the polishing unit) with up to 4.3 μg/L ClO_4^-.

Similar bench-scale tests at Clarkson University also indicate that RO was capable of removing ClO_4^- from contaminated water.[2] At an influent ClO_4^- concentration of 125 to 2,000 μg/L and an operating pressure of 20 to 90 psi, the effluent ClO_4^- concentration was found to range from 5 to 80 μg/L, giving a perchlorate removal rate greater than 95%.

The foregoing bench-scale studies demonstrate that RO treatment of perchlorate-contaminated water is technically feasible. In fact, the use of RO to remove ClO_4^- from drinking water may be practical and economical for small-scale point-of-use water treatment systems. For example, a combination of RO and granulated activated carbon (GAC) filtration was tested at Clarkson University for removal of ClO_4^- from tap water.[17] About 99% of the perchlorate was removed by the combined treatment system, of which more than 95% was attributable to the RO unit. A cost analysis suggests that use of an RO/GAC system for perchlorate removal from household drinking water may be competitive with bottled water. Disposal of RO rejectate to a household septic system is also acceptable because the anaerobic environment in the septic tank could potentially bioreduce perchlorate into harmless Cl⁻ ions, although experimental verification of this supposition is warranted. Some commercial RO units for perchlorate removal are, in fact, available for household use. Watts Water Technologies (North Andover, Massachusetts) has developed a residential RO filtration system sold under the name of Watts Pure Water, and the system has been certified for household use by the California Department of Health Services.

However, a major limitation of RO is that the presence of dissolved organic matter and silicates in water can foul, and thus plug or damage, the membrane. The elevated silicon concentration in the rejectate also can cause membrane scaling problems and shorten the lifetime of the membrane.[7] Because RO removes most of the salts in water, post-treatment – including remineralization (e.g., applications of sodium chloride and bicarbonate) – may be required to make water palatable. Obviously, large-scale application of RO for water treatment (at high flow rates) is cost-prohibitive and compares unfavorably with costs for aboveground treatment technologies such as ion exchange and fluidized bioreactors.

NANOFILTRATION AND ULTRAFILTRATION

Unlike RO, which nonselectively rejects most other salts and requires high operating pressures, nanofiltration (NF) and ultrafiltration (UF) use membranes that preferentially separate different fluids or ions. Both NF and UF do not filter as finely as RO. The semi-permeable membrane pores are typically much larger than those in reverse osmosis filters and are thus capable of concentrating salts and other constituents having MWCOs greater than 1000 daltons. UF membrane pores are usually larger than NF membrane pores and are capable of concentrating constituents with MWCOs above 10,000 daltons. Therefore, NF and UF require less energy than RO at a given flow rate. Due to the large pore size of NF and UF membranes, their effectiveness depends, among other factors, on the charge of the particles or ions being rejected. Particles or ions with greater size and charge are more likely to be rejected than ones that are smaller or uncharged.

The two most important membrane properties in determining perchlorate separation efficiency are electrostatic exclusion and steric rejection.[8] Steric rejection is primarily determined by the membrane's MWCO, whereas electrostatic exclusion is influenced by factors such as solution pH, ionic concentrations, and surface functional groups on the membrane. Amphoteric membranes may be positively charged when solution pH is less than the membrane's point of zero charge and negatively charged at higher pH, thereby influencing the separation process. Because ClO_4^- ions are negatively charged, a negatively charged membrane is desirable for the separation of perchlorate because the passage of ClO_4^- ions through the porous membrane may be inhibited by electrostatic exclusion.[8] However, compression of the membrane's electrical double layer (e.g., at high solution ionic strength) could increase the passage of ClO_4^- through the membrane and result in lower perchlorate rejection efficiency.[8,9] Thus, perchlorate removal efficiency may be enhanced by high solution pH and low solution ionic strength, especially for NF membranes.

Yoon et al.[9] reported a detailed bench-scale study of perchlorate rejection using thin-film composite NF. The study was performed using a variety of experimental conditions with an influent ClO_4^- concentration equal to 100 µg/L. Perchlorate rejection was determined as a function of pH (ranging from 4 to 10), solution conductivity (30, 60, and 115 mS/m), background electrolyte composition (KCl, K_2SO_4, or $CaCl_2$), and NF membrane type with MWCO from 200 to 1,200 daltons. Their results indicate that perchlorate rejection followed the order $CaCl_2 \ll K_2SO_4 < KCl$ at constant pH and conductivity conditions for all the membranes tested. Calcium is a divalent cation, which induces greater double layer suppression than monovalent ions. Therefore, the $CaCl_2$ solution resulted in the lowest perchlorate rejection, which ranged from 3 to 8% as pH changed from 4 to 10 at a fixed conductivity of 30 mS/m using an 8,000 MWCO membrane. On the other hand, perchlorate rejection was found to range from 16 to 42% when K_2SO_4 and KCl were used under the same experimental conditions. As expected, the membrane with a lower MWCO (400) showed a much higher perchlorate rejection rate (52 to 75%) when KCl was used. Perchlorate rejection also decreased significantly with increasing solution conductivity again reflecting higher electrical double layer suppression.

Similarly, small-scale tests performed at the Metropolitan Water District of Southern California suggest that NF can be used to remove perchlorate in contaminated water.[10] Using a thin-film composite membrane (MWCO of 300), perchlorate removal efficiency was found to vary between 87% and 95%, depending on the influent perchlorate concentration and whether or not brine solution was recycled. Brine recycling resulted in higher perchlorate levels in the permeate due to increased salt concentrations in the influent.

General conclusions can be drawn from these studies.[9,10] When a dilute ionic solution contacts a fixed-charge membrane, the passage of ions possessing the same charge as the membrane (co-ions) is inhibited. In a single component system, target ions (ClO_4^- in this case) could be excluded from negatively charged membranes with pores large with respect to the size of the ion, but this rejection capability is reduced by electrical double layer compression as the total ionic strength increases. Therefore, RO is preferred to NF or UF when contaminated groundwater with high ionic strength (i.e., high conductivity or total dissolved solids) is treated.

Because of the relatively large membrane pore size used in UF, perchlorate rejection rate is usually low,[9] as described earlier. Yoon et al.[11] investigated the use of large surfactant molecules to modify the membrane and thus to enhance perchlorate rejection and removal efficiency. The presence or sorption of surfactant molecules is expected to affect electrostatic interaction (charge repulsion) and steric (size-related) exclusion mechanisms between a negatively charged porous membrane and anionic perchlorate. Both cationic (tetradecyl trimethyl ammonium bromide, TTAB) and anionic (sodium dodecyl sulfate, SDS) surfactants were investigated, and perchlorate rejection rate was determined in the presence or absence of these surfactants in the influent water at concentrations less than their respective critical micelle concentrations (CMC). Their results indicate that SDS had little impact on perchlorate rejection because it neither influenced membrane pore size nor the surface charge. On the other hand, perchlorate rejection modified with TTAB was greater than expected, largely due to the steric/size exclusion as a result of membrane pore-size reduction (due to the sorption of TTAB). However, a significant flux-decrease occurred when cationic surfactants were used to modify the membrane.

ELECTRODIALYSIS

Electrodialysis (ED) also uses semi-permeable membranes for removing perchlorate or other ions from contaminated water. However, unlike RO, NF and UF that require pumps to force water through the membrane against the osmotic gradient, ED utilizes electrical current as the main force driving separation. The contaminated water is exposed to an electric field as it passes through the membrane, making anions such as perchlorate travel to the anode while cations travel to the cathode (Figure 2). By interposing between the electrodes a series of membranes that are selectively permeable to different ions, the electric field separates perchlorate from the stream of contaminated water.

Selective membranes are used to prevent migrating cations and anions from reaching the electrodes, where unwanted electrode reactions may occur. For example, ion-exchange membranes can be employed to concentrate process streams, separate ionic species from non-ionic species, or recover or extract charged solutes from waste streams. In commercial ED systems, many hundred cell pairs can be stacked between one set of electrodes, thus improving energy efficiency and negating the electrode effects to a great extent. A membrane configuration with alternating cation-selective and anion-selective membranes is usually used. A cation-selective membrane (cation-exchange membrane) permits only positive ions to migrate through it, whereas an anion-selective membrane (anion-exchange membrane) permits passage of only negatively charged ions. Electrodes are placed at each end of the membrane stack to induce a well distributed electrical field of direct current across the membrane stack. Spacers are emplaced between membranes to provide room for the liquid process stream to flow along the membrane surfaces.

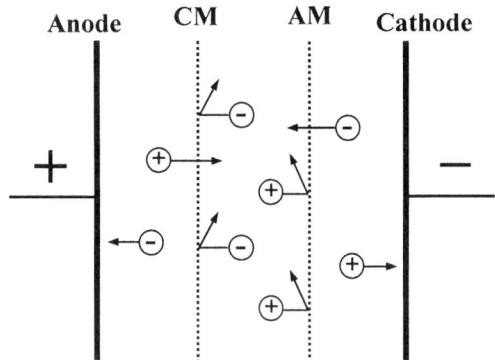

Figure 2. Schematic representation of electrodialysis with ion-selective membranes inserted. CM= cation-exchange membrane, and AM = anion-exchange membrane.

In practice, a process called electrodialysis reversal (EDR) is useful because it includes continuous self-cleaning of the membrane stack. Periodic reversal of the DC polarity allows systems to run at higher recovery rates. Polarity reversal causes the concentrating and diluting flow streams to switch after each cycle. Any fouling or scaling constituents are removed when the process reverses, sending fresh product water through compartments previously filled with concentrated waste streams. Therefore, EDR systems can operate with higher salt concentrations, or they can realize a higher concentration in rejectate streams, minimizing waste volume.

A pilot-scale test using EDR to remove perchlorate was reported by Carollo Engineers in Salt Lake County, Utah.[3] This treatment system was designed to determine the perchlorate rejection characteristics of EDR and to develop cost estimates for a range of EDR staging and implementation options. The EDR system consisted of a four-hydraulic stage EDR membrane stacked with two electric stages and operated at about 7.4 gpm with a recovery rate of 70% treated water. Under these operating conditions, the EDR process was found to effectively remove perchlorate (as well as other salts) from the contaminated water, although the removal efficiency was not sufficiently high for drinking water treatment, particularly at low influent perchlorate concentrations. At an influent ClO_4^- concentration of 130 μg/L, the removal efficiency was only 71% using a two-stage EDR configuration. Use of a four-stage EDR increased the efficiency to 94%. Removal efficiencies for sulfate (84% to 97%) and chloride (81% to 96%) were higher than for perchlorate. Additionally, the treatment efficiency appeared to depend on the influent ClO_4^- concentration. Effluent ClO_4^- concentration ranged from 11 to 17 μg/L at influent concentrations of 15 to 130 μg/L.

ED also has been utilized as a pre-treatment to purify drinking water, particularly in geographic areas where water resources are scarce; and ED has been used to concentrate seawater for a more economical production of table salt. However, ED is applicable only to separation of charged particles or ions. The separation media must be able to transfer the electrical current with relatively low resistance. ED is almost exclusively carried out in liquids, because charged particles or ions must be mobile. Like RO, NF and UF, the ED process generates two liquid streams: treated or nearly deionized water and salty wastewater.

Although RO, NF, UF and ED have fundamental similarities, the driving force and membranes employed are different, leading to different perchlorate removal efficiencies. Fouling of membranes is a major concern, especially with RO and NF technologies in which high pressure is usually used to force water through the membranes. As stated earlier, high silica concentrations in contaminated water can make RO and NF ineffective in removing perchlorate ions because silica causes scaling on the membrane and thus significantly decreases permeate flux. In contrast, silica does not cause fouling in ED/EDR because silica commonly carries no charge and therefore has no significant impact on the membrane. Nonetheless, reduced effectiveness of ED for perchlorate removal may result from membrane fouling and low selectivity of the semi-permeable membrane for perchlorate. ED/EDR equipment costs are relatively high, on the order of $3.6 to $7.8 million for a two-stage EDR treatment system, and $6.2 to $11.8 million for a four-stage treatment system.[3]

It is also worth mentioning that, although capacitive deionization (CDI) is not a membrane-based technology, its operation is analogous to that of electrodialysis. CDI uses carbon aerogel electrodes to remove ions from aqueous solutions.[12,13] Because of their porous nature and high specific surface area (400 to 1000 m^2/g), carbon electrodes provide a high electrical conductivity and high sorption capacity. It has been reported that CDI systems could potentially use 10 to 20 times less energy and achieve the same sorption capacity and treatment efficiency as ED and RO systems. Like ED, carbon electrodes may be stacked, with alternate electrodes (in the form of sheets) having oppositely polarity so that each attracts ions of opposite charge in the water. The polarization and flow can also be reversed to drive off sorbed ions after the surfaces are saturated.[14]

CDI technology has been tested in the laboratory for treatment of perchlorate-spiked water.[12,15] In this process, perchlorate-containing water enters the space between two porous carbon aerogel electrodes where an electric field forces ions to migrate into the carbon electrodes while purified water is discharged between the electrodes. It was reported that, at an influent perchlorate concentration of 80 mg/L, more than 85% of perchlorate was removed, and effluent perchlorate concentration decreased to 10 mg/L.[12] However, some difficulties were noted, such as backflushing problems, in which only 40% to 60% regeneration rate could be achieved. Additionally, high capital costs of carbon aerogel electrodes make the technology impractical for large-scale field applications at present.[16] Natural dissolved organic materials may also foul the aerogel surface and thus limit the treatment efficiency and the lifetime of CDI systems.

COMPARISONS OF TREATMENT TECHNOLOGIES

In-depth discussions of various treatment technologies for perchlorate removal have been given in Chapters 10 and 11 (ion exchange), Chapters 12 to 14 (bioremediation), Chapter 15 (modified activated carbon), Chapters 2 and 16 (chemical and electrochemical reduction), and this chapter (membrane filtration). A recent report by the U.S. Environmental Protection Agency[2] indicates that ion exchange and bioremediation are currently among the most commonly used technologies to remove or degrade perchlorate from contaminated water and soil (Table 1). There are a total of 15 full-scale ion-exchange treatment systems and 8 full-scale bioreactor or bio-reactive barrier systems installed across the United States. Obviously, these two remedial approaches are among the best developed and proven technologies for removing perchlorate from contaminated media. Other treatment options are either cost-prohibitive or still under development. However, it should be recognized that no single treatment technology is perfect or suitable for perchlorate removal under all geochemical and hydrological environments. Every treatment technology has its own advantages and limitations, as summarized in Table 2. It is important to consider site-specific hydrogeochemical conditions, water treatment criteria, regulatory requirements, human health and ecological concerns, and cost when selecting the most appropriate technology.

Table 1. Number of full-scale and pilot-scale perchlorate treatment systems using different treatment technologies[*]

Technology	Number of projects	
	Full scale	Pilot scale
Ion exchange	15	3
Bioreactor and in situ bioremediation	5	15
Permeable reactive barrier and composting	3	4
Activated carbon sorption	2	2
Electrodialysis	0	2
Reverse osmosis	0	0
Chemical reduction	0	0

*Source: EPA 542-R-05-015 report[2]

In addition, the optimum treatment technology for a given site will depend on other factors such as groundwater flow, perchlorate concentration, geochemical parameters (e.g., pH, total dissolved solids (TDS), alkalinity, nitrate, sulfate, dissolved oxygen, and metals), other water-quality parameters such as dissolved organic matter and total suspended solids (TSS), and co-contaminant presence and concentration. The best choice for any given situation will thus require a careful evaluation of options and perhaps a combination of treatment techniques. For example, in the case of drinking water treatment at low perchlorate concentrations, selective ion exchange may be preferred because of its treatment longevity, simplicity, minimal impact on water quality, and capability of running at a high flow rate with a small treatment unit (Chapter 10). Although a bioreactor may be used and is technically feasible, the technique could be ineffective or costly for treatment at low perchlorate concentrations (e.g., at tens or hundreds of ppb) because a highly reducing environment is required for perchlorate biodegradation.[4,17-19] The greatest challenge in this case will be to sustain enough biomass to create continuously reducing conditions without addition of extra amounts of electron acceptors and donors. Additionally, a much larger treatment unit will be required if the system is running at the same flow rate as an ion exchange system. This is because a relatively long hydraulic retention time (HRT) is necessary to achieve complete perchlorate reduction. Kim and Logan[20] reported an HRT of 43 to 120 min in order to biodegrade perchlorate from 24 mg/L to less than 4 µg/L. A longer HRT

Table 2. Pros and cons of various treatment technologies for perchlorate removal and/or destruction

Technology	Pros	Cons
Conventional nonselective ion exchange and regeneration	• Practical and economical • Simple to implement • Effective and able to remove ClO4- below detection limit (<4 ppb) • Fast reaction and simple operation • Can be operated at a high flow rate • Regulatory acceptance	• Ineffective in the presence of moderate concentrations of competing ions such as NO_3^-, Cl^-, $SO_4^=$, HCO_3^- • Generation of large quantities of secondary NaCl brine wastes • High capital cost • Remineralization may be needed • Subject to fouling in the presence of suspended solids and dissolved organic matter
Single-use, throw-away ion exchange	• Same as above • Relatively low capital cost	• Frequent change-out of resin beds; high O&M cost • Generation of large quantities of spent resin loaded with ClO_4^- • High waste disposal cost • Subject to fouling as noted above

Table 2 continued.

Technology	Pros	Cons
Highly selective and regenerable ion exchange	• Same as above. • No changes to water chemistry • Highly selective and efficient, particularly suited for treatment at low ClO_4^- concentrations • Resins last much longer; reduced O&M costs. • Low regenerant volume using $FeCl_3$-HCl, <1 BV per regeneration cycle • Perchlorate recovered or destroyed, and regenerant recycled; minimized waste production	• Resins cost about twice as much as nonselective resins • Handling of acidic regenerant solution • Subject to fouling, but the use of HCl in regeneration cleans up the resin bed
Ex situ bioreactor and biodegradation	• Practical and economical • Particularly effective for treatment at relatively high ClO_4^- concentrations with suspended solids • Perchlorate completely degraded • Other contaminants such as chlorinated organics also destroyed	• Highly reducing environment and a continual feed of nutrients required • Oxygen and nitrate competition • High O&M costs for treatment of water at low ClO_4^- concentrations; addition of extra e-donors or acceptors required to maintain the biomass • Slower reactions than ion exchange; large treatment unit required at the same flow rate • Post treatment may be required due to added nutrients/chemicals and potential pathogens

Table 2 continued.

Technology	Pros	Cons
In situ bioremediation (e.g., reactive barriers and direct nutrient injection)	• Same as above • Passive treatment and low O&M cost • Viable option for in situ localized soil treatment (e.g., source zone with a high ClO_4^- concentration)	• Competition by other e-acceptors such as oxygen, nitrate, Fe(III) and Mn(IV) in soil • Difficult to implement and control in the subsurface • Subject to hydrogeological and geochemical influences
Activated carbon and modified carbon sorption	• Simple to implement and low maintenance • Other contaminants such as chlorinated organics also removed • Fast reaction kinetics • Reactivation possible	• Low selectivity and inefficient in the presence of competing anions such as NO_3^-, Cl^-, and $SO_4^=$ • Low sorption capacity without surface modification • Increased cost for surface modification with cationic surfactants • Stability and bleaching of surfactants during treatment unknown • Subject to fouling by dissolved organic matter
Electrochemical reduction	• Perchlorate destroyed • Low maintenance	• High capital and O&M costs for treatment at low ClO_4^- concentrations • Energy consumption • Electrode fouling • Not fully developed

Table 2 continued.

Technology	Pros	Cons
Membrane filtration (RO, NF, and UF)	• Fast and effective • Regulatory acceptance • Suitable for household and small-scale applications	• Low selectivity and removes other salts • Fouling of the membranes and corruption • Waste concentrate disposal • High capital cost and impractical for large-scale applications • Difficult to remove ClO_4^- below the detection limit
Electrodialysis (ED)	• Same as above • Low pressure operation	• Same as above • Electricity energy consumption
Capacitive deionization (CDI)	• Same as above • No membranes used	• Same as above • Not selective and not fully developed at this time

leads to a larger reactor volume and higher capital costs. Post-treatment also may be required for a biodegradation system to remove anaerobic microorganisms and potential pathogens. Although membrane and conventional nonselective ion exchange with brine regeneration technologies also are capable of removing perchlorate below the detection limit, they are either too costly or produce large quantities of secondary brine concentrates. In other words, the selection of treatment technologies must consider potential costs of post-treatment and disposal of secondary wastes.

On the other hand, for the treatment of groundwater and wastewater with relatively high perchlorate concentrations, dissolved organic matter, or TSS, biodegradation techniques may be preferred because of their treatment efficiency and cost effectiveness under such conditions. Although ion exchange and other technologies may be used for treatment in such conditions, they may be less effective or more costly due to fouling of the treatment systems and limited sorption capacities. At many contaminated sites, perchlorate also is found in groundwater with a variety of co-contaminants, including volatile organic compounds (VOCs), halogenated solvents, and explosive compounds such as 2,4,6-trinitrotoluene (TNT), 2,4-dinitrotoluene (DNT), 1,3,5,7-tetranitro-1,3,5,7-tetrazacyclooctane (HMX), and hexahydro-1,3,5-trinitro-1,3,5-triazine (RDX). Under these circumstances, bioremediation may be used to effectively degrade both perchlorate and co-contaminants. Combinations of treatment technologies may be applicable under certain conditions. For example, a bioreactor could be used to reduce very high concentrations of perchlorate and other co-contaminants, followed by ion exchange as a polishing step. Similarly, membrane filtration may be used for some small-scale treatment systems.

REFERENCES

1. Perchlorate Contamination Treatment Alternatives *V. 1.02* 2004. http://www.dtsc.ca.gov/ScienceTechnology/TD_REP_Perchlorate-Alternatives.pdf.

2. Perchlorate Treatment Technology Update *U.S. Environmental Protection Agency, EPA 542-R-05-015*, 2005:http://clu-in.org/download/remed/542-r-05-015.pdf.

3. Roquebert, V., Booth, S., Cushing, R. S., Crozes, G., Hansen, E. Electrodialysis reversal (EDR) and ion exchange as polishing treatment for perchlorate treatment. *Desalination*, 2000; *131*:285-291.

4. Urbansky, E. T. Perchlorate chemistry: Implications for analysis and remediation. *Bioremed. J.*, 1998; *2*:81-95.

5. Urbansky, E. T. Issues in managing the risks associated with perchlorate in drinking water. *J. Environ. Manag.*, 1999; *56*:79-95.

6. Tchobanoglous, G., Schroeder, E. D. *Water quality: characteristics, modeling, modification*; Addison-Wesley: Reading, Massachusetts, 1985.

7. Forter Wheeler Environmental Corporation Perchlorate treatability studies: Use of reverse osmosis and biotreatment for removal of perchlorate from JPL groundwater. 1999:http://jplwater.nasa.gov/NMOWeb/AdminRecord/docs/NAS70984.PDF.

8. Yoon, Y. M., Amy, G., Cho, J. W., Pellegrino, J. Systematic bench-scale assessment of perchlorate (ClO4-) rejection mechanisms by nanotiltration and ultrafiltration membranes. *Sep. Sci. Technol.*, 2004; *39*:2105-2135.

9. Yoon, Y., Amy, G., Cho, J. W., Her, N., Pellegrino, J. Transport of perchlorate (ClO$_4^-$) through NF and UF membranes. *Desalination*, 2002; *147*:11-17.

10. Liang, S., Scott, K., Palencia, L., Bruno, J. Perchlorate treatment by enhanced coagulation, oxidation, and membranes. 2005: http://www.epa.gov/safewater/ ccl/perchlorate/ presentations/sun.ppt.

11. Yoon, J., Yoon, Y., Amy, G., Cho, J., Foss, D., Kim, T. H. Use of surfactant modified ultrafiltration for perchlorate (ClO4-) removal. *Water Res.*, 2003; *37*:2001-2012.

12. Wong, J. M. Treatment technologies for the removal of perchlorate from contaminated water. *Hydro Visions (http://www.grac.org/Spring_2002.pdf)*, 2002; *11*:33.

13. Ryoo, M. W., Seo, G. Improvement in capacitive deionization function of activated carbon cloth by titania modification. *Water Res.*, 2003; *37*:1527-1534.

14. Hrubesh, L. W. Aerogel applications. *J. Non-Crystalline Solids*, 1998; *225*:335-342.

15. Farmer, J. C., Fix, D. V., Mack, G. V., Pekala, R. W., Poco, J. F. Capacitive deionization of NH4ClO4 solutions with carbon aerogel electrodes. *J. Appl. Electrochem.*, 1996; *26*:1007-1018.

16. Gabelich, C. J., Tran, T. D., Suffet, I. H. Electrosorption of inorganic salts from aqueous solution using carbon aerogels. *Environ. Sci. Technol.*, 2002; *36*:3010-3019.

17. Batista, J. R., McGarvey, F. X., Vieira, A. R. In *Perchlorate in the Environment*; Urbansky, E. T., Ed.; Kluwer/Plenum: New York, 2000, pp 135-145.

18. Gu, B., Ku, Y., Brown, G. M. Treatment of perchlorate-contaminated water using highly-selective, regenerable ion-exchange technology: a pilot-scale demonstration. *Remediation*, 2002; *12*:51-68.

19. Gu, B., Brown, G. M., Alexandratos, S. D., Ober, R., Dale, J. A., Plant, S. In *Perchlorate in the Environment*; Urbansky, E. T., Ed.; Kluwer/Plenum: New York, 2000, pp 165-176.

20. Kim, K., Logan, B. E. Fixed-bed bioreactor treating perchlorate-contaminated waters. *Environ. Eng. Sci.*, 2000; *17*:257-265.

INDEX

A

Activated carbon sorption, 343-369
Ammonia tailored GAC, 352
Ammonium perchlorate, 3, 93, 257, 264
Amphibians, 158
Analysis and detection of perchlorate, 99, 111-147, 200-203, 255, 332, 346
Animal toxicity studies, 171
Anion exchange resins, 24, 210-248
  also see Resin
Atacama Desert, 34, 50
Atmospheric deposition, 37
ATR-IR spectroscopic detection, 137

B

Bacteria for perchlorate reduction, *see* Microbial reduction
Bifunctional resin, 142, 217-248, 255, 265
Biochemistry of perchlorate bioreduction, 297
Bioconcentration, 154
Biofouling, 286
Biomarkers, 199
Biomonitoring methods, 200
Bioremediation, 253, 279, 286, 311, *also see* Microbial reduction
  Pilot and full-scale treatment systems, 311, 398
Birds, 159
Blasting agents, 82
Breakthrough studies, 219-224, 257-275, 347-368,
Brine regeneration, 224

C

Capacitive deionization, 396
Capillary electrophoresis, 128
Catalytic reduction of perchlorate, *see* Chemical reduction
Cationic polymer, 352
Cationic surfactants, 345
Chemical and electrochemical reduction of perchlorate, 27-34, 378-382
Chilean caliche, 34, 51
Chilean nitrate, 72, 97
Chlorate, 21, 84, 279
Chlorate reducing microorganisms, 299
Chlorine Isotope Ratio, 99
Chlorine products, 18, 84
Chlorite, 280, 298
Chronic exposure, 185
Clinical study, 180
Coadsorption of organics on GAC, 363
Colorimetric method, 116
Colorado River, 155
Comparisons of treatment technologies, 397-403
Cost analysis of treatment, 239

D

Department of Defense, 2
Desorption of surfactants from GAC, 360
Destruction efficiency, 235-239, 274-275
Dismutase gene, 298
Dismutase Protein, 298
Disproportionation, 20, 38, 375
Distribution coefficient, 211, 257
DPRB, Dissimilatory (per)chlorate-reducing bacteria 281, 297, 301
DPRB activity, 284

LIST OF CONTRIBUTORS

Laurie A. Achenbach
Southern Illinois University
Department of Microbiology
Carbondale, IL 62901
(618) 453-7984

Todd Anderson
Texas Tech University
MS 41023
Lubbock, TX 79409-1023

Carol E. Aziz
GeoSyntec Consultants
130 Research Lane, Ste. 2
Guelph, ON N1G 5G3
Canada
(519) 822-2230

Jacimaria R. Batista
University of Nevada - Las Vegas
Dept. Civil Environ. Engineering
4505 Maryland Parkway
Las Vegas, Nevada 89154-4015
(702) 895-1585

Abelardo D. Beloso
University of Illinois at Chicago
Earth & Environmental
Sciences/MC-186
845 West Taylor Street
Chicago, IL 60607-7059

Kelly Bender
BioInsite, LLC
150 E. Pleasant Hill Dr.
Dunn-Richmond Ctr. Suite 174
Carbondale, IL 62903

Liza Valentín-Blasini
Centers for Disease Control
Division of Laboratory Sciences
Atlanta, GA 30341
(770) 488-7902

Ben Blount
Centers for Disease Control
Building 103, room 3301
4770 Buford Hwy NE
Atlanta, GA 30341
(770) 488-7894

John K. Bohlke
US Geological Survey
431 National Center
12201 Sunrise Valley Drive
Reston, VA 20192
(703) 648-6325

Peter V. Bonnesen
Oak Ridge National Laboratory
P. O. Box 2008, MS-6119
Oak Ridge, TN 37831
(865) 574-6715

Pamela A. Boss
SPAWAR Systems Center
9112 Fermi Ave.
San Diego, CA 92123
(619) 553-1603

Gilbert Brown
Oak Ridge National Laboratory
1 Bethel Valley Road
Bldg. 4500-N, Room C-250
Oak Ridge, TN 37831
(865)-576-2756

Fred Cannon
The Pennsylvania State University
Dept. Civil and Env. Engineering
212 Sackett Building
University Park, PA 16802
(814) 863-8754

John D. Coates
University of California, Berkeley
Dept. Plant & Microbial Biology
271 Koshland Hall
Berkeley, CA 94720
(510) 643-8455

Randall J. Cramer
6564 River Tweed Lane
Alexandria, VA 22312
(301) 744-2578

Shannon E. Cunniff
3616 N Upland St
Arlington, VA 22207
(703) 604-1529

Jay Diebold
Shaw E&I
111 W. Pleasant St.
Suite 105
Milwaukee, WI 53212-3939
(414)-291-2357

Baohua Gu
Oak Ridge National Laboratory
1 Bethel Valley Road
Bldg. 1505, Room 218
Oak Ridge, TN 37831
(865)-574-7286

Gregory J. Harvey
ASC/ENVR
1801 10th St
Bldg. 8, Suite 200, Area B
WPAFB, OH 45433
(937) 255-3276

Paul B. Hatzinger
Shaw Environmental
17 Princess Road
Lawrenceville, NJ 08648
(609)-895-5356

Juske Horita
Oak Ridge National Laboratory
1 Bethel Valley Road
Bldg. 4500-N, Room S-214
Oak Ridge, TN 37831
(865) 576-2750

Greta J. Orris
U.S. Geological Survey
520 N. Park Ave., Suite 355
Tucson, AZ 85719
(520) 670-5583

Andrew Jackson
Texas Tech University
Department of Civil Engineering
MS 41023
Lubbock, TX 79409-1023
(806) 742-2801

Robert Parette
The Pennsylvania State University
Dept. of Civil and Env. Eng.
212 Sackett Building
University Park, PA 16802

Namgoo Kang
Texas Tech University
MS 41023
Lubbock, TX 79409-1023

Leslie J. Patterson
University of Illinois at Chicago
Earth & Environmental
Sciences/MC-186
845 West Taylor Street
Chicago, IL 60607-7059

David R. Mattie
AFRL/HEPB
2729 R Street, Bldg 837
Wright-Patterson AFB
OH 45433-5707
(937) 904-9569

Srinath Rajagopalan
Texas Tech University
MS 41023
Lubbock, TX 79409-1023

H. E. Maupin
Naval Environmental Health
Center
Portsmouth, VA 23708

Frederick V. Sloop
Oak Ridge National Laboratory
1 Bethel Valley Road
Bldg. 4500-N, Room D-250
Oak Ridge, TN 37831
(865) 574-6716

Philip N. Smith
Texas Tech University
The Institute of Environmental &
Human Health (TIEHH)
1207 Gilbert Drive
Lubbock, TX 79416
(806) 885-4567

Joan E. Strawson
TERA
2300 Montana Ave., Suite 409
Cincinnati, OH 45211
910-692-7752

Neil C. Sturchio
University of Illinois at Chicago
Earth & Environ. Sci./MC-186
845 West Taylor Street, Rm. 2442
Chicago, IL 60607-7059
(312) 355-1182

Carey Yates
5 Euclid Avenue
Condo 142
Bristol, VA 24201

Qiyu (Jay) Zhao
Toxicology Excellence for Risk
Assessment (TERA), Suite 409
2300 Montana Ave.
Cincinnati, OH 45211
(513) 542-7475

Ping Zhou
Oak Ridge National Laboratory
Environmental Sciences Division
P. O. Box 2008, MS-6036
Oak Ridge, TN 37831